T0092978

Slouch

Slouch

Posture Panic in Modern America

Beth Linker

Princeton University Press
Princeton & Oxford

Published by Princeton University Press
41 William Street, Princeton, New Jersey 08540
99 Banbury Road, Oxford OX2 6JX

press.princeton.edu

All Rights Reserved

Library of Congress Cataloging-in-Publication Data

Names: Linker, Beth, author.
Title: Slouch : posture panic in modern America / Beth Linker.
Description: Princeton : Princeton University Press, [2024] | Includes
    bibliographical references and index.
Identifiers: LCCN 2023036266 (print) | LCCN 2023036267 (ebook) | ISBN
    9780691235493 (hardback) | ISBN 9780691235509 (ebook)
Subjects: LCSH: Posture—United States—History—20th century. |
    Posture—United States—History—21st century. | Posture—Social
    aspects—United States—History—20th century. | Posture—Social
    aspects—United States—History—21st century. | Human body—Social
    aspects—United States—History—20th century. | Human body—Social
    aspects—United States—History—21st century.
Classification: LCC RA781.5 .L56 2024 (print) | LCC RA781.5 (ebook) | DDC
    613.7/80973—dc23/eng/20230929
LC record available at https://lccn.loc.gov/2023036266
LC ebook record available at https://lccn.loc.gov/2023036267

British Library Cataloging-in-Publication Data is available

Editorial: Eric Crahan, Whitney Rauenhorst
Jacket: Chris Ferrante
Production: Danielle Amatucci
Publicity: Alyssa Sanford (US); Carmen Jimenez (UK)
Copyeditor: David Heath

Jacket Credit: a-yun / Shutterstock

This book has been composed in Arno Pro and Futura New

Printed in the United States of America

10 9 8 7 6 5 4 3 2 1

For Mark and Katie Rose

Contents

Slouch

Introduction

READERS WHO opened the *New York Times Magazine* on the morning of January 15, 1995, learned something surprising and sensational. Thousands of nude photographs, including those of prominent public figures such as George H. W. Bush, Bob Woodward, Meryl Streep, Hillary Rodham Clinton, and Diane Sawyer existed in the Smithsonian Archives, readily available for public viewing. Adding fuel to the fire, Ron Rosenbaum, the journalist who broke the story, claimed that the photos were part of a nefarious eugenics-inspired plot, research undertaken at the nation's universities that evoked "the specter of the Third Reich."[1]

The nude photos that Rosenbaum discovered were taken decades earlier, when the pictured individuals were young adults attending college. For much of the twentieth century, a remarkable number of U.S. institutions of higher education mandated that their students undergo an annual physical exam, including a posture evaluation. By the mid-twentieth century, most schools had adopted camera photography to assess human posture, requiring students to pose nude or seminude for the pictures. What had become a ritual on nearly every college campus in the United States would come to an end by the early 1970s.

Written twenty years after the closure of these programs, Rosenbaum's piece sparked outrage among alumni across the nation, placing university administrators and lawyers on the defensive. Soon after the exposé, most of those named—and even unnamed—universities and colleges ordered their archivists to destroy institutional holdings related to the practice. Within a matter of months, decades of recorded history had either been shredded beyond recognition or gone up in smoke. It was a decisive and, to many onlookers, triumphant end to an ostensibly dark chapter of U.S. history. Since then, most journalists and scholars have remained silent on the matter, seemingly content to lay an unsavory chapter of history to rest.[2]

Slouch revisits these events to gain a fuller and more nuanced perspective on the history of the posture sciences, of which posture photography was but a part. To get at this history requires close scrutiny of the ways in which culture and politics come to inform certain scientific endeavors and, in turn, how widely accepted scientific facts influence the social fabric of everyday life. It requires adopting a critical stance when faced with seemingly self-evident claims, such as Rosenbaum's portrayal of mid-century posture examinations as "bizarre," a "pseudoscience" even.[3] If Rosenbaum was right, then how do we explain the fact that to this day, approximately $1.25 billion is spent annually worldwide on posture-enhancing devices and fitness programs.[4]

It turns out that the scientific concern for human posture has a long and complicated history, attracting a wide array of professionals in medicine, education, and biology, as well as myriad health culturists and journalists. Indeed, for much of the twentieth century, Americans were told that they were living through a poor posture epidemic that, if left unchecked, would lead to widespread illness, disability, and even death.

The first recorded study to report a poor posture epidemic—dubbed the "Harvard Slouch" study—appeared in 1917.[5] University physicians found that 80 percent of students exhibited significant bodily misalignments and worried that the future leaders of America would end up chronically ill or permanently disabled. In the decades to follow, the U.S. military, secondary schools, industrial workplaces, and public health agencies would conduct similar studies, each coming to the same conclusion: that slouching was rampant in America.

This was not the first time that scientists and other social commentators expressed an interest in human posture. Since the time of Plato, naturalists, theologians, and philosophers in the West have remarked on the uniqueness of human posture and bipedalism. To many early Christian thinkers, human uprightness was understood to be a sign of divinity, an attribute that indicated proximity to angels, God, and the heavens. Enlightenment thinkers who saw the world through more secularized eyes believed that human posture was essential for rationality, right living, and self-discipline. Yet while anatomical erectness was important to the very definition of humanness, few scientists devoted their entire careers to studying it.[6]

This would change by the turn of the twentieth century, when poor posture became medicalized, in large part due to the new interest in the evolutionary sciences. After the publication of Darwin's *On the Origin of Species* (1859) and *The Descent of Man* (1871), human posture took on a new relevance. Prior to Darwin, it was widely accepted that the human brain—and thus human intellect—drove human evolution. The nineteenth-century ubiquity of skull collecting reflects the pre-Darwinian assumption that human cranial growth dictated all other advancements and development, including the acquisition of upright standing and bipedalism. Darwin, by contrast, believed that humans descended directly from simians through a gradual process of natural selection, and argued that upright human posture preceded the evolution of all other distinctive human characteristics.[7]

Darwin's posture-first theory of evolution created both intrigue and fear. If human intellect was no longer the prime mover that distinguished human and nonhuman animals, then it appeared that only a mere physical difference, located in the spine and feet, separated humankind from the apes. Human superiority seemed more fragile than ever. Though controversial, Darwin's explanation held an appeal, especially to those trained in anatomy and medicine. It was a Dutch physician-turned-paleoanthropologist, after all, who would become one of the most well-known scientists to offer material proof of Darwin's posture-first theory. In 1891, on the Indonesian island of Java, Eugène Dubois discovered human fossil remains that would come to be known to the scientific world as *Pithecanthropus erectus*, later redesignated *Homo erectus*.[8]

Applying the scientific study of nonliving fossil remains to current-day living populations of people—what would become the study of evolutionary medicine—early twentieth-century scientists began to argue that the human body was poorly adapted to the modern industrialized world, an evolutionary weakness that resulted in musculoskeletal pain, prolapses, and chronic disease.[9] While theories abounded as to why the human race seemed to be struggling with this most basic of species attributes, many agreed that the demands of modern, civilized life were to blame. From the introduction of mandatory schooling and assembly-line production to increasing urbanization and motorized travel, a large number of Americans appeared to be more sedentary than ever, hunched over workstations, slumped in school desks, and dependent on trains and cars for transport rather than their own two feet. Sedentariness did not mean leisure, but rather fixity. Contemporaries liked to use the example of a tailor: tailors labored with their hands but otherwise remained stationary, often fixed in a faulty, slumping posture.

Hence, unlike earlier eras when faulty posture was seen as low class, rude, or an indicator of a lack of civility, by the early twentieth century it became a quintessential marker of civility, albeit one that pointed to the negative consequences of too much civilization, or "overcivilization," as contemporaries put it.[10] Moving beyond manners and breeding, posture maintenance became a problem for anyone touched by modern industrialized life.

Such a dire outlook on the state of human health in the industrialized world led certain physicians and educators in the United States to create the American Posture League (APL) in 1914, an association that would grow in membership and public visibility well into the mid-twentieth century. The APL married the theoretical findings from evolutionary scientists to the practical tools of the then "new public health" movement. New public health proponents, medical historians have shown, privileged prevention over cure, and utilized top-down surveillance in order to track individuals who exhibited early-stage disease or disability. The shift in emphasis to preventive medicine made a lot of sense in the wake of the late nineteenth-century discovery that microscopic germs spread diseases such as tuberculosis and other contagious infections. In the absence of effective therapeutics such as antibiotics, many practitioners put their efforts into disease prevention since cures seemed so elusive. These new public health systems created such a level of public fear, historian Nancy Tomes argues, that "germ panic" became commonplace within U.S. culture.[11]

What is striking about the poor posture epidemic is that it gained legitimacy and public support without evidence of a contagion. The epidemic was thus defined more by lifestyle and individual behavior than germ transmission. While the germ theory brought about a reductionistic clarity concerning the spread of contagious disease, it did little to explain why certain individuals, when exposed to infectious microorganisms, would fall ill while others would not. The APL stood at the forefront of the post–germ theory medical effort that emphasized holistic health, insisting that postural health was necessary to the proper functioning of the musculoskeletal system and inner organs.[12] The scientists of the APL maintained that slumping shoulders and a protruding abdomen could indicate a wide variety of health conditions, ranging in severity from a deadly case of tuberculosis to scoliosis and generalized effects of old age.[13]

Slouch thus tracks an epidemic pathology that was defined as a disability rather than an acute disease. Unlike the majority of epidemics found in recorded history, the detection of a widespread problem of poor posture did not come about because of mass deaths. No one died due to slouching. And yet, posture experts argued, people who did not adequately attend to their own physique risked an early demise, for poor physical form made it more likely for infectious disease to take root. In short, postural defects begot disease, which could then result in lifelong disability. Though not a communicable disease itself, the slouching epidemic was built on the notion of social contagion, on the idea that deleterious norms, practices, and beliefs about bodily comportment could be passed from person to person if not corrected through proper measures.

In this context, the physical exam became paramount. And unlike height and weight measurements, posture assessments proved to be one of the quickest and least expensive tools in the new public health arsenal. With a trained gaze acutely attuned to even the slightest anatomical misalignment, health experts readily diagnosed extant and potential disease states based on a simple and quick full-body view of the examinee. In many settings, the posture exam served as a litmus test to distinguish the able-bodied from the disabled.[14]

It is little wonder that one of the first reports of the poor posture epidemic would come from Harvard University. These students, after all, were inundated by the overcivilizing influences of modern society; they were white men with the social and economic capital to pursue book learning and an advanced education rather than engage in hard labor. Yet despite their many advantages, white middle- and upper-class Anglo-Saxon men at the time lived in fear of disabling weakness. The perceived threats were many. Women were making concrete political gains, earning the right to vote with the 1920 ratification of the Nineteenth Amendment. Immigrants and Black American men donned military uniforms during the Great War, leading to a victory for the Allied forces. Talk of white race suicide filled the lecture halls and the pages of the daily newspapers. The popularity of social Darwinism and eugenics fueled the white masculinity crisis of the early twentieth century.[15]

But white men of privilege were not alone in their fear of physical weakness and disability. Anxious to dispel the notion that theirs was the "weaker" sex, women physicians and physical educators encouraged college co-eds and school girls to take up physical fitness classes, where posture work and measurement became a mainstay. African American doctors fretted about widespread posture faults among poorer Black Americans, incorporating posture health examination and education into the National Negro Health Week program. Similarly, Jewish American physicians who assessed the posture of Eastern European garment workers insisted that industrialists provide laborers with "posture-right" chairs and workspaces.

The posture crusade, in other words, cut across racial and ethnic categorizations. Men and women of all races lived with the possible fate of acquiring a defective posture. In this respect, all of these groups shared a common goal of separating themselves from the ultimate other—namely, the disabled.[16] Disability historian Michael Rembis notes that while scholars have written at length about the racism of the American eugenics movement, they have paid far less attention to how the movement was also "infused with virulent ableism." "Eugenics at heart," he concludes, "was a politics of normalization . . . and optimization."[17]

Normalization was central to the American anti-slouching campaign. The demand to stand straight carried immense moral and medical weight. Straightness signified health and youthful vitality, but also upright character and sexual chasteness. It is no accident that by the early 1940s in the United States, "straight" connoted heterosexuality, while "bent" signified homosexuality. Linguists George Lakoff and Mark Johnson contend that at the deepest levels of conception and understanding, the notion that "up is good" and "down is bad" shapes our language and the way humans make meaning out of the wider world around us. In other words, the inherent positive valuation of "up" is rooted in human bipedalism.[18]

Determining the degree to which the anti-slouching campaign was eugenic or neo-eugenic requires careful study. Posture scientists featured in this book did not speak in terms of Mendelian genetics, heredity, or controlling reproduction—the kind of "hard" eugenics associated with the American Eugenics Society and Nazi Germany, where forced sterilization and genocide, respectively, were used as methods of ridding society of "undesirable" traits and human beings. Rather, the posture crusade appealed more to a theory of "euthenics," defined as "right living," which is sometimes characterized as "soft" eugenics. Soft eugenics is often portrayed as unique to Latin American history, but the U.S. poor posture epidemic suggests otherwise.[19]

As *Slouch* demonstrates, the manufacture of posture panic served as a powerful motivator and a ready-made disciplinary tool, deployed for multiple political ends in the United States throughout the twentieth century. Arising at a time when the nation was attempting to become a global military and industrial power, posture surveillance was taken up by the U.S. Public Health Service, the military, and educational institutions as a way not only to colonize other peoples, but also to regulate "deviant" bodies and "abnormal" behavior internal to the nation.[20] Posture crusaders promised that their work would control disease and political unrest, maintain industrial supremacy, and promote physical, aesthetic, and behavioral homogeneity. In this respect, the anti-slouching campaign reconceptualized centuries-old hygiene concerns regarding dress and exercise, making human posture an indicator of national strength, population health, and fitness.[21]

Of course, the intersection of race, class, age, and gender often determined the degree of poor posture's disabling effects. A middle- to upper-class white person with a postural abnormality would have a greater chance of securing an education and gainful employment than a non-white person. Still, even for the most well off, discrimination was a constant battle. The famous early twentieth-century writer and New York intellectual Randolph Bourne, whose spine was noticeably hunched and curved from a case of skeletal tuberculosis as a child, faced repeated marginalization and rejection in lodging and employment, and from potential love interests.[22] Born with spina bifida forty years after Bourne's death in 1918, Riva Lehrer recounts being an object of scorn and unwanted stares as a teen with "a curved spine and very large orthopedic shoes" growing up in Cold War America. Or take Sunaura Taylor, who, born in 1982 with arthrogryposis, grew up being compared to animals. She writes, "I have been told I walk like a monkey, eat like a dog, have hands like a lobster, and generally resemble a chicken or penguin." A disability and animal rights activist, Taylor owns her animal identity, seeing it as integral to her humanity. And yet she also fully realizes that animal insults in a Westernized culture come from a long tradition of seeing nonhuman beasts—animals and humans who look like animals—as unworthy of care and rights, lacking full subjectivity and autonomy.[23]

While today the outsized concern for poor posture may seem quaint—or worse, deeply ableist—it is important to recognize that noncommunicable epidemics are not simply a thing of the past. The so-called obesity epidemic shares many similarities with the poor posture epidemic of yesterday; in both cases there is a sense of urgency and fear, as if a communicable biological contagion were involved. Feelings of stigma and discrimination experienced by those with nonnormative bodies are also a shared commonality.[24]

I make no claim about the realness of the epidemic or the degree to which poor posture was and is debilitating. Rather, as a cultural historian, I see past and present worries concerning poor posture as part of an enduring concern for so-called "diseases of civilization."[25] Since the Enlightenment, if not earlier, the belief that civilized society causes sickness has served a variety of political and social ends, with the medical and the moral superimposed on each other. In most if not all cases, it is the "civilized" who identify, worry, and fall victim to such diseases, while they pin their hopes for a cure on fabled peoples and places supposedly untouched by civilization. During the heyday of European imperialism and colonization, the myth of the "noble savage" became part of political criticism and commentary, a critique of the problems that plagued civilized societies as they increasingly moved away from "nature" toward human artifice. The ills caused by growing urbanization during the nineteenth century led the American educated elite to posit that cities were the main culprits of disease and that more "primitive" living, ostensibly found in bucolic countrysides or among indigenous peoples, offered the path to health. Once monogenism and the evolutionary sciences became widely accepted by the turn of the twentieth century and all of humankind became biologically tethered to the deep past, true health seemed to exist only in early human origins. In this context, surviving populations of presumed "hunter-gatherers" became model organisms for research in evolutionary medicine and public health, much as it still is today.

I depict the experts who studied, researched, and wrote about the importance of human stance as posture scientists. This is a shorthand way to describe a vast professional network of physical educators, physical anthropologists, evolutionary scientists, efficiency engineers, commercial industrialists, physicians, and physical therapists. For the first half of the twentieth century, the APL served as the professional organization where individual work and research could be shared and promoted. But since there was no official journal for the APL, most members published in academic journals keyed to their own methods and disciplines of study. After the APL dissolved in 1944, human posture research became more specialized, and, in a certain sense, siloed.

The following account pulls together some of the disparate threads of this post–World War II history by looking at the ways in which the Kennedy administration and institutions of higher education supported human posture research as a means of fighting the Cold War by proxy.

Ultimately, what bound together the professionals whom I designate as posture scientists is a nascent commitment to evolutionary medicine, of which a comprehensive history has yet to be written. The marriage of the evolutionary sciences to a concern for diseases of civilization set the stage, I argue, for the existence of noninfectious epidemics, such as poor posture, and eventually obesity, ADHD, and diabetes, along with many others. I have taken the liberty of categorizing the widespread health concerns and interventions regarding poor posture as an epidemic, even though the historical actors featured in this book do not use that precise word. In their collection of data, use of statistics, and reliance on the language of health risk, posture scientists deployed tools of tracking and surveillance to make visible a problem that was epidemic in scale.

In addition, using the epidemic frame to explain the rather sudden and overwhelming twentieth-century concern about poor posture allows me to problematize, in a more obvious way, distinctions that are often made between real epidemics and not-real epidemics. In his 1992 field-defining essay, "Explaining Epidemics," historian Charles Rosenberg maintains that "A defining component of epidemics is their episodic quality." "A *true* epidemic is an event, not a trend," he continues. "It elicits immediate and widespread response. It is highly visible." Such a definition privileges acute and deadly health events but does not adequately capture chronicity, or diseases that evolutionary scientists insist are inherent to modern life because the human body was somehow not built for it.[26]

Slouch thus opens with the "outbreak" of the poor posture epidemic, which began with the discovery of human fossil remains that, to many scientists, proved that upright posture predated human intellect. As soon as the physical record was revealed, scientists began to fret about the inherent anatomical weakness of the human upright stance, specifically the highly vulnerable abdominal cavity that, while responsible for protecting vital organs, sagged under the constant weight of gravity and modern life, a problem from which humankind's quadrupedal ancestors did not suffer. These scientists thus discovered a population-wide problem that, unlike more acute disease states, had its initial occurrence in the far distant past, when early humans first adopted bipedalism as their primary mode of locomotion. With the help of medical doctors and physical educators who developed standardized measures and tests to determine normal from abnormal human posture, the epidemic became a statistical, objective, and visually proven fact. Along the way, photography became the preferred clinical tool to both evaluate and record posture, with hundreds of thousands of Americans stripping down to the flesh so that an examiner could get an accurate read. While at first the practice raised concerns about propriety and privacy, it soon became accepted convention, making it relatively easy for scientists to continue to demonstrate the existence of the epidemic.

The middle section of the book charts the spread of the poor posture epidemic, a process that, in the absence of a biological contagion, occurred largely by way of public health awareness campaigns and the commercial marketplace. A variety of stakeholders—schoolteachers, shoe companies, clothing manufacturers, public health officials, medical professionals, beauty culturists, and the popular press—worked to convince the public to engage in various poor posture detection and improvement initiatives, wellness programs that perpetuated disability stigma by encouraging the belief that health and ableness could be purchased through various consumer goods and sustained attention to one's physique. This was not simply a top-down process. Some of those who wanted to correct their posture faults wrote letters to orthopedic physicians and magazine advice columnists, looking for a cure. In the workplace, labor unions insisted on posture health awareness programs and agitated for the purchase of ergonomically sound seating systems.

The final chapters of the book focus on the demise of the poor posture epidemic, a slow process that would begin with the mid-century discovery of antibiotics but be quickly hastened with the countercultural revolutions of the 1960s. Throughout much of the twentieth century, scientists relied on both a steady supply of compliant university students and military draftees who would pose nude (or almost nude) for the recording and compilation of posture health on a population-wide level. The tradition of performing large-scale physical fitness assessments in order to "weed out" disabled bodies arguably had its beginnings in the United States with the practice of slavery, and more specifically with the history of the slave auction block. Stripped naked, enslaved people endured dehumanizing and invasive physical assessment at the hands of enslavers who wished to determine the "soundness" of their potential human property.[27]

Though not equivalent to the slave block, the posture assessments that became routine practice for much of the twentieth century in the United States did demand that a significant number of Americans stand unclothed in front of experts who had the power and tools not only to diagnose but also to make a permanent photographic record of the exam. It was only after the civil rights and women's movements, as well as the anti–Vietnam War protests, that a majority of Americans repudiated the practice. Suspicions of the real intent behind such educational and scientific initiatives also animated the disability rights movement, activists who legally challenged the practice of employment- and school-based discrimination based on outward appearance and perceived ability. Ultimately, the largely unattainable standard of ableness set into motion by the poor posture epidemic would give way to late twentieth-century political and cultural movements that promoted and celebrated bodily and cognitive diversity.

While the history of the poor posture epidemic cuts across gender, race, and class, it is nonetheless driven by ideals of whiteness and ableness, by scientific racism as well as scientific ableism. Indeed, the very notion of diseases of civilization is a construct of the ruling class who, in creating the myth of the happy "noble savage" or healthy "hunter-gatherer," put forth an origin story that places living indigenous peoples and their presumed special connection to the "natural" world in a time and place before white ablebodiness could became a reality. "Primitive" living thus becomes an evolutionary stage necessary to the perfectibility of the white body. This worldview is not only inaccurate. It is harmful, for it attempts to strip indigenous peoples of their own agency by insisting on their immutability, treating them as biological organisms frozen in time.

This history is thus one that cannot be easily bound to one nation-state or place. Since the poor posture epidemic was rooted in the logic of settler colonialism and empire, it operated on a far more global scale, built on imaginaries that had real consequences for the colonized. For the purposes of this book, I have sought to explain how and why a wide range of U.S. middle-class professionals came to believe in the epidemic's existence, and yet I also trace how making the epidemic a reality rested on the ability and power of these same professionals to designate certain populations of people as "primitive" in the first place.

Similar to their professional peers of a hundred years ago, evolutionary scientists working today continue to express concerns about how the human body is inherently ill-equipped to meet the demands of the computer age, seeing the slouched, largely sedentary existence of those living in the industrialized world as a primary cause for disabling back pain. This has led certain researchers, whom I discuss in the book's epilogue, to conduct studies on the presumed pain-free, cardio-fit hunter-gatherers living today in the Americas and Africa. This kind of research has been taken up by many in the fitness and wellness industry, by Western entrepreneurs who then create commercial goods in the form of "paleo" diets, exercise, and, most important for this book, posture surveillance practices. What good posture means, if anything, to native peoples who are the object of current scientific researchers is a topic that I hope will be taken up in further studies.

What these modern-day research efforts demonstrate is that human posture is still believed to have predictive power concerning future health, despite the existence of compelling scientific studies that demonstrate otherwise. Take, for example, *New York Times* health columnist Jane Brody, who as recently as 2015 proclaimed: "Slouching is bad. It's bad not only for your physical health, but also for your emotional and social well-being." Or consider how, a year later, *Glamour* magazine advised its readers to turn off their "bitch switch" through posture improvement. "Good posture increases testosterone and . . . lower[s] stress," the magazine explained. "It's science, bitches. Try it!"[28] Far more disconcerting are reports of Black Americans being excluded from juries because of bad posture. Meanwhile, Silicon Valley is busy creating new apps and biometric tools that can track an individual's sitting and standing posture at home and at work, adding yet another feature to our smart phones and Fitbits.[29]

In some ways, then, the epidemic seems never to have fully resolved, or at least the fears that accompany failing human posture and its negative health effects have never died out. To be sure, talk of a slouching epidemic may have subsided, but many of the same posture-promoting practices and beliefs are still alive and well today, uncritically accepted as self-evident truths.

CHAPTER 1

The Making of a Posture Science

LEADING A PARTY of military engineers and fifty forced laborers, Dutch physician Eugène Dubois set sail in 1887 for Indonesia to explore the rivers and caves of the island of Java, looking for human skulls. After a grueling couple of years of false starts and dead ends, Dubois and his team uncovered a large collection of vertebrate fossils from the early Pleistocene epoch. The bones were like no others. With a skullcap and molar that were more anthropoid than human, along with a thigh bone that was almost completely human, Dubois concluded that he had discovered the missing link between apes and humans. He named his find *Pithecanthropus erectus*, an early species of "ape-man" that had acquired human-like erect stance before the development of higher brain functioning.[1]

The discovery of *Pithecanthropus erectus*, later dubbed the "Java man," forced the international scientific community to rethink the question of whether the human brain or upright posture was the first step toward human evolution.[2] For much of the nineteenth century, the brain-first camp dominated evolutionary thought, with most scientists assuming that higher intellect served as the foundation for all the other features that distinguished human from nonhuman animals. The upright, small-skulled Java man, however, suggested that bipedalism was actually of greater import, a necessary precondition to the acquisition of a more developed mind, language, and ultimately civilized society.

Disputes between the brain-first and posture-first camps ensued throughout Europe, the British Empire, and North America.

Although advocates of brain-first evolutionary theory would continue to hold sway over much of Europe throughout the early twentieth century, Dubois found loyal followers in Britain and the United States, paving the path for the birth of the posture sciences. Meanwhile, it was not until the 1940s, after the so-called Darwinian synthesis, that evolutionary biologists came to a consensus that bipedalism must have preceded encephalization.[3]

This chapter situates the birth of the modern posture sciences at the turn of the twentieth century within a small but influential cohort of British and American physician-paleoanthropologists who believed that evolutionary concepts could and should be applied to living populations and inform clinical practice. This was not a straightforward or immediately popular proposal. After all, the theory of evolution was still hotly contested in the United States well up to the famous Scopes trial of 1925. What's more, the posture-first thesis made human beings biologically closer to nonhuman animals than the more widely accepted brain-first theory did. According to Dubois's findings, humans and apes shared a common biological origin; intellect was no longer the prime mover that separated the former from the latter. Since antiquity, natural philosophers, theologians, and even pre-Darwinian evolutionists developed elaborate schemas and hierarchies that clearly divided humans from nonhumans based on the capacity to think and reason.

Early posture-first proponents ran the risk of severe rebuke for upending cherished and long-held beliefs about human superiority. But the opposite happened. If anything, the close proximity between humans and apes led to a kind of vigilance and glorification of bipedalism. The medically trained evolutionary scientists featured in this chapter raised the stakes even higher, for, in their observations of living subjects, they saw an abundance of postural "defects," anatomical weaknesses that led to unique ailments found only in humankind.

To convincingly apply the biological concept of human evolution to present-day concerns about disease causation required a triangulated study of fossilized bones, comparative anatomy, and health conditions of living human beings. But it also necessitated the creation of a compelling account that linked the deep past to the then-present human condition.[4] Such narratives tended to blame "civilization" for postural weaknesses, an argument that essentially pointed to evolutionary flaws and maladaptation.

Exactly when in time bipedalism became a problem and for whom was up for interpretation, influenced by the political and social context of the physician in charge of constructing the narrative. One of the most dominant narratives originated with Arthur Keith, physician-anatomist and conservator of the Museum of the Royal College of Surgeons. In Keith's estimation, the agricultural age was the turning point when chaotic forces of "civilization" began to take over the well-ordered "natural" ones. In his telling, uniquely human ailments such as flat feet, hernias, scoliosis, organ prolapse, and pulmonary disease developed because of man-made environments, the by-products of civilization. American pathologist Shobal V. Clevenger, by contrast, claimed that the problem arose from the evolutionary outset. Or, as he put it, "Man's original sin consisted in his getting on his hind legs."[5]

What is crucial to these accounts is that it placed the concern for human posture on a population level, or at least at the level of "civilized" white populations. In other words, if evolution moved toward ever-greater civilization, then all of humanity was eventually destined to suffer the consequences of maladaptive weaknesses intrinsic to bipedalism. This kind of logic, I argue, made the scale of the poor posture problem quite large, and something that would theoretically continue to spread, much like an epidemic.

Unless, of course, it could somehow be halted. While subsequent chapters will address various slouching prevention programs, the current chapter reveals the racialized and ableist underpinnings of such projects. In an attempt to apply the lessons of evolutionary theory to living populations, Keith and his colleagues upheld certain "primitive" peoples as model posture organisms to be studied and, to some extent, mimicked.[6] Influenced by trends in physical rather than cultural anthropology, these physicians made posture improvement into a preventive health movement that was, in essence, a project of upholding and preserving white able-bodiedness.

The precise way that the posture sciences became racialized in the United States is unique given the nation's history. In a quest to find a model organism of their own, U.S. physician-scientists fixed their gaze on Native Americans. After the passage of the 1887 Dawes Act and the mounting of the federal Indian boarding school initiative, American Indians, who were long perceived to be enemies of civilization, became "symbols and myths upon which white Americans created a sense of historical authenticity."[7] They also came to represent primordial, "natural" posture health, aspirational figures for white Americans to model. Of course, the image of an inert, proudly erect Native American, frozen in primordial time, was a way to also strip indigenous peoples and tribes of their agency.[8]

Meanwhile, during the height of racial segregation and Jim Crow, Black Americans were forced to resist depictions of being equated with beasts, lest they continue to be pushed down the evolutionary chain to the point where civilization seemed an impossibility.[9] To counter this, the Black educated elite insisted that they, too, faced a poor posture epidemic. After all, to have a posture problem was a signifier of civilization, and thus an indicator of evolutionary superiority distant from "savage" populations. This is one of the perversities of so-called diseases of civilization. If Black Americans were believed to be immune from postural weakness, they risked being cast as pre-civilized, pre-rational, and animalistic. Making an argument that the true ancestral origins of posture health perfection could be found among tribal Africans, Black American physicians offered a subversive human origins narrative that countered the one promulgated by white scientists.

In the end, the physician-paleoanthropologists who put forth an evolutionary account of failing human posture ended up enjoying popular appeal partly because it refracted and reified fraught race relations in the United States and beyond. Posture health and ills had very real and deep political meanings and repercussions, especially for marginalized groups who were already disenfranchised.

Posture Science and Human Origins

In the 1859 *On the Origin of Species* and later in the 1871 *The Descent of Man*, Charles Darwin postulated that upright human bipedalism had evolved from quadrupedal simians through an early shift to terrestrial existence, and that this change preceded the evolution of other distinctive human characteristics.[10] Darwin was not the first to put forth such an idea, but he was the foremost evolutionist to argue that such a change must have occurred through a process of gradual transformations and variations by way of natural selection. "Man"—a descriptor that Darwin and others often used in place of "humankind"—belonged to the animal kingdom and had originated from some lower forms. Most important, Darwin's work pointed to the possibility that humankind's far-distant ancestors could be found within the line of primates.[11]

While evolutionists in the late nineteenth century agreed that humans differed from apes, they disagreed about which of the human species attributes (i.e., upright locomotion, speech, and brain size) evolved first. Historian Peter Bowler describes this era as a time when debates about the "*process* of human evolution"—as opposed to the *fact* of evolution—began in earnest.[12] Some insisted that cranial development and increased brain capacity were necessary for any other human traits—upright posture, language, tool use—to have come about. The "brain-first" camp, bolstered by a century of skull collecting, phrenology, and craniometry, prioritized human intelligence as the measure of human beings, creating a clear distinction between them and beasts, and also between races.[13] Upright posture was but a mere consequence of higher brain development. This line of thinking persisted well into the early twentieth century, with British anthropologists William J. Sollas and Grafton Elliot Smith in the lead.[14]

On the other side of the debate were those theorists who contended that upright posture preceded significant brain development or speech acquisition. Darwin himself posited this viewpoint, arguing that upright posture "was not just an abstract sign of nobility, but a practical link in the chain that led to the increasing use of tools."[15] In 1889, Alfred Russel Wallace further advanced Darwin's theory of evolution, writing, "We must conclude that the essential features of man's structure as compared with that of the apes—his erect posture and free hands—were acquired at a comparatively early period, and were, in fact, the characteristics which gave him his superiority over other mammals, and started him on the line of development which has led to his conquest of the world."[16]

What were largely theoretical debates concerning human origins turned into more epistemological debates in the final decade of the nineteenth century. Up until that point, much of the human origins debate concerning the primacy of uprightness versus brain capacity was based on inductive reasoning. Those in the embryological sciences and comparative anatomy circles believed that they could understand the process of human evolution by looking at embryological development (recapitulation) as well as by comparing simian anatomy (living and dissected) with human anatomy.[17]

In the late nineteenth century, however, certain physicians and anatomists began to look for physical, material proof of how human evolution came about, namely through the study of human fossils. It was out of this strand of human origins research that the posture sciences took firm hold, placing even greater emphasis on the importance of human posture and the gross anatomical developments that occurred in the distant past in order for bipedalism to come about. By necessity, these scientists needed to look beyond the more customary practice of skull collecting to include the entire body, considering all anatomic features.

The Birth of Homo erectus

In 1887, the Dutch anatomist Eugène Dubois enlisted as a medical officer in the Royal Dutch East Indies Army with the purpose of traveling to Southeast Asia (now Indonesia) in order to search for physical proof of the human–ape lineage. When he was a medical student and lecturer of anatomy at the University of Amsterdam, Dubois became fascinated with the problem of human evolution, immersing himself in the works of Darwin, Thomas Henry Huxley, and, most importantly, the leading German morphologist Ernst Haeckel. A follower of Darwin, Haeckel was most well known for his devotion to the problem of what

earlier naturalists had referred to as "missing links," crucial transitional
stages of human evolution. Using evidence from fetal development,
Haeckel created a hypothetical, twenty-two-stage, monophyletic progression that began with the most primitive organism and ended with
modern humans. Most importantly, he argued that a crucial transition
occurred between the twentieth and twenty-second stage, an organism
which he named *Pithecanthropus* (Greek for ape-man). This organism
walked upright, but lacked capacity for speech.[18]

Haeckel was by no means the first to postulate the existence of a
missing link between animal and human. Since the early eighteenth
century, European explorers and naturalists often assigned the status of
missing links to indigenous peoples, whom colonists believed to be less
than human based on their aboriginal ways of life and skin color.[19] This
tradition of appointing a missing-link status to non-white or "defective"
human beings continued up through the late nineteenth century and
beyond. While Darwin himself was reluctant to speculate on the transitional episodes that led to human existence, many of his followers had
no such reservations and indeed actively engaged in digging up material
proof for how evolution happened.

Taking a cue from Enlightenment-era naturalists, some evolutionists
looked to living populations to demonstrate the veracity of evolution,
with their search largely focused on South and East Asia. Although
Darwin speculated that humankind had descended from apes in Africa,
many evolutionists, including Haeckel, subscribed to the Indo-
European thesis—or Aryan thesis—which postulated that during the
Paleolithic period, a superior race of agriculturists from Asia conquered
and ousted indigenous Europeans, who were mere hunters. This hy-
pothesis was supported by an 1878 archaeological find in India of a Plio-
cene fossilized ape that "resembled humans more closely than any of
the living anthropoid apes."[20] Anthropologists studied Australian Aborigines and other indigenous Asian tribes, believing that, while potentially ancestors to contemporary Europeans, the aboriginals were
"frozen relics of an earlier stage in the upward progress from the apes."[21]
In 1904, the U.S. Louisiana Purchase Exposition exhibited Australian
Aborigines, Native Americans, Filipinos, and three thousand other

non-white indigenous peoples in order "to trace the course of human progress and classify peoples in terms of that progress to illuminate the *origin . . .* of man."[22]

Beyond the realm of science, the search for living missing links became a booming commercial enterprise that involved human and nonhuman trafficking. One example is of a young Burmese girl, whom her captors named "Krao." On an 1882 trip to the jungles of Siam in search of "curiosities," English explorer Carl Bock and American physician George G. Shelly discovered a family of "human monkeys" that included a father, mother, and daughter.[23] Bock and Shelly captured the family, and after the father died of cholera, they made an appeal to the local tribal chiefs and the Siamese government to keep the girl. The king of Siam allowed Bock and Shelly to take ownership of the daughter—separating her from her mother—as long as they adopted her and vowed to provide her with "security and good care."[24]

Bock and Shelly first took Krao—whom they billed as a living missing link—to England, where they exhibited her at the Royal Aquarium in Westminster, London, and then to the United States, where they displayed her at various medical and popular anatomical exhibits. Describing her as a "Monkey girl," journalists reported that "unlike the usual monkey type, she has two hands and two very pretty, human-looking feet." But "a double row of teeth in her mouth; pouches in her cheeks, where she stows away surplus food; and a coating of hair over the entire body" made her, in their estimation, nonhuman. By the time Krao reached the United States, she exhibited fluency in English, German, and French. Despite her intelligence, her captors and onlookers insisted that her physical characteristics branded her as not fully human.[25]

While Krao was stripped of her human status because of her racialized appearance, other so-called missing links were excluded based on intelligence. Certain scientists and social commentators believed that microcephaly provided an ideal model for understanding the link between humans and apes. Both the scientific and commercial search for missing links perpetuated a hierarchical understanding of evolution, with white, able-bodied Europeans perceived as the highest achievement of human biological development and non-white or white "defectives" lagging behind. Despite the radicalism of Darwin's theory of natural selection, which did not support evolutionary progressivism or hierarchies, such assumptions pervaded much of the evolutionary sciences well into the twentieth century.

When Dubois set out in search of an intermediary between humans and apes, however, he was not looking for living organisms. In fact, though a student of Haeckel's, Dubois voiced dissatisfaction with morphology, in that it provided only indirect evidence of human evolution and origins. Inspired by advances in the profession of archaeology, Dubois embarked on an empirically minded search for human fossils. This was the beginning of a new profession, which would later be known as paleoanthropology whereby scientists—many of them physician-anatomists—studied human fossil remains as a means of providing "raw data" that could offer a material testimony of the past, especially concerning human origins.[26]

Dubois began his exploration in 1891 in central Java, a site that offered an abundance of caves, the customary site for hominid excavations. After a few unsuccessful months, he moved the exploration to open territory, near the River Solo, where he and his crew unearthed a large collection of vertebrate fossils from the early Pleistocene age. Among this collection, Dubois spotted a skullcap and molar that were more "humanlike than that of any known anthropoid."[27] A year later and near the same site, the crew discovered a fossilized thigh bone, a femur almost completely human in its characteristics. Dubois concluded that the three fossilized finds belonged to the same species, which, in his 1894 published reports, he named *Pithecanthropus erectus*, the same "ape-man" designation that Haeckel had used.[28]

While there had been earlier discoveries of hominid bones—the Neanderthal fossils in 1856 and the Cro-Magnon remains in 1868—Dubois was the first to claim to have discovered a transitional species that stood between human and ape. Most biologists discounted the Neanderthal and Cro-Magnon fossils as evidence for human evolution, assuming instead that they came from an era when the species had already been humanized; that both represented an earlier version of the human race—either ancient or still extant—that were less advanced than contemporary Europeans.[29] Thomas Huxley, the avid Darwinian who argued for the common ancestry of apes and humans in his 1863 *Man's Place in Nature*, contended that "Neanderthal Man" belonged to a "primitive, but completely human race, comparable with the Australian Aboriginals."[30] Rudolf Virchow thought the bones were diseased; Carl Vogt thought the remains were those of an "idiot"; Haeckel ignored the fossils altogether, at least until the twentieth century. Dubois, too, believed that Neanderthal could be "accommodated within the broad range of modern races."[31]

The discovery of *Pithecanthropus erectus* would change this, making biologists rethink the place and utility of human fossils as proof of human evolution. Dubois's discovery also rekindled the question of whether the human brain or upright posture had evolved first.[32] To Dubois, the fossilized femur proved the "gibbon thesis," which Haeckel and others had promulgated in the late nineteenth century. Contrary to the idea that humans descended from the great apes of Africa, Haeckel, Dubois, and others believed that humankind evolved from the gibbons of Southeast Asia since they had the ability, albeit limited, to walk upright. Dubois considered his fossil find to be incontrovertible evidence that the sequence of human evolution began first with upright posture, then increased brain development.

Upon his return to Europe in 1895, Dubois regularly displayed the *Pithecanthropus* fossils to the international scientific community, trying to convince his colleagues of the import of his discovery. Aside from Haeckel and Gustav Schwalbe, many on the European continent voiced skepticism. Some contested the transitional status of the find, with certain scientists claiming that they were the bones of a gibbon-like ape and others that it was a primitive human, much like the Neanderthal find. Others questioned Dubois's assumption that the bones belonged to the same species. Dubois would find his greatest support, however, in the English-speaking world, specifically in Arthur Keith, physician-anatomist and conservator of the Museum of the Royal College of Surgeons, and among American paleontologists and anthropologists.

Keith was primed to be a Dubois supporter. Before he became conservator in 1908, he, like Dubois, traveled to Southeast Asia, serving as a medical officer for a mining company in Siam between 1889 and 1891. While his primary purpose was to dissect gibbons and other native primates in order to study malaria, Keith's interest as an evolutionist was piqued. Carefully observing the orthograde gait characteristic of gibbons, Keith concluded that "the history of many of man's postural adaptations had to be traced back to the evolution of the gibbon" and that it was through this research that one could learn about the "chief postural modification of man's body."[33] Keith believed that the unique brachiating behavior of the gibbons—their arboreal locomotion, swinging entirely with their arms from tree limb to tree limb—was a precursor to human locomotion; as an anatomist, he saw a similarity in the anatomical erectness and verticality in both kinds of movement and both species. According to Keith, evolution involved four stages that roughly moved from pronograde monkeys, to orthograde gibbons, to large-bodied orthograde apes, to humans. The early stages of evolution involved the upper limbs and associated anatomy, while the latter stages required changes predominantly in the lower limbs.[34]

When Keith was asked to respond to Dubois's findings less than a decade after his own return, he found the evidence highly compelling. Remarking on his own investigation of the fossil remains, Keith wrote: "It seems to me . . . highly probably [sic] that the frame of man reached its perfection for pedal progression long before his brain attained its present complex structure."[35] He continued, saying that the *Pithecanthropus* race "were distinctly and considerably smaller-brained than the great majority of men that now people the earth's surface."[36] In conclusion, Keith wrote that Dubois's fossils "confirmed the theory of common descent." In other words, *Pithecanthropus erectus*, an upright, cranially underdeveloped ape-man, should be understood as an early ancestor of *Homo sapiens*, a category that had been reserved for wise, thinking humans.[37]

A year later, Yale University's Othniel C. Marsh, who examined the fossil specimens at the International Congress of Zoologists in Leyden, also heralded Dubois's discovery, saying that it was of "first importance to the scientific world."[38] Marsh dismissed the criticism of his European colleagues, suspecting that they did not want to believe that human ancestry began with a small-brained gibbon. Marsh concluded, with Dubois, that the "antiquity of man extends back into the Tertiary, and his affinities with the higher apes become much nearer than has hitherto been supposed."[39] Later, Aleš Hrdlička, the first curator of physical anthropology of the U.S. National Museum (now the Smithsonian), would proclaim that Dubois's discovery was "universally acknowledged as one of great importance," especially among American anthropologists.[40]

Keith had great influence on the development of American physical anthropology, especially in the early twentieth century. Immediately before assuming his position as professor of anthropology at Harvard, Earnest Hooton spent two years with Keith at his Royal College museum laboratory where Hooton learned the science of investigating and interpreting skeletal remains of prehistoric populations. From that point forward, Hooton considered himself a "disciple" of Keith's, and wished to "build American physical anthropology in the image of Arthur Keith."[41] Both men went on to publish popular and professional writings on primatology and paleoanthropology. Over the course of his career, Hooton held fast to the notion that bipedalism was the foremost trait in the evolution of humans.[42]

Keith also inspired American physician-anthropologist Dudley J. Morton, who devoted his career to researching the anatomy and biomechanics of the human foot, believing that this appendage would provide further evidence for the bipedal origin of humankind, as depicted in figure 1. Some evolutionists marveled at the human hand, which evolved to have an opposable thumb and thus could better manipulate tools and perform other fine motor activities. But for Keith and others, the foot was the last part of the body to have evolved before the brain, making it the ultimate anatomical bridge between humans and apes. Morton endorsed the view that "if 'missing links' are to be traced with complete success, the foot, far more than the skull, or the teeth, or the shins, will mark them as Monkey or as Man."[43] Human erectness, Morton declared, "was a slow development, in which great changes took place in the feet, legs, and hips."[44] The popularity of this kind of applied evolutionary thinking, especially pertaining to the human foot, pervaded U.S. medical practice and popular health beginning in the early

FIGURE 1.   A follower of the brachiating thesis of evolution, Dudley J. Morton emphasized the development of the femur bone in lower animals, culminating in the acquisition of human bipedalism. Especially important in his evolutionary staging is the close affinity between gibbons (F) and humans (G).

twentieth century. It was in this context, for example, that the U.S. military would deem flat feet a disability severe enough to preclude enlistment during the universal draft of the First World War.[45]

By the turn of the twentieth century, human posture had become a subject of scientific specialization like it never had before. Human comportment could no longer be seen as an immutable given, and more-over, those in the posture-first camp believed it was the necessary condi-tion for species differentiation and human brain development. "The evolution of our posture is not," wrote Keith, "so simple of a problem." "Such a change," he continued, "entails a *complete revolution* in the organiza-tion of the body." In his 1912 popularized guide to human anatomy titled *Man: A History of the Human Body*, Keith told his readers that "there was not a bone, muscle, joint, or organ in the whole human body but [*sic*] must have undergone a change in the evolution of our posture."[46]

Emphasizing the complete material and biological "revolution" that must have occurred to bring about erect human posture, evolutionary anatomists and physiologists began to see it as a mysteriously elaborate mechanism that required the coordination of many bodily systems to have coevolved in a harmonious and mutually supportive way. New findings from various medical and scientific specialties only seemed to confirm the complexity. In his 1906 *The Integrative Action of the Nervous System*, British neurologist Charles Sherrington determined that human beings, in order to maintain an upright stance, required a "postural re-flex," a constant muscle tone, controlled by enervation from a spinal nerve loop, which served as the "raw material of posture."[47] At the same time, physiologists extended their study of human circulation to include the vasomotor postural mechanism, a complex system for controlling human blood flow against gravitational forces.[48]

While the human origin of bipedalism was inherently a backward-looking science, its practitioners had very real and present-day con-cerns. Since most early paleoanthropologists were also trained medical doctors, they tended to emphasize the applied nature of their studies. Keith, for example, insisted that his research into posture, while intrinsi-cally of value, had a "very direct bearing on medical and surgical prac-tice." Other American physician-physical anthropologists, including Hooton, Hrdlička, and Morton, spoke of the interconnectedness of the evolutionary and medical sciences as well. Morton, who specialized in orthopedics, wrote, "The problem of evolution permeates every funda-mental branch of medical education." If the American medical profession wished to be on the cutting edge of knowledge advancement, Morton contended, it needed to turn its "attention toward the original source of man and attempt to learn the history of human development."[49]

The twentieth-century study of human posture was one way that evolutionary medicine took shape. It was a research topic that, while rooted in the study of the deep past, had pressing implications for the present. If bipedalism evolved because it was an evolutionary advantage, then it would seem to follow that human beings would still be enjoying the fruits of having developed such an advantage. Few evolutionists who studied human posture held such sanguine views.[50]

On the contrary, looking at human comportment in the modern era, most posture scientists worried instead about all the imperfections that lay before their eyes. In a lecture series devoted to the subject of posture, Keith proclaimed that "many of the obscure and distressing conditions which require [medical] treatment are in reality manifestations of a . . . derangement of the elaborate mechanism which . . . maintains the posture of the human body."[51] For Keith and his American colleagues, such "derangements" included conditions such as flat feet, hernias, scoliosis, organ prolapse, and pulmonary disease.[52]

The Making of a Poor Posture Epidemic

As the posture sciences established themselves at the turn of the century, beliefs about the evolutionary mechanism behind the acquisition of human bipedalism became more complex and varied. Commitments to different schools of evolutionary thought were commonplace as well. Certain scientists leaned more in the Lamarckian direction, believing that species change occurred through inheritance of acquired characteristics. Others inclined toward Darwin's theory of natural selection. Still others believed that they could somehow combine the two.[53] And there were even those such as Keith, who, mid-career, changed his position from espousing linear evolution (single-line ascent from apes) to variational evolution (a process of continual branching and extinction).[54]

Generally speaking, the theory of a linear sequence from ape to human fell out of favor, replaced with more racist, branching models, where certain species were believed to be wiped out by superior races.

In this context, the Neanderthals came to be seen not as a progenitor to modern Europeans but instead a "degenerate race" overtaken by fitter and more cunning races. "Given the widespread appeal and inevitability of racial conflict as a means of justifying modern empire building," writes Bowler, "it is not difficult to see this view of the Neanderthals as an unconscious extension of the imperialist theme of paleoanthropology."[55]

Race conflict became a hallmark of physical anthropology—and by extension the posture sciences. White Europeans and Anglo-Saxon Americans invariably assumed that they were culturally superior to the races that they were subjugating with their military technology and political power. Yet, at the same time, there was widespread anxiety about the ability of the white race to maintain its power and control, especially concerning physical might and biological superiority. Sociologist Edward A. Ross coined the term "race suicide" at the turn of the twentieth century, a concept that articulated a fear shared by many in the ruling class that the white Anglo-Saxon race would die out from either falling birth rates or declining physical fitness—or both. The other side of the race suicide coin, of course, was the perceived superior physical strength and high reproduction rates of non-whites.

Human posture thus became an indicator of evolutionary fitness (or lack thereof) that could not be taken for granted. Clevenger expressed this concern most pointedly when he wrote: "At present the world goes on in its blindness, apparently satisfied that everything is all right because it exists at all, ignorant of the evil consequences of apparently beneficent peculiarities, vaunting man's erectness and its advantages, while ignoring the disadvantages."[56] Working as a pathologist and professor of anatomy in Chicago, Clevenger was not concerned about just any "man," although according to language conventions at the time, the word *man* could be used to stand in for all of humankind. Instead, he had a very a particular worry: the declining birth rate among Anglo-Saxon women. The erect human posture, Clevenger wrote, "has caused the death of millions of otherwise perfectly healthy and well-formed human mothers and children."[57]

More specifically, Clevenger believed that white women had more difficulty in birthing than animals—and by extension "less evolved" non-white women—because of the former's narrow pelvises.[58] In other words, the more evolved a human, the narrower the pelvis. Quadrupeds, Clevenger contended, had "box-shaped" pelvises, which allowed for easy parturition and—essential to posture scientists—the prevention of organ prolapse. Human pelvises, by contrast, were widely flared at the top, but severely contracted at the bottom, making for a constricted pelvic outlet. The "disadvantageous" human pelvis came about from centuries of bipedalism, a position in which the organs descended farther down into the pelvic cavity, creating more demands than were intended by "nature." Whereas quadrupedal organs were protected by sturdy ribs and strong abdominal musculature, human abdominal organs were exposed and placed an inordinate burden on the pelvis. Contrary to the biblical account of the fall of Adam and Eve, whose sin was the quest for greater knowledge, Clevenger had a far more materialistic view; "Man's original sin," he wrote, "consisted in his getting on his hind legs."[59]

Clevenger's argument proved to be too radical for most posture scientists, for it seemed to undermine human superiority altogether, not to mention that it undercut both Darwin and Lamarck. Keith, for example, held a much more moderate view on the advantages and disadvantages of human posture. Posing the crucial question of whether bipedalism was, in the end, an evolutionary disadvantage, Keith wrote: "Are we to say, then, that Europeans and peoples of European origin are more liable to hernia and to uterine prolapse than more primitive peoples, such as the aboriginals of Australia, because the European body is less perfectly adapted to the upright posture?" "I think not," Keith continued. "The Australian aborigine spends his days amidst conditions very similar to those for which early man was evolved, while our conditions of living and working are of recent development, and very different from the aboriginal kind."[60] In a word, Keith blamed postural problems on civilization, rather than a defect in the evolutionary process toward bipedalism.

Despite their differences of opinion concerning the origins of the faulty human form, both Clevenger and Keith—along with many others—believed that human posture was a problem. The entire civilized Anglo-Saxon race exhibited postural weaknesses that manifested both in appearance and in disease states. While neither Keith, Clevenger, nor any of their contemporaries invoked the term "epidemic" to describe the issue, they nonetheless conceived of the problem on a population level, for, according to their logic, the entire civilized human race—or at least the Anglo-Saxon race—was affected. And hence, a poor posture epidemic was born.

The extent to which these early posture scientists were committed to eugenics and white race betterment varied from person to person and changed over time. In general, posture scientists did not speak in terms of limiting birth rates of non-white citizens or controlling heredity. But they were interested in human variation, and often spoke of such differences in terms of race. They were also committed to nationality and believed that countrywide strength would come through greater attention to physical fitness, especially among Anglo-Saxons but also among immigrants and the newly colonized. Well into the 1950s, Keith maintained that the chief driving force behind evolution was tribal and racial struggle, and he worried that "eugenicists might try to breed out the racial instinct," which he believed would stall progress.[61] Hrdlička and Hooton were early proponents of the American eugenics movement; both began their careers studying criminal and mentally ill populations, assessing whether there was a linkage between appearance and character. According to certain scholars, both had moved away from the movement by the 1930s.

What is most important to appreciate about the early posture sciences was the shared assumption that an intervention could be made concerning human evolution. Speaking about "the New American Race," Keith wrote that "no race of mankind is fixed; all are plastic." "Civilization," he continued, "never stands still; it is progressing, evolving, at a greater rate now than ever before and nowhere more rapidly than in the United States of America. Men and women have to keep pace with civilization or fall out."[62] When advising English politicians on Irish uprisings against Britain, Keith insisted on understanding the conflict in terms of evolutionary tribalism, and wrote that "we can no longer afford to be mere pawns on the chessboard of evolution; we must somehow take a hand in the game."[63] American posture scientists shared Keith's view and were even more influenced by a neo-Lamarckian belief that "living things are in charge of their own evolution."[64]

Locating the Outbreak of the Epidemic

The question of how best to intervene with failing human posture would be something that a whole host of posture crusaders would take up for decades to come. For those interested in the origins of bipedalism, however, the source of the breakdown had to be located within the mechanism of evolution itself. Or, to use the language of epidemics, the "outbreak" of the disease needed to be understood and addressed. Since the logic of epidemics is customarily applied to acute, deadly disease, the "outbreak" is often a vector or "patient zero" of the near past—an event that immediately precedes the spread of the disease. Yet the poor posture epidemic is unique in that the outbreak was assumed to have occurred in the distant past.

Of course, there was disagreement as to how far in the distant past one had to go in order to understand when human bipedalism first presented health problems. Some adopted a viewpoint close to Clevenger's— namely, as soon as human beings stood erect, anatomical and physiological problems started. This line of thinking easily lent itself to the study of human–primate comparative anatomy, and sometimes resulted in the rather radical claim that human beings were not originally intended to be bipedal. In this view, bipedalism was somehow unnatural.[65]

Reporting on the new studies being conducted in the labs of physiologists, anatomists, and neurologists in the early twentieth century, one American columnist wrote:

> The fundamental difficulty about standing up straight . . . is that the human body was never designed by nature for anything of the sort. It was originally constructed to be used on all fours, like cats and dogs. The internal organs were designed to hang freely from a horizontal backbone, like clothes from a line. The legs and arms were designed to serve as supports at the corners of a four-legged body, like a table. The spine was intended to be a girder, not a column.[66]

These kind of analogies and descriptions helped to popularize the idea that erect posture was not something that could be taken for granted, but rather something to which human beings needed to attend through self-care, exercise, and constant vigilance over the physical body.

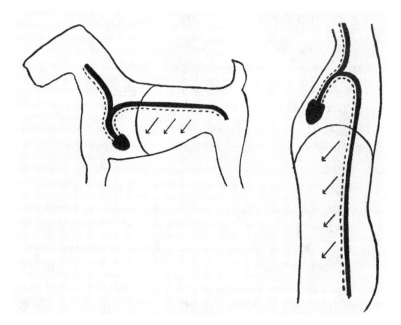

FIGURE 2.   In a university textbook, the reader is invited to compare animal and human, quadruped and biped. Of particular note are the arrows in the abdominal region, which are there to indicate how gravity affects internal organs. In the human model, the arrows point downward, suggesting a problem with descending organs, which the abdominal muscles must work to combat. The arrows in the canine's abdomen, by contrast, point in an upward direction relative to the heart. Anatomists reasoned that it was because of this that quadrupeds did not suffer from organ prolapse, hernias, and other physiological disfunctions.

The use of animals to explain the problems of human posture appeared not only in newspapers and magazines, but also in university textbooks on physiology, anatomy, and hygiene. Physical educators at Smith College learned about anatomy and hygiene from K. Frances Scott's *A College Course in Hygiene*, which illustrated the strains of uprightness through the use of schematics comparing canines to humans (see figure 2).[67]

From the outset, evolutionary theory inherently blurred the animal–human boundary.[68] It was never clear where the line between human and animal was or should be drawn. *Pithecanthropus* threw the issue of the human–animal divide into starker relief, for the fossils were part simian and part human, part ape and part man. Whatever one concluded about Dubois's find, *Pithecanthropus* strengthened the human biological ties to animals, and led certain anatomists and physiologists to contend that animals—in particular quadrupeds—had certain anatomical advantages over humans.

Those who made such arguments insisted that the weaknesses left from bipedalism needed to be strengthened and that humans had to make a conscious effort to do so. While the early twentieth century marked the beginning of myriad movements aimed at enhancing physical strength, fitness, and the outdoor life—organized sports, muscular Christianity, the Playground Association of America, the Boy Scouts of America and Girl Scouts of the United States—posture scientists and proponents saw themselves as addressing fitness at the most rudimentary, primordial level. Rather than rely on elaborate gym apparatuses or the complexity of sport, they focused on the "fundamentals" of movement, or "functional exercises," which required not only an intimate knowledge of evolutionary anatomy and physiology but also a critical eye for the (usually poor) physical habits promoted by everyday modern life.[69]

But what did foundational exercises look like? One approach was to mimic nonhuman animal movements so as to strengthen the trunk and shoulder girdle and ultimately relieve the increased intra-abdominal pressure that came with an upright stance. This presumed solution was quickly picked up by experts outside of the evolutionary sciences—by physical educators, medical doctors, and health culturists, many of whom preached the gospel of good posture through the popular press. Readers of white popular beauty and parenting magazines were told that "children enjoy imitating animals in their [posture] practice and so do grownups."[70] Other columnists reporting on the posture sciences advised that exercises such as the "panther crawl," "crab walking," "frog hopping," and "horse trotting" could strengthen key muscles needed to maintain an upright stance.[71] Engaging in such activities, however, was largely intended for white, able-bodied Americans. For non-whites and the disabled, adopting animal-like movements and behaviors risked furthering prejudicial treatment that they already faced, much of which rested on the ruling class belief that they were less than fully human.[72]

Because of the theory of recapitulation—the idea that infant and toddler development mimicked the evolutionary stages toward upright posture—there was a particular emphasis on children and exercise. In a cross-country scientific tour of America at the turn of the twentieth century, Hrdlička studied infant development and movement of all races, native and non-native to the United States. During a stint in Mexico, he noticed a two-year-old native Indian girl running around on all fours, but in a manner closer to a "monkey" because she did not progress on her hands and knees (which Hrdlička associated with non-poor white babies), but rather on her hands and feet.[73]

When he assumed the post as curator of the Smithsonian, Hrdlička continued to collect observations on these "quadruped progressors" from other anthropologists working in the field across the globe. He also received letters from mothers who described how their hands-and-feet-walking children progressed quickly, arriving "at where they want to go in half the time taken by those who crept on their knees in the usual way."[74] Hrdlička found that these children were not only faster in getting around but also "developed muscles that made them . . . more competent walkers."[75]

Infants and toddlers who stood and walked early, Hrdlička maintained, were at a disadvantage, having missed a crucial opportunity to "rehearse" the evolutionary process, with quadrupedalism serving as an essential stage of musculoskeletal development that led to a body better adapted to bipedalism. This same kind of logic was used when Progressive-era playground experts installed so-called "monkey bars"— in hanging on these bars like a brachiating gibbon, human children could rehearse and strengthen earlier evolutionary stages. Of course, evolutionary rehearsal could not persist throughout life or outside the bounds of prescribed exercise. Older children and adults who moved on all fours often were deemed to be "savage"-like, examples of postural atavism, humans who were not quite fully human, stuck in an evolutionary stage before bipedalism.[76]

Other posture scientists blamed civilization and modern life for the poor posture epidemic. Keith was a proponent of this view. "Beyond a doubt," he claimed, "civilization is submitting the human body to a vast and critical experiment."[77] Keith believed that the first humans to arrive in Western Europe "were hunters; their bodies unaccustomed to either manual labor or an indoor life." Under the stress of civilization, which Keith believed to have begun eight thousand years before his time, somewhere in Egypt or the larger Mesopotamian region, the "hunter's body" had to operate under "modern needs."[78] Working for the Ministry of National Service, assessing mandatory physical exams of all British Army recruits, Keith found numerical proof for the physical strains of modern life. According to early results, nearly one-third of recruits suffered from hernias and only a slightly smaller fraction from flat feet.[79] Keith blamed both the demands of industry—repetitive muscular labor—and the sedentary lives of city dwellers, conditions that affected the working poor, the new white-collar professionals, and even the monied elite. By contrast, he insisted, "such breakdowns . . . do not occur with this frequency among hunting peoples."[80]

Keith, like many anthropologists of his era, believed that we could see the unfolding of human history among living populations, especially in regions of the world that had been colonized by Europe and the United States. Physical anthropologists often looked to the Australian Aborigines or "bushmen" as representative of "primitive," pre-civilized life. But myriad other such populations existed around the globe—in Africa, South America, and in North America where Native Americans and Eskimos piqued the interest of U.S. anthropologists. Many historians have demonstrated the racialized underpinnings of the concept of civilization, an attribute that white imperialists believed they possessed and their non-white subjects did not.[81]

It is important to note, however, that those in the posture sciences believed that so-called primitive peoples had a physical advantage over their civilized brethren. This did not mean, of course, that white scientists such as Keith believed that aboriginal populations were racially superior in all regards. Civilization was a sign of cultural advancement, and thus the white man was placed at the top of the evolutionary scale. Still, Keith and his colleagues worried that the unnatural physical demands of civilization could also be the undoing of white superiority.

The worry that civilization led to physical weakness was pervasive in the early twentieth century—it can be found in scientific writings, news dailies, literature, and film. The articulation of the fear as it affected human posture arose in several specific ways. First, modern dress and lifestyle was blamed as a culprit that promoted unnatural strains on the human body. In this regard, posture scientists took up the mantle from earlier physical culturists who were part of the anti-corset and physiologically living movement. But posture advocates took the movement even further, criticizing footwear, trousers, chairs, homelife, and workplace conditions that impacted all classes of people. They worried not only about an increase in overall sedentariness brought about by mechanized travel, mandatory schooling, and the rise of a leisure class, but also about the muscular demands placed on industrial workers. "Of the present manhood in Britain," wrote Keith, "half earns its bread by muscular labor; the other half lives sedentary lives. . . . Our forefathers," by which he meant primitive hunters, "were unaccustomed to either manual labor or an indoor life."[82]

Visions and Politics of Primitivism

In criticizing the so-called modern life, posture advocates simultaneously promoted idealized visions of "primitive life," men and women who, living in otherwise adverse environments, exhibited admirable physical strength, muscular form, and agility. Dr. J. Albright Jones of Philadelphia told his colleagues that "primitive Man, living an outdoor life, not hampered by clothing and securing much more physical exercise, usually manifested good posture." Americans, he believed, could witness such primitively ideal human comportment among Native Americans. As Jones put it, "The Indians have always been symbolic of grace and beauty of form." The problem with civilization was that "human spines were not evolved to withstand the monotonous and trying posture entailed by modern environment and activity."[83]

Writing for the *Ladies' Home Journal*, Dr. Barbara Beattie advised her readers that the easiest way to accomplish good posture was to "go native." The indigenous woman of West Africa, she wrote, "carries her household belongings poised on her head, burdens weighing sixty pounds or more." As a result, she continued, "the primitive woman has a carriage so beautiful that she might well become the envy of her more cultivated sisters."[84]

Posture experts appropriated the head-carrying customs of indigenous peoples, turning the practice into a prescribed exercise as opposed to a labor of necessity. Books, instead of food, water, kindling, and other basic necessities, became the preferred objects for prescribed head-carrying. Such an exercise reinforced the appearance of whiteness, education, and civilization while simultaneously correcting the biological weaknesses wrought by such an existence. The media capitalized on this imagery. Parenting magazines, women's beauty guides, and popular health journals of the time are filled with photographs and images of children and adults carrying books on their heads in a concerted effort to promote posture perfection.[85] Joseph Pilates, the famed physical culturist who would go on to establish an eponymous exercise regime of improving "nature's corset" (i.e., the abdominal musculature), told his clients that, when they had leisure time at the beach, they should "sit crossed-leg in the sand . . . like those people of the East," who in his estimation "had the best of all sitting postures." "Sit upright on the beach, legs crossed" he advised, "with a book or sandbag on your head as you read or sew, pulling in the abdomen."[86]

Middle-class African Americans themselves engaged in a variation of this cultural movement as well. For example, a 1913 article in the *Chicago Defender* held up "the Zulu women" as having a "noble carriage, as the result of carrying burdens on their heads."[87] Such reportage could have been taken on by the Black press as an element of race pride, but this would not happen until much later in the century.[88] Race could cut both ways in the posture sciences. As Jones indicated above, Native Americans, in their primitiveness, served as models of good posture, just as the "head-carrying Africans" or "cross-legged Asians" did. But while such peoples, presumably untouched by civilization, exhibited strength in form, they also represented "savage" inclinations in mind and behavior, which white scientists—in their barely disguised fear of being physically dominated by such other races—construed as evolutionarily regressive. This was particularly true in the scientific establishment's judgment of African Americans who, throughout the early twentieth century, were often depicted as ape-like in museum exhibits, popular magazines, and other graphic illustrations.[89]

The Black elite did little to combat such portrayals, and indeed seemed to confirm such prejudices, especially when addressing posture and race.[90] Dr. Algernon Jackson, a prominent African American physician who, with Booker T. Washington, helped establish the National Negro Health Movement, complained that young African Americans "too often slump along, stoop-shouldered and walk with a careless, lazy sort of dragging gait."[91] A regular health columnist for Baltimore's *Afro-American* newspaper, he believed in "race building through body building."[92] He depicted poor posture among Black Americans as a "racial habit" that, if left unchecked, would lead to the extinction of his race. "A slumping, stoop-shouldered hollow-chested people," he warned, "are destined to failure in meeting opposition and developing that un-conquering [*sic*] vigor which present day civilization demands."[93] Jackson was not an outlier.

Concern about African Americans' posture was a regular feature in the Black press throughout the early twentieth century.[94] At the urging of Booker T. Washington, the U.S. Public Health Service (USPHS) established the "National Negro Health Week," a preventative health campaign targeting Black Americans. One goal of the health week, as depicted in figure 3, was to get Black households to encourage daily hygiene rituals, devoting as much time to teeth-brushing as to posture checks. The stakes for African Americans were quite different from those of the white scientists engaged in the same pursuit, but it is important to note the complexity of the intersectionality of race and class in the history of the posture sciences. Middle-class African Americans, interested in "uplift," had plenty of motivation to distinguish themselves from undereducated Blacks, who were, in certain circles, depicted as evolutionarily closer to apes than to white men.

As the poor posture epidemic became increasingly popularized, both explanations for the outbreak—whether it started at the moment of the

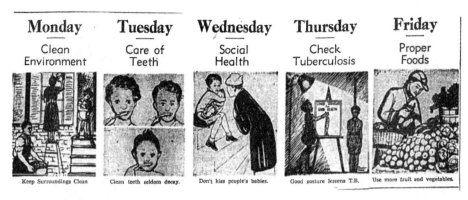

FIGURE 3.   At the urging of Booker T. Washington and the Tuskegee Institute's findings of the poor health status among Black Americans, the U.S. Public Health Service instituted a "National Negro Health Week" beginning in 1915. Posture checks became an integral part of the campaign. According to this comic strip, which assigned one health-promoting behavior to each day of the week, Black Americans were supposed to pay particular attention to cultivating an upright posture on Thursdays.

evolutionary leap from ape to human or later with the birth of ancient civilization—became largely conflated, and for the most part did not really matter. To be sure, the distinction mattered among evolutionary scientists, for the two explanations pointed to very different problematics concerning human evolution, natural selection, and adaptation. But outside the circle of experts who focused exclusively on the particulars of evolutionary science, the question of exactly when and how the epidemic began eventually fell by the wayside.

The shared belief that modern human bodies—Anglo-Saxon men, civilized women, and also non-whites—on a population level were somehow failing, ill-equipped to meet the demands of modern life, was enough to bind together a whole host of Progressive-era professionals from medicine, engineering, physical education, and the actuarial sciences. These professionals made themselves instrumental in the effort to combat the poor posture epidemic not only by medicalizing the effects of poor posture, but also by denaturalizing the human form. Since the new posture sciences challenged the long-held belief that human beings were naturally endowed with uprightness, some form of artifice was needed to correct poor form. This assumption opened up a space for professional, material, and commercial interventions of all kinds, for the responsibility for maintaining a proper erect attitude now fell on the individual, who needed to be educated and disciplined to consciously work at adopting good physical form.

CHAPTER 2

Posture Epidemic

AT THE TURN of the twentieth century, the United States was an expanding nation. Although Fredrick Jackson Turner pronounced the closing of the western frontier, the nation's eastern and western ports teemed with immigrants from Asia as well as Southern and Eastern Europe. This was also an era marked by mass internal migration, from country to city as well as from South to North, with African Americans looking to earn a living wage and to escape the violence of Jim Crow. At the same time, the U.S. foreign policy establishment began to utilize military and economic power to acquire new lands beyond the contiguous states. Between 1890 and 1920, the United States took control over Hawaii, the Philippines, Guam, Puerto Rico, and the U.S. Virgin Islands; the federal government would also grow a standing army of nearly 4 million men during the Great War, sending drafted and volunteer soldiers overseas to defend foreign soil.

Becoming a global power required many things—money, a strong military, natural resources, and political stability. But it also necessitated healthy and fit citizens, or "national health," to use a favored terminology of political leaders. By the turn of the twentieth century many industrialized nations accrued and publicized vital statistics, including the metrics of infant mortality, life expectancy, and disease rates. These data served both geopolitical ends (a yardstick to compare nation-states) and domestic ones. The quantification of the nation's populace normalized the physical examination, making it a mandatory ritual for any citizen who wished to participate in the public sphere. By the early twentieth century, U.S. citizens (including would-be citizens) were forced to undergo bodily inspections in order to obtain and retain employment, to receive an education in the nation's public and private schools, to fulfill the requirement of military service, to secure life and health insurance policies, and, in the case of the newly immigrated, to gain entry into the country. It was within this context that posture screening became a regular part of American life.

From the time of the First World War until the early 1930s, most large-scale posture studies found that anywhere from 60 to 80 percent of Americans suffered from faulty form. This chapter reveals the myriad material, financial, intellectual, and political investments that went into making this number an objective fact. Establishing a high incidence rate of faulty form required not only a population of docile bodies, to invoke Michel Foucault, but also standardized tools to measure and build consensus on what constituted good and bad human form. It required trained professionals who could conduct the exams, convert visual findings into numerical data, and communicate those results to the wider public.

A notable feature of this normalization effort was that the standard of "good" posture applied equally to all races, genders, and ethnicities. Considering the history of the anatomical sciences and anthropometry, in which the actors in this chapter were trained, it is surprising that they did not create different posture scales according to racial and gender types. Additionally, the field of physical education, which also played a significant role in the establishment of posture norms, was known for its adamant sex segregation in theory and in practice. The anatomical and physiological differences between the sexes and races, it was believed, were many, from respiratory capacity and pelvic shape to overall bodily stature and muscular strength. Such dissimilarities became the rationale for sex-specific sporting events and competitions. The professionals interested in establishing posture standards, however, understood upright standing to be the most basic of human features, and thus universal.[1] On the one hand, this belief could be (and was) used to challenge sexist notions of ability. On the other, it denied human anatomical variety or even the fluctuations in an individual's posture over the course of that person's day, let alone a lifetime.

To get the public and fellow professionals to see the prevalence of poor posture was a significant challenge. In a commissioned report to President Herbert Hoover in 1932, Dr. John Carnett, surgeon and vice dean of the medical school at the University of Pennsylvania, wrote, "Faulty body mechanics, being present in the great majority of children and adults, is so very common that the average observer does not appreciate its abnormality."[2] Carnett's argument—that a serious condition threatened the nation's health, if only people could see it—fit neatly with contemporaneous efforts in public health, preventive medicine, and bacteriology. Educating the public on how to control the spread of invisible germs and microscopic organisms required a campaign replete with visual media to convince laypeople of the threat.[3] With the introduction of germ theory also came a new category of patient, the "healthy carrier," who posed a danger equal to—if not greater than—a person who felt and looked sick.[4]

While posture experts never used germ theory as a causative reason for the existence of bad posture (they appealed to the evolutionary sciences for this), they did believe that poor posture made individuals more prone to acquiring diseases such as tuberculosis. Posture surveillance and improvement, proponents maintained, thus promised a way to *prevent* both contagious disease and disability, the latter of which was particularly appealing to industry owners who wished to steer clear of the newly legislated workman's compensation and to the Republican-controlled government of the 1920s and early 1930s, which looked for ways to minimize federal spending.[5]

Creating programs of prevention necessitated agreement that a problem existed in the first place. This chapter looks at the health experts who created a system to quantify posture, which they then used to convince employers, school administrators, and various branches of the federal government that poor bodily alignment was not only a widespread problem, but also a health hazard that could lead to America's demise. This was not a simple top-down process, for what made the new posture sciences so compelling was that they tapped into already existing social and cultural anxieties about post-Victorian modern life. The sight of depleted, slumping bodies amplified a sense of alarm about changes affecting large swaths of the population, from universal schooling and mechanized forms of travel to women's suffrage, assembly-line production, labor strikes, urban overcrowding, and military readiness.[6]

Of course, large-scale population health studies require human subjects who agree or are coerced to undergo such examinations.[7] Only a few sources from this time period offer glimpses into the perspective of what it was like to undergo such physical scrutiny. The subjective experience of the new forms of posture surveillance is varied and situational. Some of the earliest clinical posture studies were conducted on laborers and "pink-collar" workers (nurses and teachers) who came to treating physicians with complaints of foot and back pain.[8] In a 1921 study conducted by the U.S. Women's Bureau, Nelle Swartz noted how factory workers modified their own physical environments so as to minimize bodily ailments and strain; describing the self-made seating accommodations of one factory, she wrote, "Backs have been put on or knocked off, arms or cushions have been added, a box or barrel has been made into a seat, a chair seat has been raised by using a block or strip of wood."[9]

For the New York City textile worker who suffered from chronic back strain or the reluctant military draftee who was found to have flat feet and was thus exempted from military duty, a diagnosis of poor posture could mean improved work conditions or a welcome deferment. Others, however, had no idea that they had a health-threatening posture problem until they were told so by the examining expert and then mandated to undertake body alignment training, or worse, barred from employment or refused access to the country.[10] For the symptomatic, a poor posture diagnosis could legitimize suffering, but for the otherwise healthy, it could bring about unwanted policing and medical intervention.

Defining Good Posture

Physical educator Jessie Bancroft founded and named herself president of the American Posture League (APL). Like many female physical educators of her generation, Bancroft did not have a formal degree in the subject. She was first exposed to posture exercises in her adolescent years. A self-proclaimed invalid who grew up in a remote region of the upper Midwest, Bancroft sought the help of Mrs. Jenness-Miller, an anti-corsetry reformer who held parlor classes on hygiene and recumbent exercises, providing instruction on how to attain good carriage.[11]

Having benefited from this form of therapeutic exercise, Bancroft devoted her young adulthood to pursuing physical fitness as a career path. Following Jenness-Miller's example, Bancroft provided traveling lectures at private schools, churches, and homes throughout the Midwest. Entirely self-taught, she based her lesson plans on medial books (*Gray's Anatomy* and H. Newell Martin's *The Human Body*) and popular health culture journals. Wishing for a more formal education, Bancroft moved to Boston in 1891 to take a summer instruction course with Harvard's Dr. Dudley A. Sargent, the leading U.S. authority on physical training at the time. There, Bancroft would become acquainted with orthopedic surgeons, gymnastic trainers, YMCA leaders, industrial physicians, and public school educators, many of whom would later join her in the posture crusade.[12]

Although Bancroft would go on to enjoy great professional success in the field that would come to be known as physical education, she regularly portrayed herself as an outsider, a frontier "Westerner" (she was born and raised in Winona, Minnesota) who did not abide by any one school of thought. In the late nineteenth century, America's fitness culture—especially in schools and universities—was dominated by European systems of gymnastics on the one hand and military drill on the other. In the wake of the U.S. Civil War, when early anthropometric studies suggested that Union Army recruits were weaker, shorter, and sicklier than European men, many public and private schools hired military officers and trained experts in the German or Swedish systems of physical calisthenics to strengthen young citizens. Regardless of orientation or country of origin, many of the nineteenth-century systems of physical training required mass drill formations, with regiments of students performing choreographed physical stunts in response to the vocal commands of a teacher or officer.[13]

When she assumed the newly created position of assistant director of physical education with the New York City public schools in 1904,

Bancroft became a leader in the "new physical education" movement. Influenced by John Dewey, William James, and G. Stanley Hall, the new physical education movement emphasized the psychosocial importance of play, swimming, games, and outdoor life.[14] Dewey undertook postural work himself, taking private lessons from F. Matthias Alexander, an Australian Shakespearean actor, who created a method to help clients develop a better kinesthetic awareness in order to enhance upright form (what is today known as the "Alexander technique").[15] Meanwhile, Harvard philosopher William James insisted that "posturegymnastics," as he called it, had the effect of waking up "deeper and deeper levels of will and moral and intellectual power."[16] James came to this conclusion after undertaking training in hatha yoga, a decision inspired by his meetings with Swami Vivekananda.[17] Proponents such as Dewey and James denounced the artifice of military drill and European systems of calisthenics in favor of a more "natural" form of exercise.[18]

Bancroft too understood posture health to be foundational to a more naturalistic approach to physical fitness. Persuaded by the findings of her colleagues in evolutionary medicine, she insisted that the human spine was essentially a "quadrupedal spine set on its end," and that to be a healthy biped required attention to and support of the muscles surrounding the spine.[19] Citing the work of Sir Arthur Keith, Bancroft maintained that the maintenance of proper form was not something that could be left to reflex or innate ability. Posture was not, she wrote, "an inherent characteristic for which nature cares without any special assistance."[20]

While she advocated the undertaking of posture work for an individual's entire life span, Bancroft emphasized the particular importance of training the nation's youth. In addition to having more pliable muscles, bones, and nerves, children, she argued, had the advantage of being temporally closer to the evolutionary process, the phases of which were recapitulated from infant to toddler, when upright stance was attained for the first time.[21] In other words, if children were trained early, faulty posture habits could be avoided altogether. Appealing to nature, Bancroft emphasized the acquisition and repetition of childhood developmental markers, incorporating activities such as crawling, jumping, standing on one foot, and climbing into her lesson plans.

Despite the attempt to create homogeneity in human stance, posture experts such as Bancroft claimed that the goal was to focus on the individual, emulating clinical work more than collective military drill. The interest in individualism reflected larger social concerns in the Progressive era. Physical educators argued that compulsory military exercises or European-style gymnastics would undermine American democracy and injure young bodies if adopted in the public schools.[22] Such mass activities, one educator contended, would "develop unquestioning obedience" and thus a student's "sense of right" and "conscience" would be left to languish.[23] Individualized posture work, on the other hand, was understood to be compatible with creating independent-minded citizens.

Bancroft and her physical educator colleagues understood themselves to be vital to the Americanization process, molding the bodies of would-be citizens in the new U.S. territories of the Philippines and Puerto Rico, at boarding schools for Native Americans, and in urban public schools where the children of immigrants predominated. Physical educators worked alongside general teachers, providing lessons in proper handwashing, table manners, English language instruction, and posture work.[24] Recounting her early days in New York City, Bancroft saw herself much like other Progressive-era reformers at the time. "The 'gilt-edged' neighborhoods where chauffeurs and nursemaids waited to take children home after school did not interest me deeply," Bancroft writes; she was drawn instead to "the crowded tenement sections, where undeveloped standards of living, ethical sense, personal ambitions and ideals, and an embittered attitude toward all government, formed the basic problems for our country."[25] To the new immigrants from Southern and Eastern Europe, the public school system, according to Bancroft, "stood for all law, order, courts, and even the Federal Government itself."[26] She felt that public schools, and specifically her work in physical education, was "democracy in the making."[27]

But what kind of posture was "good" for a democracy, and more specifically American democracy? What should the public face—or, in this case, stance—of the United States be? As I indicate above, Bancroft and her colleagues associated military postures with "Old World" authoritarianism. Accordingly, such bodily comportment came to be seen as harmful not only to American politics, but also to the physical health of the nation's citizens. Dr. R. Tait McKenzie, physician and professor of physical education at the University of Pennsylvania from 1904 to 1930, believed that military postures "tended to make students angular in their movements."[28] A student of Sargent (and Bancroft's peer), McKenzie learned from his mentor that the "constrained positions and closely localized movements" demanded by military formations did "not afford the essential requisites for developing the muscles, and improving the respiration and circulation"; military drill, in other words, increased defects rather than ameliorated them.[29]

An activity conventionally taught to men by men, military drill enforced gender segregation and promoted sex-based biological determinism, both of which Bancroft and her closest female APL colleagues resisted. When in the late 1880s, female students at the University of Nebraska fought to gain access to courses in military drill, the ROTC commandant, Lt. John J. Pershing (who would later become the commander of the American Expeditionary Forces in World War I), agreed to teach co-eds fencing and light marching, but not the masculine-coded activities of push-ups, chin-ups, and weighted marching.[30] Bancroft wished to dismantle this approach to physical fitness through posture training. For starters, she contended that human posture standards applied equally to both men and women. In this regard, she stood in opposition to professional contemporaries such as Havelock Ellis, who claimed that erect posture was, for evolutionary reasons, more difficult for women than men to attain.[31] Bancroft also advocated for gender parity within the posture science profession, a fact best exemplified by the equal representation of professional men and women in the APL.

The other primary threat to American democracy was the monied elite, aristocrats who indulged in new, youth-driven forms of leisure and dress, behaviors that the professional middle class associated with loose mores and ill health. The educated middle class detested the "debutante slouch," a term used to describe a high-class, fashionable person who telegraphed their leisured status by standing and walking with a slight stoop and a jutting of the hips.[32]

Striking a relaxed pose was not an entirely new custom of the monied class. Aristocrats of the eighteenth and early nineteenth centuries, for example, often held themselves in a more languid stature, with one foot forward and a graceful backward lean of the upper body. John Hamilton Moore, author of America's most popular etiquette guide of the early republic and Jacksonian era, insisted that only "the awkward and ill-bred . . . sit bolt upright." In his book titled *The Young Gentleman and Lady's Monitor, and English Teacher's Assistant,* he advised his reader that "A man of fashion makes himself easy, and appears so, by leaning grace-fully."[33] While the upper crust of society had long flouted common rules regarding bodily discipline, the debutante slouch represented a particular kind of menace in the early twentieth century. The newly established department stores and mail-order catalogues all tended to glamorize the debutante slouch, using it to sell a range of commercial products. In a way, the fashionable slouch threatened to become a commodity in itself, a cost-free way to climb the social ladder.

Along with concern about the influence of the debutante slouch was a worry about the slumping, disorderly bodies of the working poor and non-white Americans. To reformers like Bancroft, overworked European immigrants bore the scars of long hours of labor and poor living conditions in their outward appearance. Historian Daniel E. Bender contends that Progressives fretted over the debased lives of European immigrant workers, fearing that such work and living conditions would lead to an "overall racial degeneration among workers who shared at least a visible whiteness with the reforming class."[34] With the second wave of immigration, whiteness became a fractured, heterogeneous category where other characteristics besides skin color came to determine racial categories.[35] "Race was etched on immigrant workers' bodies," Bender writes, "visible in their stooped, diseased frames."[36]

In other words, posture served as a marker of social status similar to skin color. Both attributes—human carriage and hue—contributed to Progressive-era schemas of racial belonging, hierarchies, and able-bodiedness. Posture surveillance was an intellectual and material space where scientific racism and scientific ableism met. One did not exist without the other.

The hierarchies based on color and carriage reflected multiple social, cultural, and political concerns. Take, for example, how the APL propped up Native Americans as posture ideals. Much like Arthur Keith, Aleš Hrdlička, and Dudley J. Morton, who argued that colonized aboriginals should serve as physical models for white Anglo-Saxon slouches to mimic, the APL did much the same. Native Americans were often assumed to be, as one physical educator put it, "naturally endowed with good posture," because they "lived active, outdoor lives," even though in reality many indigenous children were being removed from their lands and families, with children forcibly sent to boarding schools.[37] Bancroft's mentor, Dudley A. Sargent, articulated the idealization of the "primitive" other and posture in the following way: "Many of these bodily weaknesses and imperfections . . . have arisen largely from civilized man's neglect to care for the form and strength of his bodily mechanism as an African Zulu or Sandwich Islander would do."[38]

This belief became materialized in an award that the APL would bestow on students with grade-A posture: a posture pin made in the image of a Lenape man, as pictured in figure 4. The APL commissioned McKenzie, who, in addition to his career in medicine and physical education, created sculptures, medallions, and reliefs of athletes, many of which still adorn the University of Pennsylvania's campus. Although most of McKenzie's work celebrated the white male body, he chose the symbol of an indigenous man for the posture pin. Scholar Shari Huhndorf writes of the many ways during this time period when men like McKenzie looked to Indians as a source of rejuvenation in an increasingly degenerate white society. "Since most observers believed that 'authentic' Natives no longer existed as a historical presence," she argues, "Indians came to be regarded as disembodied symbols and, ironically, symbols of white American identity."[39]

While primitive African postures—particularly in reference to head-carrying—would be glorified, the American Black body was often vilified. One physical educator teaching at a private all-girls school admonished slouching students, saying, "You American girls walk like a lot of slaves. There is nothing of the queenly pose, the power, the upright carriage which ought to mark freeborn women."[40] Black elites

COPYRIGHTED, 1917, BY
AMERICAN POSTURE LEAGUE

GOOD POSTURE PINS

ISSUED BY

THE AMERICAN POSTURE LEAGUE

· 1 MADISON AVENUE, NEW YORK

FIGURE 4.   The American Posture League's (APL) letterhead included a photograph of a metal pin with a Native American man standing in profile and exhibiting a strong, erect, upright posture. The pin, which the APL bestowed upon schoolchildren and college students who demonstrated grade-A posture, was made and designed by APL member R. Tait McKenzie, physician and respected sculptor. This artifact speaks to the ways in which the APL glorified "primitivism" and hoped to commercialize it for the betterment of "civilized" peoples, who were thought to alone suffer from poor posture.

worried about the same habits among poor, working-class African Americans, fearing that the entire race would be judged by the behavior of their lower-class brethren.[41] Famed African American vocalist E. Azalia Hackley implored Black Americans to "conquer [their] habits of laziness, untidiness and extravagant gesture."[42] "We are a poor people," she wrote, "but we can be quiet, clean, becoming, and fittingly dressed."[43] The Great Migration enhanced such concerns among both white and Black elites, who feared that the supposed slovenly traits among Southern blacks would travel and be picked up by Northerners. A social commentator living in New York City in the 1910s wrote:

How often have I watched the dear old mammies in our Dixie Land home, as they planted their flat feet carelessly, indifferently, step by step along the road. A well-born white child prancing by mammy's side would fairly fly along the road like a breeze. . . . [Mammy] would still slop on her way and it is not alone colored people, but some white folks who never know success because they slop along instead of walking briskly.[44]

To such onlookers, voluptuous, curvy bodies indicated either laziness or hypersexuality. Jewish immigrant women came under similar scrutiny, often criticized for not wearing undergarments to properly support their breasts and hips.[45] In the post-Victorian era, when the slender, athletic build of the Gibson Girl increasingly became the standard of beauty in fashion houses and magazines, the full-figured woman, once heralded for maternalism and fertility, became reinterpreted as dangerously sensual.[46]

In her work with New York City immigrant children, Bancroft found a similar dichotomization of body types among the near-white, working poor.[47] Similar to what she imagined of African natives, Bancroft witnessed head-carrying among immigrant child laborers and idealized such bodies in her published materials (see figure 5). Yet these same working-class bodies could just as easily collapse under the weight of industrialized work, as Bancroft and reformer Lewis Hine both documented through photographic images reproduced here (see figures 6 and 7).[48]

What exactly was being carried on the head, however, was crucial when it came to social status. A basket or a basin on the head (as the young, working-class boy pictured in figure 5 is doing) became a sign of a lower class existence. This is why Bancroft and others insisted that, in order to signify educated middle-class whiteness, students should carry books on their heads during posture training. The head-carrying of books was not a necessity, nor, as objects, were books particularly conducive to remaining on the head given their material and shape. In other words, the very impracticality of the head-carrying of books is what made it a middle- to upper-class activity, intended for people who had the leisure time and resources not only to engage in posture training but also to read books.

FIGURE 5. APL President Jessie Bancroft used photographic images such as this to demonstrate the ways in which the white, urban, working poor could serve as models of good posture, similar to Native Americans and other non-white cultures where head-carrying was more the norm. She and her colleagues would appropriate the practice of head-carrying, but with books as the objects of choice. The impracticality of this practice was precisely the point, for it emphasized that with a certain degree of learnedness (symbolized by books), one would never need to actually head-carry as a form of labor.

FIGURE 6.   Bancroft used images such as this, in contrast to figure 5, to depict faulty posture practices among laboring children. This young boy carries the weight of his load entirely on one shoulder. Bancroft argued that such asymmetries increased the risk of developing permanent posture deformities.

Amid this elaborate classification of body types dictated by the intersectionality of race, class, and environment, Bancroft and her colleagues agreed that the scientific postural ideal was the biomechanical "plumb line," a positional mean between the angular soldier on the one hand, and a slumping or overly curvaceous individual on the other. The fixation with the plumb line was by no means new, nor unique to physical educators or physicians at the time.[49] Derived originally from ancient static

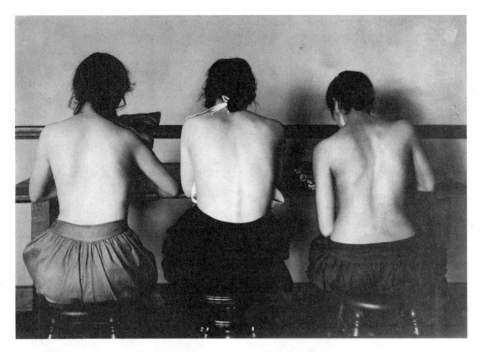

FIGURE 7. Lewis Hine, famed Progressive-era sociologist and documentary photographer, fought to bring an end to child labor. He did so partly by focusing on the perils of "work scoliosis," and the ways in which certain work practices contributed to unhealthy postures.

mechanics, the "aplomb" was further theorized in Renaissance art, and finally integrated into academic architectural design training in the nineteenth century.[50] In the late nineteenth and early twentieth centuries, art history and physical education were often taught alongside one another, with both placing an emphasis on the "classical" beauty of Greek sculpture, body symmetry, and ideal proportions.[51] This preference for linear sleekness would go on to influence industrial designs in the 1920s and 1930s, an era known for streamlined commercial goods.[52] As several historians have noted, anatomists looked to machines to describe the workings of the human body, and, at the same time, engineers saw the human body as a model to explain and improve the workings and designs of early twentieth-century buildings and manufactured items.[53]

In many ways, the appeal to a plumb line postural ideal and neoclassicism was an outgrowth of nineteenth-century physiognomy, phrenology, and race science. The Dutch physician-anatomist Petrus Camper used a "facial line" to establish a racial hierarchy, putting Greek sculpture at the top.[54] Early anti-corset reformers relied on the visual imagery of the linear *Venus de Milo* alongside depictions of the compressed and deformed torsos of the modern corseted woman.[55] All of these attempts to appropriate the aesthetics of "classical beauty," writes art historian Robin Veder, "encouraged the adoration and emulation of the 'white' classical body."[56]

Bancroft's attraction to the plumb line was also inspired by her contemporaries working in industrial manufacturing. Frederick Winslow Taylor, Henry Ford, and other industrialists and efficiency engineers aimed to harmonize the human body with assembly-line manufacturing in order to conserve energy, minimize sick days, avoid worker unrest, and ultimately increase productivity.[57] Efficiency engineers worked firmly within industrial capitalism, attempting to fix the injurious results of the modern workplace rather than questioning why such disabilities occurred in the first place. In Bancroft's mind, schools faced similar problems, as evinced by the reported rising absentee rates, a metric made possible by mandatory schooling laws.

Rather than adopting a physiological approach to the problem of fatigue—a tactic used by most efficiency engineers who conducted vascular skin-reaction tests, urinalysis, and respirometry tests on workers to assess energy expenditure and conservation—Bancroft proposed a more biomechanical solution to the problem. "Fatigue comes less readily in correct posture," Bancroft wrote, "and the energy spent through unconscious muscular action in maintaining a bad position is available in good posture, for other uses. In good posture, also, better circulation, respiration, and digestion keep the stores of energy and sense of wellbeing at a higher level, and the efficiency and even the spirts of the individual are thereby placed on a loftier plane."[58] Unlike physiologists, who saw the body as a motor—a complex system of chemical combustion and the production of energy—Bancroft and her colleagues relied more on the metaphor of the body as a machine, an organism of muscles and bones that functioned on the same principles as levers and pulleys; it was an ergonomic approach that relied on assessments made by the naked eye rather than a chemical one that required instrumentation to probe deep inside the biochemical processes of the human body.

When she took over as assistant director of physical education with the New York City public schools, Bancroft thus sought to establish the plumb line as a scientific standard that could be used by all school districts, regardless of the demographics of the student body. It was a goal within reach, for she had at her disposal a large and diverse student body upon which she could develop a system of measuring and training plumb line verticality. It was in this setting that Bancroft became convinced—and, in turn, convinced others—that a posture problem existed in America on a population-wide level.

The Vertical Line Test

The job of taking over physical education in the New York City public schools proved to be a daunting task. "The incredible number of children," Bancroft would later recount, "children by the hundreds, by thousands, by tens of thousands, yes, and by hundreds of thousands. It was overwhelming."[59] Most of Bancroft's days were spent traveling from one school to another, many of them with enrollments of 5,000 students, providing instruction to classroom teachers who would then demonstrate Bancroft's lessons to their students. Since few schools had a dedicated space for physical education (gymnasiums would not become a regular feature in most U.S. schools until the mid-twentieth century), most instruction took place in the classroom, or, if they existed, school playgrounds.

When not occupied with teacher instruction, Bancroft devoted her time to conducting and standardizing student physical examinations. Overseeing a small "corps of assistants," she established a system whereby each student would be "stripped for the study of contours."[60] While relatively new to secondary schools, medical inspections (the more common terminology for physical examinations) had become routine in the actuarial sciences, in certain industries, and in the evaluation of immigrants at the nation's ports of entry. Anthropometry— detailed measurements of the human body and its proportions—along with height, weight, and respiratory capacity became the bedrock of life insurance medical inspections, where virtually every detail of the human physique was statistically translated into life expectancy numbers. While examiners for life insurance agencies had the time and material resources to conduct elaborate inspections, the same was not true of the U.S. public health immigration inspectors. With only a few examiners stationed at every port and thousands of people entering every day, the physician had to make a "snap diagnosis," assessing current and future health in a matter of seconds.[61] Moreover, immigration officials were concerned not only with future health, but also with detecting current cases of disease, especially contagious infections.

In terms of resources and rationale, medical inspection in schools resembled immigration stations more than life insurance companies. A primary focus of the physical exam in schools was contagious diseases detection. Health officials trained in the new science of bacteriology worried that the new mandates of public school attendance could end up making the classroom a breeding ground for the spread of germs. A colleague of Bancroft's and future member of the APL, Dr. S. Josephine Baker took over New York's Bureau of Child Hygiene in 1908. Baker directed approximately 400 medical inspectors and nurses who examined around 3,000 children per week, visiting homes and classrooms on a daily basis, lining up children to examine eyes, hands, throats, and hair.[62]

Bancroft developed her own exam method, however, one based on the belief that the quickest and most failsafe method of making a "snap diagnosis" was through posture assessment. So as to create a standardized method for evaluating and grading posture, she invented a "vertical line test," which she developed with the help of Dr. Robert Lovett, a Boston orthopedist who conducted extensive research on the effects of gravity on the human body. In his clinic, Lovett used a scale-like apparatus to determine the location of the human center of gravity when standing. In the interest of educating teachers, parents, and the public who were not trained in the technicalities of the posture sciences and who, unlike clinicians, had to assess "the posture of large numbers of children with clothing on," Bancroft simplified Lovett's work, defining the ideal posture by assessing readily observable bodily landmarks.[63]

Ideally, according to Bancroft, the head, neck, and trunk would form a straight line so that the front of the ear would fall on the same vertical axis as the ball of the foot. If classroom teachers needed help visually with such assessments, Bancroft advised that they use a window pole and place it vertically next to a standing student and look in profile for bodily deviations from the ideal vertical line represented by the pole. This method was simple and easily adaptable to any classroom or domestic space. Indeed, it was so simple that the method became adopted for the purpose of self-surveillance and peer surveillance, as well as top-down surveillance.

With the intent to standardize the vertical line test, and to legitimate its need, Bancroft tested her method on some 250,000 New York City schoolchildren (in over 5,000 classrooms) in the early 1910s.[64] In addition to the vertical line test, she incorporated a five-minute march test, with the goal of assessing a student's postural endurance, seeing if the student could retain good posture after cardiovascular exercise. If the student passed both tests, she had him perform deep knee bends and arm swinging exercises to test the strength of the student's postural muscles, specifically the extensors that run along the spine. A student was then given a grade of A, B, or C and retested monthly. Paperwork and record-keeping were vital to Bancroft's vision of classroom efficiency and, more importantly, to translating the problem of poor posture into numbers.

On the basis of her findings, Bancroft concluded that 60 percent of schoolchildren could not pass the baseline posture test.[65] After three months of targeted exercises, the number who failed decreased to 20 percent, proving the benefits of school-based education. While most anthropometric studies at the time classified human types by skin color or country of origin, Bancroft did not create posture hierarchies according to race. If anything, disability dictated her posture hierarchy more than race, class, or gender did; "the most marked characteristics of idiots and mental defectives," she wrote, "are their collapsed posture and imperfect carriage, their slouching gait, and in extreme cases their inability even to stand erect and their undeveloped capacity for movement."[66]

Failing a posture test had great import, for according to Bancroft, erect carriage served as "a sensitive barometer of fatigue and illness."[67]

Aligning her interests with the newly formed National Association for the Study and Prevention of Tuberculosis—the model for the "new public health" effort that focused more on individual surveillance and health behaviors than on larger-scale sanitary and environmental interventions—Bancroft contended that "the germs of tuberculosis and other infectious and contagious diseases will find lodgment and will flourish in a body whose general tone and power are reduced."[68]

The belief that poor posture was both a sign of and a precursor to infectious disease was held by several leading medical scientists in the early twentieth century. The germ theory brought about a reductionistic clarity concerning the transmission of contagious disease, but it did not explain why certain individuals, when exposed to infectious microorganisms, would fall ill while others would not. Harvard orthopedist Joel E. Goldthwait stood at the forefront of the post–germ theory effort within medicine to emphasize the importance of posture health, an approach that took account of how musculoskeletal bodily alignment affected inner organ functioning. Conducting research with X-ray machines and clinical observations, Goldthwait contended that slouching posture crowded the internal organs, especially of the abdomen, and thus predisposed a person "to tuberculosis, nervous disease, acute mental disorder, as well as hyper-glandular disturbances, . . . and many intestinal disorders."[69] McKenzie, the first to hold a full professorship in physical education in the United States, persuaded university officials to mandate posture exams and training for all University of Pennsylvania undergraduates based on his research that showed that "the best time to treat tuberculosis was ten to fifteen years before it begins."[70]

This medicalized view of posture health was taken up in popular culture and writings as well. In her 1916 etiquette manual, *The Colored Girl Beautiful*, Hackley advised the following: "proper poise prolongs life because pressure on certain organs is evenly distributed and no strain is placed on any particular muscle to cause abscesses or tumors. Improved circulation of the blood results and good circulation spells health."[71] Future APL member and Yale economist Irving Fisher shared Hackley's view of the matter, writing that "one of the simplest and most effective methods of avoiding self-poisoning [i.e., disease] is by maintaining an erect posture." A follower of John Harvey Kellogg's biological living and a leading voice of the eugenics movement, Fisher blamed his own faulty posture—what he called a "consumptive stoop"—for his physical demise in 1898 when he contracted tuberculosis.[72]

Public health officials promoted the medicalized view of posture as well. In its promotional materials, the National Association for the Study and Prevention of Tuberculosis endorsed the view that good posture could combat infectious disease. Take, for example, the poster titled "Posture and Tuberculosis" reproduced in figure 8. Public health experts used the biblical David and Goliath story to represent how the small act of good posture (David) could defeat a looming problem like tuberculosis (Goliath).

In this context, Bancroft's findings of poor posture in over one-half of New York City schoolchildren had potentially dire consequences. Accordingly, the New York City school system—followed by many other districts across the country—began to mandate posture training, and a passing grade on posture tests, for graduation, even from elementary schools.[73] As each school made posture training compulsory, the studies and statistics of the nation's children and their (inevitably failing) bodily comportment accumulated.

While Bancroft did not actually use the term "epidemic" to convey the extent of the poor posture problem, she nonetheless employed contemporary language and methods needed to prove a population-wide health threat. The linkage of poor posture with tuberculosis—arguably the most popularly publicized epidemic in the United States during the Progressive era—helped to emphasize the scope and depth of the risk of poor comportment. In this sense, poor posture became an epidemic by association, co-constructed alongside tuberculosis.

But it is also true that Bancroft knew how to build a case for taking the poor posture problem seriously, mainly through her emphasis on numbers. "Percentage figures," she wrote after presenting her research findings, "have roused an interest on the part of all concerned."[74] All numbers mattered, she insisted, from the overall district averages of the number of students suffering from poor form to the grade point

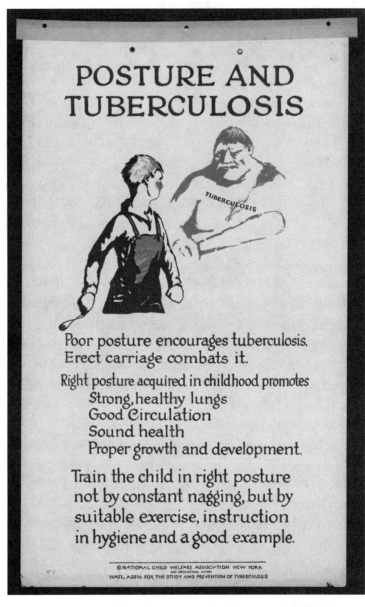

FIGURE 8.   A public health propaganda poster that demonstrates the interconnectedness between the anti-tuberculosis and anti-slouching campaigns. The illustration depicts how the biblical David (good posture) vanquishes Goliath (tuberculosis).

averages of individual students who underwent monthly posture testing. In her teaching manual, she emphasized that every student and classroom should know their percentages and make them public to the rest of the school and the entire district, for *"to omit this is to fail [to] use one of the most potent psychological elements in the situation."*[75] If the data were public, teachers and children would be compelled, Bancroft reasoned, to compete against each other, especially when posture prizes and other honors were awarded. It was a multilayered arrangement of posture surveillance that promoted policing at all levels, a constant feedback loop of seeing slouching, correcting it (in oneself or others), and inevitably seeing it again later in someone else or the same person. In this system, poor posture was an ever-present danger that required constant monitoring.

Nationalizing Posture Standards

With the intent to expand the scope and reach of her work beyond New York City's youth to the nation, Bancroft formed the APL in 1914, inviting her closest colleagues to join as executive board members. She first turned to several preeminent orthopedic surgeons who had devoted their careers to posture research, namely Goldthwait, McKenzie, Henry Ling Taylor of New York, and Elliott G. Brackett of Boston. She also enlisted her fellow Brooklynite friend and colleague, Dr. Eliza Mosher, who had been publishing on the topic of posture and women's health since the 1880s.[76] Within a year, the APL grew to include nearly seventy professionals, all of whom volunteered to serve in some sort of administrative capacity. A majority of the members had advanced degrees in medicine, while others had doctorates in economics and law, and still others were graduates of teaching colleges that specialized in physical education and physical therapy. Regionally, most APL leaders came from Boston or New York City, but some worked in the Midwest and the West Coast. The APL had roughly equal involvement from male and female professionals from various ethnic backgrounds.[77]

Though Bancroft had developed a way to grade the posture of public schoolchildren, her method did not appear entirely scientific according to early twentieth-century standards. Within medicine, clinicians increasingly turned to camera photography, X-ray machines, and motion pictures to both see and record disease and illness. At the same time, the mass circulation of magazines, mail-order catalogues, and the advent of motion picture films contributed to a "frenzy of the visible," with a particular emphasis on the portability of visual artifacts and entertainment. Bancroft readily used visual aids in her 1913 teaching manual on posture and in her teaching; *The Posture of School Children* is replete with photographs of classrooms and individual students, with reproductions of classical paintings and sculptures, and with anatomical drawings. But her vertical line test did not lend itself to graphic reproduction; rather, it was based on a subjective one-on-one interaction, a test that required a passive subject and knowledgeable examiner who had been trained to see postural deviations.

APL member Dr. Clelia Mosher, cousin of Eliza and director of women's physical education at Stanford University, would solve this problem. In 1915, she announced the invention of her "schematograph," a tool used to assess, record, and quantify human posture (see figure 9). Mosher developed her device in the Stanford school gymnasium with the assistance of Everett P. Lesley, a professor of mechanical engineering and later known for his work in airplane propeller design between the two world wars. Clelia's device consisted of a reflecting camera equipped with tracing paper, which allowed an examiner to outline a mostly disrobed person's figure. The final product was known as a schematogram, an outline of the human form on a single sheet of paper (see figure 10).[78]

A self-proclaimed physiological feminist, Clelia had specific reasons for engaging in posture research. In essence, she wished to prove that women were just as biologically capable as men in the workplace and institutions of higher education. The schematograph was an extension of her earlier work in respiration studies, wherein she proved that women—who had been portrayed by male scientists as shallow, costal breathers—actually breathed diaphragmatically, just as men did. She also believed that upright posture would mitigate the disabling effects of menstruation. At the time, it was commonplace for women to take up bed rest during their monthly flow; Clelia believed that women

𝒯𝒽𝑒 MOSHER-LESLEY
SCHEMATOGRAPH

The Schematograph is an instrument for easily recording posture outlines of the human figure

FIGURE 9.   Dr. Clelia D. Mosher operating her schematograph, a device used to trace the postural silhouette of examined subjects.

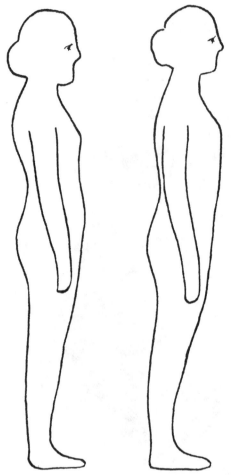

Fig. 20.—Schematograph records of posture.

FIGURE 10.   An example of a schematograph
record, known as a schematogram.

could and should work while menstruating if they were to achieve true
equality with men. Last but not least, Clelia was concerned about the
aesthetics and presentation of the new, modern, educated woman.
Knowing that dress and gesture had political and social import, Mosher
urged her students to remain ever circumspect about their appearance,
instructing them to strike a "pleasing" silhouette that did not attract too
much sexualized attention—as the new high-society debutante slouch

was doing—but also did not veer toward the plain or masculine so as not to jeopardize their marriageability.[79]

Using her device, Clelia measured the posture of every freshman woman entering the university. Clelia made it a practice to produce two identical copies of each tracing. She placed one copy in the student's file, using it as both a baseline against which to measure an individual's future progress (or lack thereof) and as a data point when making a class-wide assessment. Clelia gave the other copy of the schematogram to the student herself for the purpose of guiding home exercise practice. In this regard, Clelia found her device to be an ideal teaching aid. Because the record was made on tracing paper, she could take a schematogram that exhibited ideal posture and lay it on top of a schematogram of an examinee with bad posture, shading in the difference between the two so as to demonstrate more clearly the nature and degree of the subject's posture deficiencies. The silhouette served as a kind of disease map in its own right, a topography of both obvious corporeal defects (slouching) and imagined inner organ damage.[80]

Clelia insisted that her technology produced not only "hard facts," but also communicable information that the wider public could easily access. The tracings, she maintained, could be read and understood by the "uninitiated."[81] Embarking on a science that had not yet achieved widespread professional or public acceptance, Clelia's schematograph succeeded in giving the posture sciences a recording device that fixed the human form in time, making a static yet highly manipulable and re-producible record that could be used to advance both research and public outreach efforts. It was also a fairly easy technology for the examiner to master. A skilled examiner, such as Clelia, could complete up to a dozen schematograph exams in an hour.[82]

In 1916, Bancroft and the APL endorsed the schematograph as one of "the most practical and valuable contributions that has been made to corrective and educational work on the subject of posture."[83] To facilitate sales, the APL, in concert with the New York Posture Standards manufacturing company, arranged for the production, sale, and distribution of Clelia's schematograph. Compared to other popular visual recording technologies, especially camera photography and film, the schematograph was remarkably affordable, making it accessible to money-strapped public schools, state universities, and small private schools. Within a matter of years, dozens of physical education departments, clinics, and even the U.S. Navy adopted the schematograph.[84]

The most widely publicized schematograph study took place in early 1917 at Harvard under the direction of Dr. Lloyd T. Brown, a Boston orthopedic surgeon. Popularly known as the "Harvard Slouch" study, it found that 80 percent of university freshmen exhibited faulty posture.[85] Unlike Mosher, who largely kept her assessments at the individual level, Brown manipulated the visual data to a far greater extent. Brown took the hundreds of tracings from his study and created composites that other educators and clinicians could use for the purpose of assigning grades to examinees (see figure 11). The result was a one-page chart that divided human posture into four categories of "good" (A grade), "fairly good" (B grade), "bad mechanics" (C grade), and "very bad mechanics" (D grade). These composites served the very important function, as Bancroft and others had already learned, of creating percentages and statistics, a move that could, with one number, demonstrate the widespread extent of the poor posture problem. Brown also created posture-image aggregates to improve inter-examiner reliability and to train the eye of his students, teaching them to distinguish good from bad posture.[86] If optic agreement concerning posture norms could be achieved, Brown reasoned, then the schematograph could help standardize the posture sciences, controlling for an otherwise wide range of visual and subjectively experienced interpretations of what constituted good bodily form.

Published a month before the United States' official entry into the First World War and at the height of military preparedness, the Harvard Slouch report hit a nerve. If the Harvard man was to represent the typical man, then the nation appeared to be woefully ill-equipped for military engagement. Early reports from medical draft boards confirmed such worries. Within the first week after President Woodrow Wilson's declaration of war, the state of Arkansas had already rejected over 8 percent of its volunteers for foot and posture weakness.[87] By August 1917, the Texas National Guard had rejected 1,125 out of approximately 13,000 volunteers because

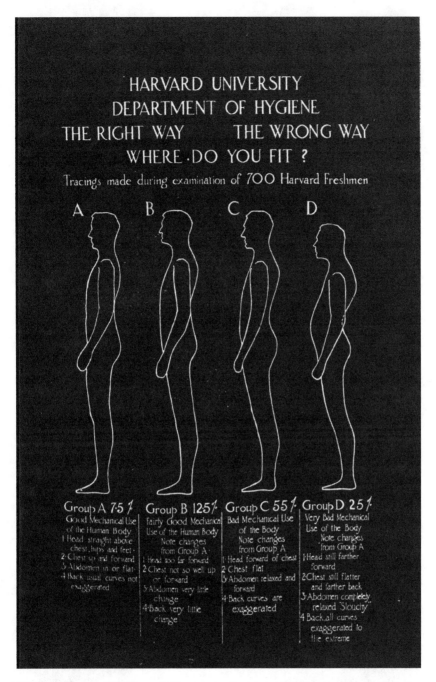

FIGURE 11. Lloyd T. Brown's posture grading system with accompanying statistics, the basis of the "Harvard Slouch" study. Only 7.5 percent of Harvard students had an A-grade, or "good," posture. 12.5 percent exhibited B-grade, or "fairly good" posture; 55 percent showed C-grade, or "bad," posture; and 25 percent had a D-grade, or "very bad" posture. He publicized his classification system in scholarly articles and also on large-scale posters that the APL sold to be hung on the walls of school gymnasiums and clinics across the country.

of postural defects, ranking it the third highest reason for rejection, just behind the diagnosis of "underweight" and "hydrocele and varicocele" (swelling and varicosities of the scrotum).[88] In Philadelphia, one examining surgeon found that up to 25 percent of the first group of draftees had to be turned away because of flat feet.[89]

In response, the U.S. Armed Forces called upon Goldthwait, McKenzie, Brackett, and other APL members to establish and administer "Flat-foot Camp(s)," where draftees who were found to have poor posture or weak feet underwent a specialized exercise program to make them ready for service.[90] Foot breakdown was of particular concern to the military; soldiers were expected to be carrying heavy packs, to march long distances, and to "be at the ready" in the theater of war.[91] To treating physicians, the primary cause of flat feet was biomechanical, that one or multiple deviations had occurred along the plumb line. The army flat-foot camps thus served as posture training camps, reorienting the military ideal to be one that matched the biologically proven posture standards promoted by the APL rather than those of an authoritarian state. On a day-to-day level, camp operations looked similar to Bancroft's classroom work and Clelia's gymnasium activities where students—or in this case draftees—were examined, instructed in targeted exercises, and taught the "art of observation" so that everyone involved could distinguish good from bad posture.[92]

The Postwar Posture Campaign

After the Great War, the effort to conduct large-scale posture studies intensified, particularly within the newly expanded, yet fiscally conservative federal government. Since the early days of the APL, Bancroft saw posture improvement as a public health intervention that was both foundational and affordable. During the prewar years, Progressives poured time, money, and energy into playground construction, free lunch programs, proper ventilation in tenements and workshops, and industrial efficiency. But, as Bancroft was quick to point out, "the good that all these things ought to accomplish is being discounted by the fact that few [Americans] stand or sit in correct posture." "Pure air," she continued, "will not invigorate the body the way it should if the lungs are cramped . . . pure food will not give its entire nourishment to the body if the stomach and digestive tract are not allowed normal activity."[93]

Simple posture training, by contrast, promised to prevent and cure a whole host of diseases, improve the health of the impoverished, and also enhance the productivity of schoolchildren and industrial laborers. It was a win-win proposal for businesses and governmental public health officials alike. Outside the cost of basic surveillance technologies and the salaried professionals who conducted the exams and produced educational materials, the time and effort of posture improvement fell largely to the individual, who bore the responsibility of daily self-monitoring and self-improvement.

Much of the postwar federal spending concerning posture research targeted school-aged children. Ever since the 1918 "Children's Year," there was a concerted effort to improve the health of the nation's youngest citizens, readying them for future wars and colonial expansion, for economic productivity, and for a better tomorrow with improved national health statistics.[94] With this goal, several federal agencies were able to extend their reach, albeit in more conservative, less politically charged directions. Rather than pushing for more progressive policies that addressed the social determinants of health (e.g., housing, universal health care) or structural changes, the Harding and Coolidge administrations favored studies and efforts in preventive health, much of which centered on distributing educational pamphlets and collecting biometric data, such as height, weight, and posture measures.

One of the first postwar federal efforts to address the nation's youth and posture was the 1919 anti-venereal disease campaign, known as the "Keeping Fit" series. Aimed at middle-class, adolescent boys, the Keeping Fit series counseled its readers to maintain a routine of regular physical fitness, to eat healthful foods, and to attend to one's posture and clothing. There were many "handicaps revealed by the war," read one pamphlet, with the principal causes of draft rejections being "defective eyesight, poor teeth, bad feet, and venereal disease." To conquer the third, these pamphlets aimed to teach American boys how to attend to their foot health and overall physique. Through educational programs such as this, the federal government simultaneously appealed to both medical science and conservative mores, linking the two together in the name of improved hygiene. An upright citizen would abstain from sexual relations until marriage, and would improve his own health and serve as a model of fitness for others with his plumb line stance.

The U.S. Public Health Service (USPHS) was no stranger to posture education and research. USPHS medical inspector Joseph Schereschewsky, the son of a Jewish-turned-Episcopal bishop and missionary in China, garnered praise as one of the most efficient immigration officers, possessing the rare ability to examine 1,200 immigrants a day, an average of five people a minute.[95] The secret to his success, he claimed, was his ability see the whole body "at a glance," an art of "observation and deduction" that required three years of "constant application," inspecting the human form. He described his process, when the immigrant was still clothed, the following way:

> I begin as soon as [the alien] begins to walk toward me. His gait, his carriage, his color, the shape of his body . . . all give me indications of the normality or abnormality of his condition. . . . As soon as the immigrant is within range of my eyes I look first at his feet. I notice whether they are symmetrical, one with another . . . whether they are deformed in any way. . . . Then my eyes keep on up [sic] the man's figure. I notice the knees and the shin bones. I see whether the man is knock-kneed. . . . Then I notice how the legs are joined to the body . . . then I look to the region of the stomach and see if it seems to be placed correctly . . . then I look at the chest and notice particularly if one of the lungs is smaller than the other. . . . I notice whether the chest is flat or too round or if it is misformed [sic] in any other way. . . . Then I look at the shoulders. Is one shoulder higher than the other? . . . As the alien walks away from me, I notice his carriage and look to see whether he has a spinal disorder.[96]

Later in 1915, when Schereschewsky was in charge of the USPHS's Office of Industrial Hygiene, he conducted a study of 3,000 garment workers (both male and female) and found that apart from tuberculosis, the most common defects were poor vision (74 percent), faulty posture (50 percent), and flat feet (26 percent).[97] To counter such problems, Schereschewsky claimed that the nation's schools needed to require posture training and inspection of all of its students. "Children are the potential capital of the State," he wrote, "and it is upon the subsequent efficiency of these citizens in embryo that the future prosperity of any body politic depends."[98]

The USPHS continued its research into posture standards and health throughout the 1920s and early 1930s. Dr. Louis Schwartz—who like Schereschewsky began his career in the USPHS as an immigration medical inspector—used both Mosher's schematograph as well as camera photography in an attempt to establish a more numerically exact posture norm.[99] To that end, he recruited exclusively native-born white boys and men who were free of disability and had no "marked postural or other deformities, including marked kyphosis and scoliosis."[100] To his delight, he found many willing participants who volunteered to be part of his study.[101] Because his subjects represented a wide age range, from two and a half to seventy years old, Schwartz's aim of finding a standardized plumb line proved more difficult than expected, for, as he discovered, a person's center of gravity fluctuated throughout the human life span. "At the beginning of life," Schwartz noted, "there is a protuberance of the abdomen that disappears by about 6 years of age. It is interesting that this tendency appears again late in life among quite a proportion of persons, but in old age the abdomen appears more relaxed or sagging."[102]

To assign posture grades according to the classificatory scale developed by Brown in the Harvard Slouch study, Schwartz relied on the expertise of Dr. E. Blanche Sterling, a USPHS colleague with a specialized interest in child hygiene. Although the subjects of Schwartz's study were to represent the white—and thus human—ideal, Sterling found that less than 20 percent exhibited excellent or good posture, reporting nearly the same percentage breakdown of posture grades as the Harvard study.[103] Sterling discovered similar numbers among non-white and working white populations. In a 1925 investigation of over 5,000 Black schoolchildren in Atlanta, Georgia, she found an 80 percent poor posture incidence rate among African Americans as well.[104] Motivated by such numbers, Dr. A. Wilberforce Williams, a prominent Black physician who worked in Chicago's South Side and served as health editor for the *Chicago Defender* in the 1910s and 1920s, regularly penned articles reporting on the prevalence and evils of poor posture among African Americans, urging them to engage in regular posture training in the interest of self-help, disease prevention, and racial uplift.[105]

The U.S. Children's Bureau produced findings similar to those of the USPHS. Chronically underfunded and increasingly under physician control after the Great War, the Children's Bureau turned to university researchers to conduct studies and produce educational pamphlets on its behalf.[106] Dr. Armin Klein, Lloyd T. Brown's successor at Harvard and orthopedic surgeon at Massachusetts General Hospital, and Leah Thomas, a physical education instructor at Smith College, served as the Children's Bureau posture experts throughout the 1920s.[107] In addition to producing a posture education film and several other promotional pamphlets, Klein and Thomas conducted the first controlled study at a public school in Chelsea, Massachusetts, just north of Boston across the Mystic River.[108] The students were predominantly Russian Jewish, with a minority of native white children, and even fewer non-white. Klein and Thomas set out on a two-year study (1923–24), dividing approximately 2,000 students into two groups: a control group that would not receive posture training and a treatment group that would make weekly visits to a posture clinic. At the outset, Klein and Thomas found that 92 percent of the students exhibited poor posture. After one to two years of posture training, however, 62 percent of the treatment group showed marked improvement, while the control group numbers were largely unchanged.[109]

Klein and Thomas's study confirmed, once again, the high prevalence of poor posture, but it also provided scientific evidence for the need to establish posture clinics nationwide, hire more physical educators, and redouble the effort to educate parents and schoolteachers on the importance of good posture. School directors across the country wrote to Grace Abbott, head of the Children's Bureau, requesting copies of the posture standards charts included in the Klein and Thomas study.[110]

Concerned mothers and teachers of infants and preschoolers also wrote for advice about which toys to purchase and how to structure play in the toddler years in order to promote upright carriage.[111] In a 1928 pamphlet, "Your Child's Posture," the Children's Bureau encouraged parents to have their youngsters make shadow pictures of themselves at home, and then compare the wall outline to the posture standards chart, purchasable through the Children's Bureau.[112]

By the time the third White House Conference on Children was held in 1930 under the Hoover administration, it had become accepted fact that a large proportion of the nation's youngsters suffered from poor posture. Dr. Ray Lyman Wilbur, chair of the conference, appointed a Subcommittee on Biomechanics to survey the field of posture sciences and to assess training in the nation's schools. The subcommittee—made up of five physicians, including Brown, Klein, Carnett, and the only female and nonphysician, Thomas—held firm in its assessment of the extent of the problem, claiming that "over two-thirds of the young children of the United States exhibit faulty body mechanics and usually continue to exhibit faulty body mechanics in adult life."[113] With the Great Depression as a backdrop to the 1930 conference, health experts expressed concerns about malnutrition and other childhood depravations. Accordingly, the Biomechanics Subcommittee emphasized the posture–disease connection, claiming that poor body mechanics in children could result in "failure to gain weight, . . . insufficient defecation, and cyclic vomiting."[114] This emphasis fit with Hoover's belief in individual self-determination rather than socially caused problems. "The ill-nourished child," he pronounced in an address to the participants of the conference, "is in our country not the product of poverty; it is largely the product of ill-instructed children and ignorant parents."[115]

By the 1930 conference's end, forty-six U.S. states required physical education of its school-aged children, incorporating posture training and exams into regular school operations.[116] The same trend occurred in higher education, as an increasing number of universities and colleges established departments of physical education where graduate students could earn PhDs in the field while also using undergraduate student bodies to advance posture research.[117] At the same time, many medical doctors and nurses who specialized in student health, public health, and industrial health increasingly made posture exams a regular part of the physical exam, referring patients with defects to orthopedic surgeons and nurses, physical therapists, and gym teachers for remedial treatment.

Americans wishing to buy life insurance were also expected to have their posture assessed. Believing that "posture is one of the chief determining factors in general health," the Metropolitan Life Insurance Company and its team of home health nurses distributed health education pamphlets such as *Standing Up to Life, The Importance of Posture,* and *Posture from the Ground Up.*[118] Whether the public at large subscribed to the existence of a poor posture epidemic held little weight as the country entered the mid-twentieth century, for the professionals involved had succeeded in persuading enough legislators, health officials, and businesses of the import of posture health to have it become a normalized part of several key facets of everyday American life.

With the support of the federal and state governments as well as private corporations and industry, posture experts had succeeded by the middle third of the twentieth century in making slouching a problem of epidemic proportions. The primary achievement was getting other professionals as well as the general public to *see* the problem. The APL, which led the way throughout the 1920s and 1930s, produced posture charts, measuring devices, and methods for quantification, which all aimed to train the eye of every American to detect postural deviations. In a post–World War I era of fiscal conservativism, the posture campaign appealed to educators, public health officials, and industrialists who wanted to quell worker unrest through the least expensive means, avoiding prohibitive workplace adaptations or the creation of a social welfare state. With virtually everyone in the professional class on their side, posture scientists gained legitimacy, not simply as clinical experts but as scientific managers of the workplace, as guardians of child health, and as protectors against an epidemic that appeared to threaten national strength.

CHAPTER 3

Posture Commercialization

SURVEYING EPIDEMIC disease from Greek antiquity to the recent past, medical historian Charles Rosenberg identifies three primary ways that people have, regardless of time period, tended to conceptualize epidemics. Before the widespread acceptance of germ theory, the "configuration" view of disease was more common, a holistic worldview that understood widespread illness to be a "a disturbance in a 'normal' arrangement of climate, environment, and value-imparting relationship," the latter of which included God-ordained diseases and plagues as punishment for individual or collective sin. At other times, people in the past believed that epidemics came about through "contamination," the "transmission of some morbid material from one individual to another." "For many laypeople throughout history," Rosenberg writes, "the terms epidemic and contagious were synonymous." A third form that often accompanied both the configuration and contamination views was the concern for "predisposition," the view that certain individuals, by virtue of their lifestyle or constitution, were more susceptible to the disease than others.

What all epidemics seem to share, according to Rosenberg and other scholars, is a "logic," a "formulaic plot," a dramaturgy that begins with an outbreak, followed by a search for the vector of the disease (whether it be physical or metaphysical), leading to a collective response to the deaths and destruction until a dénouement is achieved, when the incidence of the disease gradually declines to the point where the public begins to feel safe again.[1] A defining component of epidemics, Rosenberg maintains, "is their episodic quality. A true epidemic is an event, not a trend. It elicits immediate and widespread response. It is highly visible."[2]

The poor posture epidemic—a plague that did not kill, per se, but that was believed to contribute to deadly diseases—does not fit into the standard dramaturgy. As a scourge defined by physical defects rather than mortality rates, it was an epidemic almost solely defined by risk, animated by visions of the distant past and the far-off future with the intent to intervene in the present. Through the efforts of certain evolutionary scientists, the American Posture League (APL), and the U.S. Public Health Service, the poor posture epidemic became a statistical fact, with the number of those affected hovering around 80 percent of the population. Since the outbreak occurred in the distant past, the vector, theoretically speaking, was a flaw in human evolution, thus making the epidemic widespread from the outset. Likewise, the duration of the epidemic was not episodic, but rather long-standing, and would, according to posture scientists, only be brought to an end if individuals became more conscious in their efforts to maintain plumb line verticality through adequate musculoskeletal exercise and fitness.

Posture experts worried that they would never be able to control the already high incidence of poor posture if they could not convince the wider public of the problem, getting the American people to see biomechanical defects in themselves and others, and to understand such deviations as deleterious to individual and national health. Not trusting that the wider public would have the fortitude or clear-sightedness to live up to the ideal posture standards, these experts engaged in public health measures to facilitate proper form through body technologies.

Posture improvement body technologies took many forms. Specially designed exercise programs were one line of attack (the subject of the next chapter). These fitness programs aimed to enhance human posture through abdominal strengthening, or what we call today "the core." To supplement such training—or, in some cases, to offer an alternative to it—posture experts engaged in a second line of attack through the creation and manufacture of posture-enhancing technologies, such as specially designed clothing, undergarments, footwear, and seating systems, the subject of the current chapter.

In her discussion of how germ theory transformed early twentieth-century domestic spaces through new technologies such as sanitary sewer systems, plumbing, and everyday customs of domestic cleaning, Nancy Tomes demonstrates how the American public health movement was very much interested in "the private side of public health," of reforming individual behaviors including habits of purchasing and hygiene. To that end, many reformers borrowed from and worked with advertisers and manufacturers to sell the message of health and fitness, while also promoting material goods to help achieve that end.[3] Just as inventors, plumbers, and engineers saw the movement to germproof houses and public buildings as lucrative lines of business, so too did many of these same actors see the profit potential in improving the nation's posture.

The marriage of science with capitalistic entrepreneurship ended up solidifying the reality of the poor posture epidemic, making it visible and known not only through traditional public health agencies but also by means of commercial billboards, goods, advertisements, and corporate-sponsored competitions. Looking primarily at the years between the two world wars—in many ways the golden era of the American posture crusade—this chapter will explore how posture scientists became deeply involved in American manufacturing, both in the products produced and in the laborers and workers responsible for the assembly of goods. Several sites are particularly fruitful for this kind of study.

One of the primary industries of concern for posture crusaders was the clothing industry. On the heels of New York City's Triangle Shirtwaist Factory fire in 1911, labor unions as well as hygiene professionals pressed for reforms in the built environment of textile manufacturing. The effort to improve hygiene and safety in the workplace led to multiple scientific studies on chairs for industrial laborers. Prior to World War I, very little attention had been paid to workplace seating systems. Posture crusaders saw themselves as embarking on a new and neglected science of "applied anatomy" (later known as ergonomics and kinesiology) in order to combat worker fatigue, extending the work conducted by industrial physiologists prior to the war. Their applied practice also found a market in the burgeoning public school system, where certain educators hoped to mitigate the deleterious effects of long, mandatory school days on young American bodies.

But the interest in the workplace did not end with seating systems. Posture crusaders were equally concerned with the *kind* of clothing that manufacturers produced. Of particular importance to the APL were undergarments and footwear. In a certain way, the clothing reforms mounted by the APL were an extension of the hygienic clothing initiative of the nineteenth century, a movement that led to bloomers and union suits. But some crucial differences exist between these two movements. First, because the APL took up clothing reform squarely in the era of "the new woman," they targeted the "debutante slouch," the Jazz Age woman who insisted on new freedoms in politics and fashion. Whereas many nineteenth-century clothing reformers held an anti-corsetry stance, posture crusaders of the early twentieth century advocated for "scientific" corsetry in order to give women assistance in overcoming the inherent evolutionary challenges of bipedalism. This twentieth-century medicalization of clothing extended to both men's and women's wear, creating a degree of gender fluidity in the construction and design of both abdominal supports and footwear.

Taken together, the case studies demonstrate how posture scientists took an interest in body technologies as a means of preventive health, of reducing the health risks that they believed resulted from the poor posture epidemic. At the same time, these experts believed that such technologies could have a powerful moral and cultural influence, molding the attitudes, behaviors, and bodies of Americans from all classes, genders, and creeds. A properly designed chair, they argued, had an educative function, combating "Old World" customs and hygienic ignorance through the sensory input provided by such a seat. Likewise, a scientific corset or belt could cue an otherwise slovenly or slouching person to sit and stand straight. Similar to Michel Foucault's observation that discipline in the twentieth century would become ever more geared toward coercive manipulation of individuals, with professionals

managing and attempting to bring order to otherwise unknowing subjects, posture scientists aimed to instruct and inform individual bodies
without requiring such persons to be necessarily cognizant of it.[4]

Ergonomic Chairs and Lumbar Supports

Presenting a lecture on the importance of posture to a room full of New
York society types and professionals at the Hotel Astor, APL cofounder
Dr. Eliza Mosher captured the audience's attention when she accused
her onlookers of sitting poorly. "Looking around this room, I find almost everybody is sitting incorrectly; certainly not in posture," she re-
buked. According to a journalist covering the event, everyone became
painfully aware of their own posture and sat up straight. Mosher continued: "It is not so much your fault as it is the chair. Manufacturers
make chairs to appeal to the eye, and not to the shape of the human
form. If this league doesn't do something to produce better chairs for
the health of the public," she continued, "then the league should go out
of existence."[5]

While erect standing had become biologically and evolutionarily
fraught, the act of sitting caused deep concern as well, for it too pointed
to the inherent weakness of bipedalism. The two-legged human body
presented a continuous problem in maintaining postural balance, becoming easily fatigued and requiring respite, which could be found in
sitting, kneeling, or lying down.[6] As with standing, sitting was a culturally suffused act. Early twentieth-century French anthropologist Marcel
Mauss divided human societies and cultures into "squatting mankind
and sitting mankind."[7] Like Mauss, posture scientists saw the chair as
both an emblem of civilization and its very undoing, interpreting squatting and cross-legged sitting found among native "savages" as a sign of
admirable physical strength even if, from a cultural perspective, primitive. Or as Dudley A. Sargent wrote, "Instead of priding himself upon his
ability to sit straight without support for his spine and legs, as shown by
many of the savage tribesmen, civilized man luxuriates in upholstered
chairs and lounges molded to his physical defects—and then wonders
why he has a weak back and cannot stand in a vertical position."[8]

Much like the effort to create posture standards and norms for the purposes of examining the upright stance of human beings, APL members and other posture crusaders wanted a similar scientific universal for sitting, and they wanted manufacturers to help them in this venture. One culprit of widespread poor posture, according to the APL, was the unregulated American marketplace that favored the production of goods that satisfied the whims of fashion without regard to science, functional anatomy, and health. To force greater attention to the latter attributes, the APL developed consulting teams to provide industrialists and store owners with scientific data on material goods such as chairs, foundational garments, shoes, and outerwear.

Describing the neoliberal impulse of the APL, George J. Fisher, physician and deputy chief scout executive of the Boy Scouts of America, praised the APL for being "a group of idealists and skilled technicians [who] combined their science with business." In Fisher's mind, it was a "welfare movement" that introduced "humanitarian ideals" into the realm of business.[9] To the APL and the manufacturers who participated in the effort, it was a mutually beneficial arrangement. The market of educated consumers who wanted the latest in what the medical sciences had to offer and the purchasing power to acquire it was growing. The APL helped manufactures and store owners to tap into this market by offering a "seal of approval" for the goods it deemed to be scientifically sound.[10]

Significant changes in schooling, the industrial workplace, urbanization, and motorized transportation meant that an increasing number of Americans spent more time sitting than ever before. Though the APL focused its efforts on public seating systems rather than home furnishings, it still hoped that the former would influence the latter. APL leaders reasoned that if they provided large swaths of the population with the bodily experience of "posture-right" seating systems in schools and the workplace, they could promote innate sensory learning through scientifically engineered material objects. Building on new scientific research into proprioception and muscle memory, APL leaders believed that occupants of posture-right chairs would learn, on an innate level, what proper form should be, and that this subconscious lesson would carry over into the home, where consumers would begin to desire similar seating systems in the domestic sphere. Moreover, unlike conventional household chairs and lounges that nurtured comfort, leisure, and personal affection, "posture-right" chairs were to be depersonalized objects that would fit all body types, and ultimately recede into the background.

Some of the earliest research in scientific seating occurred in schools. The goals of preventing bodily deformities while also maintaining classroom discipline often went hand in hand. Certain German schools, for example, adopted the *Geradehalter*, a yoke-like straitjacket attached to chairs that prevented children from leaning over their work while reading and writing. American schools, concerned about preserving notions of individualism and freedom, looked to other, less overtly visible means of restriction and control. The softer, less visible forms of control would become more prevalent in the twentieth century.

The Boston school system was one of the first in the country to request extensive research on school seating, calling on medical doctors to lend their expertise.[11] So-called "school scoliosis" had long been a concern. In the latter half of the nineteenth century, physicians studied desk design, focusing primarily on the angle and placement of writing surfaces. A properly designed desk that allowed for both upright sitting and good penmanship proved to be a challenging problem. Before the late nineteenth century, students were taught to write on a slant. To produce the proper slope in lettering, students had to angle their paper and pen on the right side of their bodies (left-handed writing was unacceptable), which inevitably led to asymmetrical sitting, with spines rotated and curved to one side. To combat a high number of cases of scoliosis that resulted from school penmanship lessons, certain educators and physicians campaigned for a new type of script—namely, the vertical script—which increasingly became the standard adopted by most schools in the early twentieth century. These same health reformers worked with seating manufacturers to develop desk units that had built-in writing surfaces directly in front of seats, thereby encouraging students to maintain straight spines, even when lettering.[12]

Once vertical script became the norm, physicians and educators began to blame poor school chair design for posture defects. Harvard's

Dr. Edward H. Bradford concluded in 1899 that chair backs "remained the weakest point in even the best of the modern school furniture built in this country."[13] Dr. Tunstall Taylor at the University of Maryland complained that school chairs failed to follow the physiological curves of the spine, leaving, most crucially, the low back unsupported. Without proper support, he maintained, "weak children were too prone to try to 'lie down' in their chairs," reclining so that their heads rested on the chair back.[14] In response to this professional outcry, Drs. Frederick J. Cotton and Robert Lovett of Boston worked with a local furniture manufacturer to develop the "Boston school seat," a movable wooden chair (many desk chairs were bolted to the floor at the time) with a supportive back.

In many ways, early school chair design was a direct outgrowth of the boom in Victorian-era patent furniture. In America—where, since its founding, the patent system rewarded individual innovation and mechanical dexterity—artisans (often with physician consultation) began to think of seating systems more in terms of engineering than fashion, resulting in chairs that could swivel, incline, and recline all with a hand control mechanism that could be manipulated by the occupant.[15] Design historians mark the end of the U.S. patent furniture craze with the 1893 Chicago World's Fair when "European ruling taste flooded America," and people began to "be ashamed of" mechanized, nonartistic furniture and exhibited preferences instead for upholstered and overstuffed chairs, objects that signaled status, gentility, and fine taste.[16]

Another difficulty the patent chair industry faced was that the general public tended to associate mechanized seating systems, such as rocking chairs and other adjustable recliners, with invalidism, disability, and aging. Normalizing these features, making them mainstream instead of signifiers of illness and debility, required a new kind of streamlined design to satisfy the emerging tastes of twentieth-century Americans. These design changes did not erase the disability history inherent in these seating systems, but rationalizing these objects did manage to mask such histories.[17]

Papering over the original intent of adaptive chairs made it possible for the interest in physiological seating systems to persist. Indeed, the same designs that were originally intended to support nonnormative bodies became reformulated as features of able-bodied enhancement. Take, for example, the work of efficiency engineers Frank and Lillian M. Gilbreth and their "Fatigue Museum" in Providence, Rhode Island. The museum featured a range of devices, with a primary focus on the industrial workplace. The Gilbreths believed that of all efficiency-enhancing technologies, better chairs were "needed immediately and pressingly in all industries."[18] Compared to the patent chairs of the nineteenth century, early twentieth-century posture seats were sparse and simple. The seats on display in the Rhode Island museum varied in types and styles, and included several re-purposed chairs that workers themselves had modified.

The Gilbreths could not recommend any one single chair because of the sheer variety of machinery and industries that they surveyed. Each workplace required a multitude of sitting and standing positions among operators. But since the couple was supposed to control costs, saving money for business owners and industries, they had to find an affordable solution. To that end, the Gilbreths suggested that every employer, college, and office set up their own fatigue museum so that workers and students could help determine (and even tinker with, if need be) the seating system according to each context. In a rather utopian statement on the matter—a statement that was meant to anticipate resistance among employers, while also attempting to appease overworked laborers—the Gilbreths concluded that "fatigue elimination does not demand a large expenditure of money, nor depend upon . . . highly skilled mechanics to make the devices."[19] The one recommendation that the Gilbreths did make was that seating systems should allow workers to both sit and stand, and that chairs be built to provide support to the worker's back, legs, and feet.

The promise of a low-cost fix to human fatigue that could also potentially stave off worker unrest gained tremendous traction during the business-friendly, post–World War I era. New York City's Joint Board of Sanitary Control offers one of the best examples of how workplace seating would come to dominate reformers' attention in the 1920s, a decisive move away from the more radical agenda of shortened workdays, wage increases, and child labor laws that defined the Progressive era. Made up of physicians, social workers, and other white-collar professionals, the Joint Board was a third-party negotiator between industry and some of the most militant labor unions of the day, especially the International Ladies' Garment Workers' Union, a collective mostly made up of Jewish and Italian immigrants. Many of the leaders of the Joint Board were Eastern European Jewish immigrants themselves, and thus shared a common tongue—and at times a similar political bent—with the laborers.[20]

During its first decade of existence, the Joint Board focused its energies on safety reforms, inspecting workplaces so as to ensure that garment makers would be protected against the dangers of overcrowding, poor ventilation, and fires, a pressing concern after the 1911 Triangle Shirtwaist fire.[21] In 1914, at the urging of its medical director Dr. George Price, the Joint Board opened a clinic and provided free physical examinations for all garment workers. The union health center, according to Tomes, was the first union-sponsored health care facility in the United States; during the 1920s, it was a major health care provider for the Lower East Side of Manhattan.[22]

In the tenth annual report of the Joint Board, Price heralded his colleagues' successes, claiming that safer workplaces demonstrated the fruits of "true industrial democracy." Looking to the future, Price urged his colleagues to look not only at the work premises, but the workers themselves, claiming that the success of the Board depended on teaching the worker "the precepts of right living and the conservation of health."[23] To that end, the Joint Board recruited Theresa Wolfson, who would later become a well-known economist, to head up the organization's health education initiative. "The next job of this organization," she wrote in 1922, was "to acquaint the worker with the simple principles of health, the necessity of cooperation in keeping the workshops clean, and their own bodies healthy."[24] Wolfson offered noontime public health lectures in Yiddish and Italian, Friday night "health nights" with moving pictures and lantern slides, and a question box where workers could express any concerns, queries to be answered by management or a physician. Seating and posture, Wolfson found from her outreach work, was a matter that needed immediate attention in order to settle any "future problems in industry between the workers and employers."[25]

Wolfson teamed up with Edith Hilles from the newly formed Bureau of Women in Industry, a branch of New York State's Department of Labor, to survey the seating arrangements of 250 clothing factories in New York City. Studies in factory seating were a relatively new endeavor. A few years prior, the Pennsylvania Department of Labor and Industry reported on a munitions plant that hired an efficiency engineer to adjust the chairs of every new employee, hoping to speed up worker output.[26] In a similar vein, the California Industrial Welfare Commission observed female workers in the vegetable canning industry. The commission concluded that if manufacturers wished to decrease worker absenteeism, they needed to provide their employees with stools that had seat backs. The norm, as many industrial studies found, was that workers were given the option of backless stools or no seating system at all.[27] The Bureau of Women in Industry found that one-third of manufacturing sites—such as those that produced paper goods, electrical supplies, paper boxes, rubber goods, clothing, camera supplies, and hosiery—supplied their workers with only backless stools or benches.[28] A similar figure was found in a survey conducted by the Board of Trade of the Laundry Industry.[29]

By 1920, forty-seven states passed legislation that required employers to provide chair backs for their workers.[30] In response, many industrialists purchased the cheapest chairs on the market. Frances Perkins, head of the New York State Industrial Commission, observed that most of these seating systems were so ill-equipped that they actually "handicapped" the worker. Additionally, as one of Perkins's colleagues reported, "even if there are chairs, workers are discouraged from using them."[31] "The old idea still stands," one Department of Labor investigator noted, "that to be seated is to be lazy."[32] Some workers reported that they feared they would be fired if they were found sitting on the job.[33]

To further investigate the quality and need for workplace seating systems, Wolfson and Hilles visited 250 clothing shops in New York City, assessing the work postures of 2,528 machine operators, 1,774 hand finishers, and 734 pressers. They paid special attention to New York City garment workers since it was believed that by virtue of their work, these laborers were particularly susceptible to "spinal curvature, gastric ills, liver congestion, and prone to diseases of the chest, bronchitis, pneumonia, and tuberculosis."[34] Indeed, the association between disability and textile manufacturing led to eponymous disease categories such as "cobbler's chest" and "tailor's bunion."

The first step in curing the biomechanical ills of textile workers was a scientific seat. Certain industrialists resisted, claiming that the chairs would go unused since tailors from the old country would sit "Turk fashion," a racialized slur for cross-legged sitting on the floor. By the mid-twentieth century, this form of sitting would come to be called "Indian style," connoting the image of uncivilized Native Americans. Contrary to this myth, Wolfson and Hilles found that most laborers wanted chairs, but often did not use the ones offered because the seating options were crude and unsuitable. Over a quarter of the workshops they visited offered nothing more than makeshift wooden benches, where workers sat hunched over their work.[35]

To demonstrate the necessity of proper seating, Wolfson and Hilles outfitted two clothing shops with scientific "work chairs," designed and tested by a team of experts at the Massachusetts Institute of Technology (MIT). They also got the Joint Board to provide anatomy lectures and APL pamphlets so that laborers would appreciate the importance of proper comportment and begin to monitor it in themselves and their family members and coworkers.

The MIT work chair (see figure 12) was designed by Joel E. Goldthwait, Harvard orthopedist and vice president of the APL, along with his colleagues Dr. Lloyd T. Brown and Dr. Arthur Emmons. The university hired these men to create a scientific seating system for its students; the chair was eventually mass-produced and purchased for use at MIT and Smith College, where Goldthwait, in addition to having his own daughter enrolled there, served as consulting posture expert to university President William Allan Neilson. The Boston furniture manufacturer Plimpton Scofield Company eventually secured a patent for the chair, calling it a "posture chair," producing it in four different heights, including a reduced size for elementary schoolchildren.[36]

While the chair was originally intended for university students, Goldthwait and his team tested early chair models, with good results, on workers at Joseph and Feiss, a large clothing manufacturer in

FIGURE 12.   The ideal work chair, designed by
Joel E. Goldthwait, head of the Massachusetts Institute of Technology's 1921 "chair committee."
The slight bend in the bottom of the chair back is
an early iteration of building lumbar support into
workplace seating systems. This item would go on
to become the American Posture League's endorsed "posture chair."

Cleveland, Ohio, known as the "Ford Motor Company of clothing."[37]
They also researched the sitting posture of city clerical workers to determine the means of support for prolonged desk work.

The intended user of the MIT chair was thus relatively universal.
Apart from the four different heights, the chair was standardized, with

the intent to support seated work regardless of the sex, age, class, or race of its occupant. Although it looked, in many ways, like an ordinary wooden side chair of modified Shaker design, the work chair provided lumbar support with a slight bend in the lower back slat and the upper slat reclined farther back. The sturdiness of the solid wood construction without any cane work or fabric was also believed to lend the chair durability and the sitter a constant level of support. The ultimate goal of this chair was to get the sitter's muscular system to retain a kind of firm yet resting muscular tone in the torso rather than being at complete rest, as happened when the body sagged into an overstuffed, upholstered chair.[38]

The textile workers who had access to the MIT chairs for a period of four months, Wolfson and Hilles found, experienced less fatigue and pain than their coworkers who were sitting on kitchen chairs or benches.[39] The key, they argued, was the chair's lumbar support. Trained as orthopedic surgeons, Goldthwait, Emmons, and Brown paid particular attention to the biomechanics of the waistline in sitting. Poor chair design, they observed, led the seated worker to droop at the waist, creating a slump. (Examples of this can be seen running along the bottom of the poster in figure 13.) If a worker had to bend over to conduct work, the "forward incline should take place at the hips, not the waist," keeping the weight of the torso directly over the hip joint. Chairs with lumbar support, they argued, kept the line of the torso intact, leaving the arms and head free.[40]

Based on the results of their investigation, Wolfson and Hilles produced informational posters on seating systems that the New York City's Joint Board of Sanitary Control hung in workplaces throughout the city. The posters were intended to educate and persuade both industrialists and workers of the benefits of proper ergonomic seating. The message of the poster reproduced in figure 13 makes clear that supportive seating, with a lumbar support, would keep the worker in proper body alignment, assuring labor efficiency and pain prevention. Seating systems with lumbar supports became the gold standard at work, and eventually in other public places. In the early 1920s, the Brooklyn Rapid Transit Company outfitted its new streetcars with a seating system devised by the APL.[41] Likewise, the New York City park commissioner installed park benches with angled seat backs similar to the work chair.[42]

This demand for lumbar supports and other posture-enhancing features led to new markets throughout the 1920s and 1930s. Furniture companies, such as the American Seating Company in Chicago (originally the Grand Rapids School Furniture Company), began hiring seating engineers well versed in the latest in the posture sciences. Such companies funded large-scale studies in student posture, such as one conducted at the University of Chicago Laboratory Schools in 1925, where 5,000 elementary school–age children underwent schematographs and photography measuring both standing and sitting postures.[43] These studies, like those outlined in the previous chapter, often concluded that a majority of students exhibited poor posture, thus confirming the need for interventions such as improved seating.

Meanwhile, Boston area retailers looked to Harvard University to help with worker absenteeism among salesclerks. The Harvard Mercantile Health Work, housed in Harvard's Medical School, grew out of an interest in maximizing business profits, with merchants realizing that they could use university researchers to achieve this end. Headed by Emmons, the Harvard department interviewed and observed sales workers and noted an estimated 60 percent employee turnover rate. Rather unsurprisingly, Emmons blamed ill-fitting shoes and inadequate seating systems for poor worker outcomes. The Mercantile Health Work developed specialized footwear and stools to assist salesclerks who were on their feet, standing at counters or assisting customers, for most of the workday.[44]

Hand levers that controlled seat heights and seat back angles, first seen in patented furniture, became reincorporated into certain high-end posture chairs during the interwar years. But, as a rule, these chairs were reserved exclusively for the managerial class. Goldthwait and Emmons both expressed misgivings about offering such features to individual laborers or students, arguing that if the masses were given the freedom to control seat settings, the chairs would be "rarely correctly adjusted" to meet individual posture needs.[45] Industrialists and school superintendents also worried that adjustable chairs would be financially prohibitive, both because of the base cost but also due to expenditures on maintenance

FIGURE 13.   New York City's Joint Board of Sanitary Control used posters such as this to educate workers on the virtues of using the right chair. The main, top image is an exemplar of a woman seated in a straight chair with lumbar support, a contrast to the bottom four depictions of unhealthy posture and seating. The versions of poor seating include, from left to right, a saucer stool, a Windsor "stick chair," a cane chair, and finally a wooden bench.

and upkeep of the movable parts. Hence, most mass-produced posture seating systems for factories and schools prior to World War II were fixed structures much like the MIT chair and APL's streetcar seats. Similar to the German effort at providing energy-saving posture seating, as historian Jennifer Alexander writes, such seat design treated users "as deeply passive."[46]

By World War II, posture chair manufacturing expanded to include government contracts as well as sales to private homes. One company, aptly named Do/More, created an entire line of health seats aimed at both markets. Based on advice coming from the company's research division, Do/More salesmen were encouraged to give customers the "blood and guts posture story," a pitch that warned potential buyers of the deleterious illnesses that came with prolonged slouching, such as hemorrhoids, kidney trouble, constipation, dyspepsia, and hardening of the blood vessels. Do/More went on to become a supplier first for Bell Telephone Company and then State Farm Mutual Insurance, whose management praised Do/More chairs for bringing about a 99 percent employee attendance rate.[47] After World War II, Domore (they removed the slash from their name) won a government contract from the Federal Aviation Administration to produce posture chairs for air traffic controllers. The success of Domore prompted other chair makers to specialize in the new "ergonomic explosion," as historian Edward Tenner calls it, that permeated the U.S. military, schools, and workplaces for the remainder of the twentieth century.

Expanding the Market for Orthopedic Shoes

Posture crusaders believed that human feet, like the abdominal viscera, were under constant strain, especially prone to the problems created by bipedalism. This concern, after all, was backed by the latest discoveries in evolutionary science. As discussed in chapter 1, physician-anthropologist Dudley J. Morton, among others, believed that the human foot represented the final stage of human physical development before encephalization, making it a key feature separating humans and apes. Similar to slouching, flat-footedness carried with it comparable unvirtuous connotations of laziness and moral lassitude.[48] One way to overcome flat-footedness was through health-conscious footwear. This belief was spurred on by the APL as well as legions of professional nurses, physicians, educators, and savvy consumers, all of whom believed that the clothing and shoe industries should provide scientifically sound wearables that still looked tasteful.

A disability imaginary played heavily in the campaign for posture-right footwear, for high-fashion shoes, posture crusaders maintained, created crippled feet. And yet neither the crusaders nor manufacturers advised outfitting the nondisabled public with orthopedic shoes, footwear that was and still is strongly associated with disability. Instead, these experts attempted to find a compromise, outfitting the American foot with a "sensible" shoe that satisfied the anatomical needs of a bipedal body while also attending to style.

As soon as the League formed in 1914, Bancroft appointed orthopedic surgeons Elliot G. Brackett and Zabdiel Boylston Adams, both affiliated with Harvard, to lead a technical committee on footwear.[49] The task of the committee was to standardize foot measurements, provide a classification of foot "types," and to solicit footwear samples from shoe manufacturers in order to determine which products promoted good posture. As with other wearables, shoe manufacturers at the time were moving toward ready-made shoes that could be sold and distributed to customers through mail order as well as in the new retail department stores frequented by urban shoppers. This was an era, however, before shoe and width sizes had been nationally regularized. The well-known Brannock device, a linear, metal plate instrument with dual heel beds and adjustable slides to measure both the length and width of a foot, was not invented until 1925.[50]

Boston orthopedists were known for their research interests and studies on foot health and posture, an interest aided by the fact that the city was a leader in shoe manufacturing and home to the premier trade journal in the United States, the *Boot and Shoe Recorder*. Lovett conducted some of the earliest posture and foot health studies at the request of the Boston City Hospital. Hospital administrators were unhappy with absenteeism among the nursing staff.[51] One of the reported reasons that nurses consistently missed work was because of foot and back fatigue. Lovett conducted foot exams on approximately 500 nurses over the course of eight years at the turn of the twentieth century, and even collaborated with Dr. Morton Prince, who supervised thousands of foot exams on men applying for positions as police and firemen at the city's Civil Service Commission.[52]

Lovett found that at least a quarter of the nurses had significant "foot disabilities," especially flat feet. To prevent such conditions, he designed supportive, rubber-soled boots for the entire nursing staff, giving each employee a script to be fulfilled at a shop of her choosing.[53] Years later, the hospital's superintendent of nurses, who made regular boot and foot health inspections, reported that the orthopedically minded shoe wear had solved the problem of absenteeism, with "no loss of work time from foot trouble."[54]

Faced with the broader goal of standardizing foot "types" and determining posture-healthy footwear that could accommodate men, women, and children of all classes, Brackett and Adams solicited shoe samples from hundreds of commercial manufacturers, informing them that the APL was looking to endorse footwear brands that proved to be scientific and "anatomically minded." Hundreds of manufacturers, mainly in the Northeast, responded to the solicitation, sharing their wares and views on the League's effort. Frank P. Aborn of Lynn, Massachusetts—an industrial town near Boston known since the nation's founding as the "greatest shoe town in the country"—sent samples, guessing that the physicians would be interested in their line of low-heeled shoes with a broad toe box.[55] With advances in mid-nineteenth-century mechanization and the rationalization of production, Lynn specialized in producing women's shoes, and the town of Brockton, just south of Boston, was a leader in men's shoes. "The great majority of our customers," Aborn warned, referring to women, "prefer shoes . . . with high heels; in fact a very small percentage of our business is done on the lasts which are represented by the shoes which we send you."[56] A retailer at William Filene's Sons Company (the department store later known simply as Filene's) told Adams that the clerks in his department "refused to make sales to women who have the determination of buying shoes too short or otherwise not right for the feet."[57]

For decades, commercial shoe manufacturers and retailers had been negotiating two often opposing customer demands: comfort and fashion. The latter of the two brought the greatest criticisms in the trade. Shoe "evils," as they were called, included "toothpick shoes," a derogatory term for footwear with pointed toes, a fashion trend popular with both men and women at the time. Heel height, especially for women, was also a cause for concern. Attempting to professionalize and standardize the making and selling of shoes, manufacturers and retailers adopted the rhetoric of science in order to establish their expertise.

In order to move their products, some shoe and foot experts mimicked the work of scientists such as Morton, creating images of evolutionary foot hierarchies that would instill fear and disgust in potential consumers, motivating them to seek out expert help in their shoe fashion decisions (see figure 14). A regular feature of the *Boot and Shoe Recorder* was photographic images of "primitive" barefooted people, who were at once heralded for their "natural" strength and foot health and admonished for their lack of civilization (see figure 15). The intended audience for such images were those who suffered from the overcivilizing forces of modernity.

Overcivilization had many guises. In certain contexts, social commentators used it to mean luxury and sedentariness, which promoted physical weakness and disease. In other contexts, it connoted an excess in whims and fashion tastes, a kind of irrationality that became associated with the monied class. To demonstrate the potential harm of such overcivilizing forces on foot health, posture crusaders often pointed to the practice of foot binding in China. Bellevue Hospital chiropodist M. J. Pullman, for one, deemed footbinding to be a barbaric practice of a less enlightened Asian race whose own tendencies toward overcivilization resulted in degeneracies. Pullman, like many of his peers, used the practice of footbinding and its ills to warn of a similar potential fate for the newly liberated American women and their preference for high-heeled fashions (see figure 16). As such, American footwear became a site where middle-class professionals could vilify both primitive "savages" and the monied, decadent elite. Each in their own way posed a threat to evolutionary progress.[58]

EVOLUTION OF THE FOOT

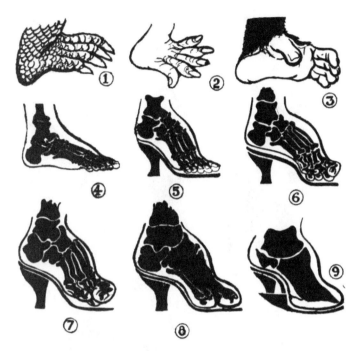

FIGURE 14.   A foot specialist provides a vivid illustration of foot evolution and devolution, the latter caused by civilization and artifice. Numbers 1–3 depict the progressive stages of foot development beginning with reptilian, moving to simian, ending with numbers 4 and 5, human foot ideals, unshod and sensibly shod, respectively. Numbers 6–9 illustrate ever-increasing stages of foot degeneration, caused by women who insist on wearing fashionable footwear. The end stage, number 9, represents an atavistic return to the reptilian stage.

Making footwear and shoe-fitting more scientific was not simply a matter exclusive to the pages of trade journals. Retailers began incorporating orthopedic departments into their commercial shoe stores in the 1910s so as to increase profit margins. Joe Marr of the Scholl Manufacturing Company in Chicago (the originator of "Dr. Scholl's" arch supports and other podiatric devices) pronounced in 1916 that since "ninety percent of the people suffer from some sort of foot trouble," the public

FIGURE 15.   A colonizer's view of ideal posture health and model
feet. This photograph was used in trade journals and advertisements
to depict the importance of the "primitive" shoeless foot, instilling
fear among civilized white people to buy anatomically correct shoes
lest they lose their evolutionary edge over other, non-white races.

wanted merchants who could advise them on shoes to relieve pain and
improve posture.[59] The Kahler-Bryant Shoe Company in Des Moines,
Iowa, opened its orthopedic department that same year, a "modern operating room . . . finished completely in white enamel," decorated as
such to give shoppers the feel of a controlled clinical environment, free
of germs and the whims of fashion.[60]

In order to advance their knowledge in foot anatomy and postural
biomechanics, retailers hired orthopedic surgeons to conduct evening

American girl reverting to
ancient Chinese custom—
deforming her feet with
"stylish" shoes.

FIGURE 16.   An unflattering caricature of the early
twentieth-century "New Woman." The exaggerated features are a political comment on how new freedoms in
politics, wage earning, dress, and sports for women inevitably leads to deformity and degeneration. Certain
politically and socially conservative experts insisted
that the foot fashions common among a certain class of
European and American women were akin to the custom of Chinese foot binding. Conflating the two was a
way to denigrate both demographic groups as racially
backward.

classes and author educational articles for their professional publications
and conventions.[61] Orthopedic surgeons, in turn, requested that shoe
manufacturers supply products that could assist patients who complained of foot and back pain. Orthopedic surgeons made shoe manufacturers and retailers look more scientific, while the "orthopedic
shoe" moniker gave the otherwise little-known medical specialty of

orthopedics visibility as both health care experts and purveyors of consumer products that promised greater health and well-being.

After assessing hundreds of shoe samples, Brackett and Adams opted for two specific kinds of commercial shoes that would earn the ALP's seal of approval. The first, the "Ground Gripper," was produced by the E. W. Burt and Company of Lynn, Massachusetts. With this shoe, E. W. Burt and Company had developed a highly adaptable design, a shoe mold that could fit children, women, and men. The Ground Gripper represented a significant shift in shoe manufacturing. By disrupting the convention that linked shoe production to sex, it served as a harbinger for gender-fluid footwear.

Billed as a "medical cure for flat foot," the Ground Gripper utilized the latest in rubber and welting technology. Rubber-soled shoes, such as the "Posture Foundation" one pictured in figure 17, rose in popularity at the turn of the twentieth century, made possible by the colonial extraction of raw materials along with the rise of American sporting and leisure culture. In 1917, the U.S. Rubber Company produced a new line of shoes which they called "Keds," one of the first all-canvas, rubber-soled sports shoes on the market.

A distinguishing feature of athletic and other health-oriented shoes, such as the Ground Gripper, was their relative gender neutrality. E. W. Burt and Company, like other manufacturers of comfort and sports shoes, believed that a thick projecting sole; a low, sensible heel; and a broad footbed—all of which characterized men's shoes—should be applied to women's and children's wear. The demand for such shoes was relatively high among female laborers and the leisure class. This is not to say that Keds suddenly replaced the felt need among women for high-heeled dress shoes, mainstays for churchgoers and formal occasions.[62] With new political freedoms and purchasing power, monied and working-class women began to buy multiple shoes, each pair keyed to a particular occasion: a pair of heels for the dance halls, a pair of boots for the workplace, a pair of oxfords for hikes, golf, and tennis.

The second shoe endorsed by the APL came from the Churchill and Alden shoe company in Brockton. Like E. W. Burt and Company, Churchill and Alden was known for its interest in health and comfort.

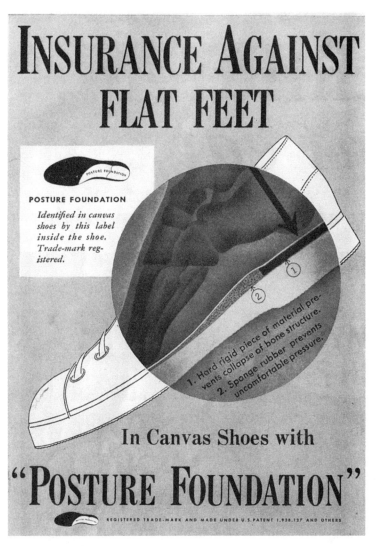

INSURANCE AGAINST FLAT FEET

POSTURE FOUNDATION

Identified in canvas shoes by this label inside the shoe. Trade-mark registered.

1. Hard rigid piece of material prevents collapse of bone structure.
2. Sponge rubber prevents uncomfortable pressure.

In Canvas Shoes with

"POSTURE FOUNDATION"

REGISTERED TRADE-MARK AND MADE UNDER U.S. PATENT 1,938,127 AND OTHERS

FIGURE 17. To posture scientists, flat feet caused slouching. Shoe manufacturers saw this as an opportunity to create a market for more medically minded footwear that aimed to correct and prevent the twin problems of flat feet and poor posture. With the colonial extraction of rubber from Africa and South America, canvas uppers with rubber soles became a popular commercial product and a staple of the burgeoning sportswear market.

At a sponsored shoe convention in the spring of 1918, the company announced the production of a new line which they called the "Trupedic."
A laced, leather upper for men, the shoe came in three standardized fits
according to foot "types"—"inflared" flat feet, straight feet, and outflared
feet. Dr. Percy Roberts, treasurer of the APL, became a devoted supporter of this line of footwear, and he frequently delivered public lectures
promoting the anatomical correctness of the shoe.

In its marketing, Churchill and Alden assured potential buyers that
fashion did not need to be sacrificed in a health-conscious shoe, pitch-
ing the Trupedic as "not a corrective shoe, but a trim, good-looking
shoe."[63] The company worked hard to disrupt the popular association
between disability and orthopedics, but not entirely so. The shift in
marketing strategy was a subtle one. Rather than focus exclusively on
consumers with already existing foot pain and deformities, Churchill
and Alden wished to expand the market to include health-conscious
buyers who could be persuaded that a future of disability could be prevented with proper footwear.

In an attempt to normalize what otherwise would have been considered an orthopedic shoe, Churchill and Alden claimed in one advertisement that the Trupedic was "a real anatomical shoe without the
freak-show look."[64] The reference to a "freak-show" spoke both to overly
stylized footwear, such as high heels and narrow toes, but also to shoes
prescribed to accommodate disabilities, such as clubfoot, gout, and
other deformities.

A "real" anatomical shoe also signaled white, heteronormative masculinity. By pitting "real" shoes against "freak-show" footwear, Churchill and
Alden imagined a white, male, able-bodied consumer as the norm, while
lumping together non-white fashions, female irrationality, and homosexual effeminacy as the dangerous other. Overly ornate or pointed shoes
were considered to be the product of defective tastes and people.[65] The
rational, scientifically minded man could thus distinguish himself from
"dandies," "fairies," and disabled "freaks" by wearing Trupedics.

Infant and children's footwear was yet another commodity that scientists and commercial shoe manufacturers used to market the gospel
of good posture. In an attempt to establish brand loyalty among the

youngest of consumers, shoe manufacturers and retailers turned to psychologists to help in the creation of window displays and advertisements. More coercive types of incentives were used as well. For example, Henry B. Scates, head of the William Filene's Sons department store in Boston, established a system of biannual visits in the children's section so that growth could be closely monitored. The "Filene method," as this practice would become known, significantly raised profit margins while posing as a humanitarian effort intended to "save the coming generation against prevalent foot ills."[66] Other successful posture footwear campaigns geared toward child health included Goodrich sports shoes, Keds for kids, and Buster Brown, the latter of which instituted a health exam involving a "triple check" of shoe fit, conducted by a "scientifically trained" store clerk who made detailed foot measurements and whole-body posture exams.[67] As is seen in Buster Brown's "Tread Straight" advertisement—a tagline suffuse with sexual, moral, and ableist overtones—parents at the time were promised nothing less than a happy and healthy future for a child who wore these shoes (see figure 18).

Female workers, relative newcomers to both shoe manufacturing and retailing, benefited from the rise in department store children's shoe sales. By the early twentieth century, boot and shoe manufacturing ranked a close second after textile manufacturing in the employment of women.[68] In 1922, the *Boot and Shoe Recorder* estimated that there had been a 100 percent increase over the ten years prior in the number of women shoe buyers and department managers in large department stores and specialty shops.

Women writing for the journal demanded equal pay for their work, insisting that, because of their "nature," they had more patience and could thus produce more exacting fittings than men could. As one female fitter put it: "this wonderful work of fitting feet properly will call for more women leaders to take charge of children's sections where seemingly one needs . . . great love for children, and unusual pride in analyzing for mothers the greatest need for health-building and beautiful posture in the growing child." Women workers, it was argued, could both empathize with and protect fellow female consumers, attending to

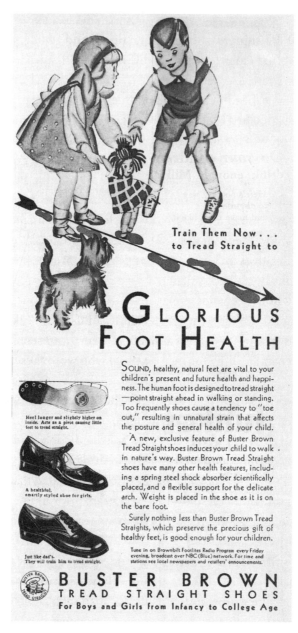

FIGURE 18. While the public face of Brown Shoe promoted sensible shoes for healthy, precocious children (the comic's sidekick, Mary Jane, led to the popularization of the namesake shoes for girls), its factories were known for their deplorable conditions, hiring underpaid child and women laborers, and also a team of entertainers with dwarfism who would dress up as young Buster and tour the country. Despite being a site for repeated labor strikes and antitrust violations throughout the mid-twentieth century, the Brown Shoe Company retained its public image as a maker of wholesome, health-promoting shoes, most apparent today in its flagship Naturalizer line for women.

the details of quality construction for young feet, assessing the inside of a shoe, "where a mother does not think of looking."[69]

During the interwar years, shoe wear and fitting became an object of scientific investigation well outside the APL and manufacturing. The First World War had normalized standard-issue footwear among not only draftees, but also scores of nurses and female factory workers. After the war, universities continued to be involved in researching footwear, but with the added goal of outfitting their own students. Under the presidency of Arthur E. Morgan, an engineer and future chair of the Tennessee Valley Authority, Ohio's Antioch College conducted extensive studies during the 1920s on footwear and posture (see figure 19), with particular concern for its female students. The most pressing matter, it seemed, was determining the best heel height for good posture. Although posture scientists insisted on supporting the growth and maintenance of the "natural" foot, they also retained biases concerning sex and shoes, especially for women.

Higher heels remained a signifier of femininity, a marker that even the most scientifically minded of individuals did not want to upend. The sexual politics of shoes informed the very research design at Antioch, where posture silhouettes were taken of women in various sizes of heel heights, dismissing the notion that the best shoe may be a flat-heeled shoe, similar to that conventionally worn by men. In the end, Antioch recommended two shoe types—a 1½-inch heel intended for dressier occasions, and a 1-inch heel for work and physical activity (see figure 20). Heels of 2 inches or higher, the researchers found, caused significant posture "distortion."[70] One outcome of the study was that the college began to franchise its own footwear, the "Antioch Shoe." These heels could be purchased at major retailers throughout the country well into the mid-twentieth century.[71]

By the time of the Second World War, both scientific and athletic shoes, with a particular emphasis on foot hygiene and posture, had become mainstays in the commercial marketplace. Shoe manufacturers, retailers, and posture scientists capitalized on the fears and concerns of a public who were being told through advertisements, public health campaigns, and in the clinic that ill-fitting shoes could lead to foot disabilities and chronic back pain. For day laborers, such a message struck a chord of

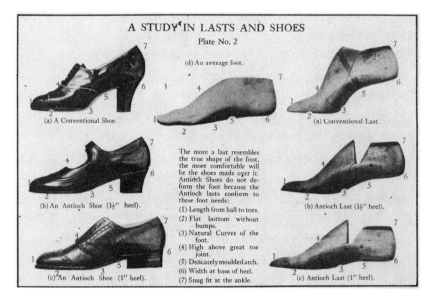

FIGURE 19.   The progressive, coeducational Antioch College in Ohio conducted extensive studies during the 1920s on women's footwear and posture. While not progressive enough to advocate that men and women would benefit from the same shoes, Antioch recommended to its female students a 1½-inch dress heel and a 1-inch everyday shoe. This image shows the different shoe models and lasts (wooden shoe molds) that the researchers used in the study.

fear concerning employability and the reliability of their bodily health, necessary for wage earning. For the professional classes—but also for the marginalized—occupational success and public acceptance increasingly became tied to looks and appearance. Putting your best foot forward, as the saying goes, required time, attention, and money.

The Sports Coat

"Clothes may not make the man," wrote one physician-posture advocate in 1916, "but we all conform more or less to our clothing."[72] While the concern for hygienic clothing dates back to the advent of bloomers, if not earlier, the belief that clothes could help combat a nationwide poor posture epidemic attracted the attention of a wide array of professionals and commercial manufacturers in the first half of the twentieth century.

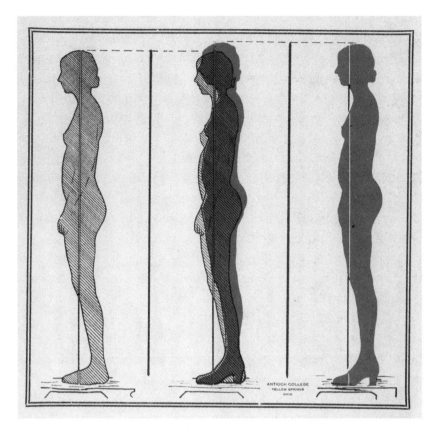

FIGURE 20.   A colorized schematograph triptych, demonstrating the posture of heel and non-heel wearing of the same research subject. In the middle panel, the right image is superimposed on the left so as to compare the difference between the two plumb lines. Since heel wear changes a person's center of gravity, Antioch researchers aimed to determine which heel height brought about the most ideal plumb line.

As with footwear, posture crusaders saw themselves as fighting against several cultural and political trends.

U.S. posture crusaders toed a line between the extremes of social conservatism on the one hand and cultural radicalism on the other. They did not condone the calls among radical feminists to banish corsets, nor were they swayed by the political cultural utopia of nudism, as was common among physical culturists in Germany.[73] Nor did they embrace free-wheeling capitalism, for they believed this fed into the indulgent tastes of the elite upper class. The wealthy were believed to set fashion trends based on irrational desire rather than science, dangerously influencing the commercial choices and desires of lower- and working-class Americans.

Much like shoes, children's clothing was a concern for the APL since growing bodies were thought to be more malleable than adults. Jessie Bancroft, president of the APL and head of its technical committee on children's clothing, focused her attention on jackets in particular. A poorly constructed jacket, she contended, contributed to the slumping of shoulders. After years of research, she gave the APL seal of approval to the Philadelphia-based wholesale clothing manufacturer Snellenburg Clothing Company for its modified Norfolk jacket for boys, known as the "Right-Posture" jacket. A style that originated with the hunting attire of the Duke of Norfolk in the 1860s, the Norfolk jacket became widely popular among the middle class by the turn of the twentieth century, having been adopted for use in sporting activities, including the new sport of cycling. Thanks to its adoption by the British Navy and other militaries, the jacket came to be seen as a crossover between formal and leisure wear, a precursor to the modern "sports coat," and eventually hybrid clothing such as athleisure. A single-breasted jacket buttoned high to the neck, the Norfolk offered support, allowed just enough movement, and looked tidy.

Other types of leisure jackets, Bancroft noted, were made with too much cloth between the shoulders and too little for the chest, a cut that, she believed, dragged the head and shoulders forward. Suspenders added to the problem, with elastic bands crossing just below the neck, placing force on the cervical spine, causing boys to thrust their heads forward. With the help of Bancroft, Snellenburg's patented its jacket with a "device built into the coat that continually remind[ed] the boy to stand straight."[74] With a trim cut across the upper shoulders, the wearer had little room to roll his shoulders forward, thus prompting a broad, erect chest.

The presumed additional benefit of the Right-Posture jacket was that it would keep boys in line behaviorally. Historian Julia Grant details how a close relationship between schools and juvenile courts grew in the early twentieth century. While ordinary mischief was often dismissed when seen in middle-class boys, the same behavior could land poor immigrant boys in a penal institution. Meanwhile, so-called coded-male troublemakers—categorized as "subnormal, incorrigible, truant, backward, and maladjusted"—would be sent to programs in special education. In many ways, slouching was seen as the first step down a slippery slope of more serious infractions.[75]

Norfolk jackets continued in popularity throughout the interwar years, with variations worn by men, women, and children alike. The jacket, according to one fashion historian, was a "multi-faceted garment, becoming a symbol . . . of democratic fashion."[76] After Brown Shoe bought the rights to the *New York Herald*'s Buster Brown comic strip in the 1910s, the company outfitted Buster with a Norfolk jacket to symbolize the company's commitment to posture health, from head to toe.[77]

In an era when advice columnists and child psychologists urged parents to do more than nag children about posture, clothing became a way to control form through material rather than verbal or instructive means.[78] Eventually the Norfolk jacket would give way to the sports jacket in the 1930s, which became a popular form of dressy but nonformal wear for men and boys. In both forms of attire, men and young boys were to look casual, yet athletic and well-poised. As the century progressed, and athletic and other types of "performance" wear became more mainstream, the concern for upright posture, while not overt, became something built into the construction and design of clothing.

Scientific Corsets, Belts, and Girdles

While Bancroft wished to oversee the standardization of ready-made clothing according to the dictates of anatomical erectness and correctness, she also wanted greater control over "foundational" garments, items such as corsets, waist supporters, hose supporters, and belts. Taking a cue from architectural engineers, she spoke of these articles of clothing as providing the "ground level" of support. Since undergarments sat nearest to the skin and muscles, they were also the most intimate and influential, in her mind.

When the APL set up a committee to investigate undergarments, the corset industry was in the midst of a highly publicized culture war. On the one side were "corset-burning" suffragists and slouching debutantes, fighting for feminism and also the right of freedom in dress. On the other side were cultural conservatives who wished to preserve the laced-up Victorian-era fashions. The APL helped the corset industry navigate these divisions by taking a middle road, turning undergarments into a medically necessary device rather than a material object rife with personal or political values.

Before the twentieth century, medical garments—including back braces and trusses—were often produced and sold by device manufacturers with the intent to correct a deformity or prevent pain. By contrast, mainstream undergarment manufacturers focused on aesthetics, shaping body lines so as to improve the drape and lay of outerwear, enhancing a person's outward appearance. The fabric and construction of both medical back braces and high fashion corsets, however, were oftentimes nearly identical. But the imagined user was very different. In the case of medical garments, the user was assumed to be disabled, either permanently or temporarily. Mainstream undergarments, by contrast, assumed an able-bodied user who had the means and wherewithal to be concerned about propriety and good looks.

But with the rise of the poor posture epidemic, linked as it was to the evolutionary sciences, bipedalism itself appeared to be a disability, or at least a potential anatomical weakness primed to create future disabilities. As a result, the categories of medical device and fashionable undergarments began to blur. To signal a new era marked by a happy co-existence between fashion and cure, corset manufacturers began to rename many of their products, and they did so by borrowing the language of medical device makers. The expanded offerings of commercial corsetieres included "belts," "waists," and "girdles." Such a move had the added benefit of expanding the consumer base to include both men and women.

Waists, belts, and girdles thus became comparatively androgynous articles of clothing, with only slight variations made for each category of wearer. Prior to the twentieth century, only a subculture of "effeminate" men donned corsets. But as more and more women began to adopt breast supporters and brassieres in the early decades of the twentieth century, women's undergarments moved from one-piece to two-piece items.[79] This move made it possible for manufacturers to sell separates, with each garment focused on a particular body part. Early bust supporters, historian Jane Farrell-Beck demonstrates, were rather simple garments compared to corsets and did not require industrial machines and fabrics in their construction. As such, industrialists and consulting posture scientists could thus focus their energies on the construction of garments that targeted the diaphragm, pelvis, and lower back, the region that was believed to be the origin of most postural deficiencies, not only in women, but also in men and children, as well.[80]

To promote the virtue and modernity of scientific corsetry and girdles, manufacturers hired physicians as spokespersons and educators. Dr. Alice S. Cutler, a physician at the Westborough State Hospital, traveled the country to give lectures under the auspices of the Royal Worcester Corset Company. Her job was to train classrooms of mostly female corset fitters in anatomy and physiology. "Anatomically," she told one group of fitters, "we are so constructed that we ought to navigate on 'all fours' and it is only through the different stages of evolution that we have come to walk in an upright position, thus causing the abdominal muscles to relax and bringing on a pernicious condition known as enteroptosis." Echoing the same statistics cited by public health officials, Cutler maintained that scientific corsets were of utmost importance, since it had been "conclusively proven that 90 percent of all women and . . . men are suffering from one or more troubles caused by this abdominal relaxation."[81]

To have a physician promote the wearing of corsets and girdles was a notable departure from earlier hygiene reformists, many of whom appealed to the physiological sciences in their adamant opposition to corsetry. While an older generation of health reformers used illustrations and eventually X-rays to educate the public on the ways in which tight lacing created horrific anatomical deformities of a women's rib cage, the new generation of physicians like Cutler used similar graphic technologies to argue nearly the opposite. In her lectures, Cutler replaced the nineteenth-century pictures of corset deformities with equally alarming depictions of uncorseted women and men, their abdominal and pelvic organs sagging due to a lack of muscular control and insufficient undergarment wear.

Arguments from nature also shifted in kind. Whereas the earlier generation appealed to the "natural" body, devoid of artifice, the new biomechanically oriented experts believed that "nature" needed some manmade assistance in order to compensate for the imperfections of human evolution. To deny someone such support, Brooklyn physician Albert Judd maintained, was akin to taking a cane away from a disabled person who needed an assistive device to walk. "LET THE CRIPPLE HAVE A CRUTCH," he wrote at the end of a medical treatise on the importance of scientific corsetry to body posture, poise, and health, lumping cane users and sufferers of ptosis and abdominal sagging into the same disability category, at least insofar as both populations needed assistive devices.[82]

The increase in abdominal surgeries during the early twentieth century facilitated the growth of support garments as well. Many posture experts—including leading orthopedic surgeons and gynecologists such as Goldthwait and Eliza Mosher—opposed the surgical approach to curing ptosis, preferring instead specific exercises, wearables, and seating systems to remedy the sagging of the body over time. Their opposition, of course, could not stem the tide of a lucrative and quick-fix approach of surgical correction. Historian Regina Morantz-Sanchez contends that women were by far the largest subscribers to and subjects of surgical experimentation, being assured that a whole host of symptoms ranging from fatigue and endometriosis to constipation and nervousness could be cured with the knife.[83]

At the same time, hernioplasties and appendectomies—abdominal surgeries performed on men, women, and children—were also being performed at a far higher rate in the early twentieth century than ever before, due to advances in antiseptic practices and techniques.[84] When such surgeries failed to alleviate the original complaint, or when they brought about further medical problems, patients often turned to posture specialists. Abdominal belts remained the standard of care for postoperative patients, and with the explosion of the market, providers had even greater access to a wide variety of products. To promote brand loyalty, certain manufacturers offered financial incentives to treating physicians.[85]

Looking to stem the tide of declining sales that came to define the corset industry at the turn of the century, manufacturers and retailers welcomed the partnership with medical doctors. After all, the most difficult task was to convince wayward women—feminists and those who had youthful, boyish figures that conformed to the fashion trends of the day—that they needed corsets. Because of this, several large corset manufacturers targeted educated white women, and more specifically the "college girl." Manufacturers and retailers believed that "college girls set the style"—for other women as well as their own progeny—and thus required the greatest attention, commercially and instructionally.[86] Whereas in an earlier generation, a corsetless woman was often associated with prostitution and loose mores, after World War I she was further condemned for being too independent and emotionally swayed by fashion trends, ignorant of the latest developments in science and hygiene.[87]

Merging aesthetics with the scientific, the Michigan-based Jackson Corset Company developed its own line of "College Girl Corsets," assuring customers that "wearing these corsets is actually like taking well-directed, healthful, systematic exercise for figure development."[88] The company distanced itself from old-fashioned corsetry by emphasizing the healthfulness and ease of their new line of step-in girdles, made with elastic insets instead of hard boning. The girdle, Jackson Corset Company argued, perfected the science of "redistribution," molding muscles, organs, and flesh into proper alignment.[89]

Similar to its work with footwear and jackets, the APL set up a committee to survey women's corsetry, a group led once again by Bancroft. An aspiring professional woman herself, Bancroft found a fellow traveler in Bessie Lazelle. Until the early twentieth century, corset manufacturing was dominated by men who owned, operated, and sold such items to female consumers. Lazelle bucked this trend, taking over the design, manufacture, and sales of her own line of corsets. Lazelle portrayed herself as being better suited to appreciate the science behind foundational garments, and also more able to connect with her consumer base.[90]

Owner and operator of her own factory, as well as salons where her products were sold by women sales agents, Lazelle was known for her innovative use of rubber elastics, specifically in bathing corsets.[91] At Bancroft's urging, Lazelle began to use rubber elastics to produce lightweight, flexible athletic undergarments. Together, Bancroft and Lazelle designed "waists" and girdles specifically targeted at young girls and slender women. Like the College Girl Corsets, Lazelle promised its wearers healthful support as well as youthful grace and beauty. In the "Posteur" girdle campaign, seen here, the image of an uncorseted woman, with a dowager hump and protruding abdomen, is juxtaposed with an attractive, perky girdle wearer (see figure 21). The preservation of youth, Lazelle insisted, was something that resided in a woman's backbone.

Lazelle was not the only manufacturer attempting to corner the posture corset market for young girls and women. The H. W. Gossard Company and its head designer, Ethel Lloyd Minific, created a "Stand Up Straight" line of corsets in the 1920s, with accompanying booklets targeted at girls entering puberty, pamphlets that could be handed to young daughters to peruse while their mothers were being fitted for their own undergarments. Corset companies zeroed in on a young girl's first experience in the corset shop in the hopes that she would become a lifelong customer.[92] Similar to menstrual promotional materials, the Gossard booklet came "in the form of a heart-to-heart talk," wrote one contemporary, "such as a wise mother or counselor might give, yet so disguised that it appears to be a 'beauty talk' rather than a bit of propaganda encouraging the wearing of corsets."[93] Having spent time in Paris, London, and with Boston-area orthopedic surgeons, Minific, like Lazelle, believed that the modern woman "who thinks for herself" would be persuaded to see science and beauty as one and the same, that the "importance of standing, sitting, and walking correctly [was] essential to beauty and good health."[94]

During the interwar years, many of the major corset companies employed physicians to provide lectures, bringing the medical and retail sciences in line with one another.[95] Department stores devoted significant space and resources to corset sales, providing customers with trained staff who had completed courses in anatomy and scientific corsetry fitting. Over time, corsetieres began to adopt the same language

FIGURE 21.   An advertisement for Lazelle's line, "Posteur." This particular foundation garment was intended for the mature, yet slender, women. At the far left of the advertisement is a shaded silhouette of a women who ostensibly did not wear proper garment supports, leaving her with a dowager hump.

as medical posture scientists. Alphonsus P. Haire, managing editor of *The Corset and Underwear Review*, wrote in 1921, "When it is realized that the right corset, correctly fitted, can and does alleviate cases of backache, indigestion, palpitation, pelvic disorders, ovarian and uterine troubles, intestinal disturbances and other peculiarly feminine ailments . . . the importance of the corset and consequently of providing corset fitters with more than a superficial knowledge of anatomy and physiology becomes at once more apparent."[96]

Profits for department store corsets remained strong through the 1920s, 1930s, and 1940s, often outpacing all other store departments. Retailers attributed such strong sales to the scientifically trained corsetieres who were able to sell the higher-priced undergarments, wearables that used surgical grade fabrications. By virtue of their trained expertise, which the new corsetieres emphasized in their interactions with customers, buyers were less inclined to return already purchased items. In this new setting, customers were also less likely to request alterations, "the bane of retailers," as historian Jill Fields notes.[97]

Medical and health professionals benefited from a booming business as well. By cooperating with industrialists, posture scientists acquired a large commercial outlet both to promote their own services and to warn against the dangers of the poor posture epidemic. While the APL made it a policy to not profit from any of the official endorsements that they granted, these professionals could nonetheless charge lecture fees, and also increase their patient base if more people began to see their own bodily deviations as medicalized problems that warranted physician treatment. To be sure, the orthodox medical profession and the American Medical Association (AMA) enjoyed a kind of public trust they had never had before. Yet the relatively new specialties of orthopedics and physical education—while endorsed by the AMA—still had to guard against chiropractors, osteopaths, and other "quacks" who specialized in spinal health, as well as electrical devices and braces that promised to be a quick fix to poor posture and back pain.[98]

Retailers promoted medicalized corsetry in the Black press as well, and the National Association of Negro Tailors, formed in 1920, worked with experts to create supportive undergarments that followed the

APL's guidelines.[99] One corsetiere writing for the historically Black newspaper *Philadelphia Tribune* implored her "indifferent sisters" to purchase the proper foundational support. "You are going to be very lonely this Spring," she wrote, "with your uncorseted look. Since gone with the styles of yesterday is the woman with Ptosis bulge, spreading hips, bulging diaphragm and sagging breasts."[100] Another advertisement for Spencer Corset—a British company that made significant inroads into the American retail market, in part due to its employment of Black female corsetieres—proclaimed that "good posture will release strain on back, abdomen, legs and feet. You won't tire so easily, and you'll have brand new energy."[101]

Ready-to-Wear Abdominal Supports

The retailer to emerge in the 1930s as the leader in abdominal supports was Camp, a company that blended the medical marketplace with the fashion trade, creating undergarments and other wearables that sold in large commercial retail houses as well as local drug stores. At the age of twenty-one in 1900, Samuel H. Camp bought the Bortree Corset Company in Jackson, Michigan, the first major corset company west of New York. By 1908, he renamed it the Camp Corset Co., and by the early 1920s had patented a unique "block-and-tackle system of lacing," which he used for a wide range of undergarments including corsets, abdominal belts, girdles, and "ptosis" supporters for men and women. Following the ready-to-wear trend, Camp produced his undergarments in several standardized sizes, yet the block-and-tackle system allowed the wearer to make personal adjustments according to individual need and comfort.

Camp used his patented lacing system in the popular "Camco" corset, which he advertised in wide variety of local newspapers and many leading women's fashion magazines (see figure 22). A step-in garment that provided "expert adjustment with one movement"—namely a quick tug on the straps at both sides of the hips—the Camco did not rely on fitters or individualized tailoring as much as other undergarments did. Additionally, in an era when women purchased several different corsets specific to various activities—one for work, one for evening wear, one for

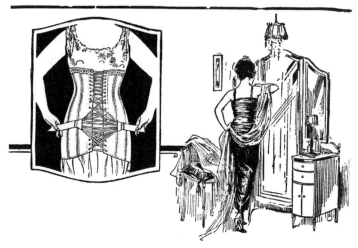

Expert Adjustment With One Movement

Every woman is her own corsetiere when she wears a CAMCO Corset! One gentle pull of the adjustment straps – and the laces move together automatically and in harmony. The straps are instantly secured and held flat against the corset.

The block and tackle system of lacing is only one of the many CAMCO features now being nationally advertised. The ease with which the CAMCO reduces the hips, supports the abdomen and brings natural grace to every figure, makes it a leader in sales.

We shall be pleased to give you further details about the corset which is building business for hundreds of retailers. Write to us.

Chicago office
57 E. Madison St.

S. H. Camp and Company
Manufacturers
Jackson, Mich.

New York office
373 Fifth Avenue

FIGURE 22. With a unique block-and-tackle system of lacing, the Camp Corset Company manufactured one of the first and most popular ready-to-wear foundational garments, making it possible for consumers to avoid visiting the corsetier or surgical device maker for an expert fitting. The wearer also had control over how form-fitting they wished the garment to be. Camp sold this feature as "Expert Adjustment with One Movement," connoting freedom and liberty on both the practical and ideological levels.

leisure—the Camco served as an all-in-one undergarment that could be worn and adjusted throughout the day according to a woman's activity.

Unlike the Lazelle and College Girl undergarments, Camp aimed to accommodate all figure types, including "stout" women, pregnant mothers and men who suffered from organ prolapse, hernias, and pendulous abdomens (see figures 23 and 24). While the design and fabrication of the undergarment was essentially the same, Camp marketed each as solving different "problems" for distinctive conditions. Worn by a slender woman, his garment was a "support" corset. Worn by a "stout" woman, it became a "reducing" garment. Worn by a pregnant woman, it would be called a "maternity" belt. Worn by a man looking to improve his figure, it sold as a "ptosis" support. For example, in an advertisement that appeared in a 1929 Lane Bryant catalogue, Camp claimed that its garment could correct posture, create a slimmer figure, and improve energy, among other benefits.

At a time when over three-quarters of corset manufacturers in the city of Jackson, Michigan, closed due to declining sales, Camp Company continued to thrive during the first half of the twentieth century, due in large part to its ability to satisfy market demand for undergarments that solved newly medicalized problems. As mentioned above, the drastic uptick in surgical operations in the early twentieth century created a need for postoperative supportive wear, an entirely new market created by the surgicalization of a variety of perceived medical conditions at the time. For every new operative technique—appendectomy, hernia repair, gallbladder removal, rib resection—came the opportunity to market a concomitant postoperative garment. At the same time, the medicalization of obesity, aging, and, of course, poor posture offered companies like Camp further opportunities to extend their market base and reach.

Similar to pharmaceutical companies, Camp hired a team of nurses and physicians who served as consultants in the research and development of his garments and devices. Some of these same nurses and physicians worked as traveling consultants and lecturers for the company. Samuel Camp himself wanted others to see him as a public health educator, a contributor to medical knowledge and its

FIGURE 23.   Camp appealed to both slender and "stout" women. Although Camp's undergarments differed very little in construction and design, the company created separate lines with different demographic groups in mind. For example, the company sold "reducing" garments for stout women and offered "supporting" ones to slender women.

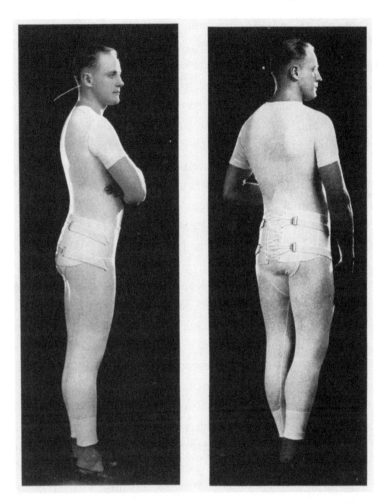

FIGURE 24.   Camp produced an entire line of supportive undergarments for men. The construction and design were nearly identical to the corsets and girdles they sold to women.

dissemination. In the mid-1920s, he hired a University of Illinois professor of medical illustration, Tom Jones, to produce a pictorial atlas, *Anatomical Studies for Physicians and Surgeons,* a booklet first published in 1930, with eleven subsequent editions running until 1943. The atlas featured anatomical drawings of men and women, emphasizing, in particular, sagging organs and other body parts, with the explicit message of selling Camp's remedies.

Although the Camp Company sold gender-specific garments for both sexes, women remained the company's primary concern and clientele. In an attempt to further legitimate his role as public health educator while also remaining one of the top sellers of supportive undergarments, Camp commissioned the engineers and artists at the German Hygiene Museum in Dresden to make a "Glass Woman," a companion and replica of the famous "Glass Man." When Camp commissioned the statue in 1935, the Dresden museum had already produced four glass men—one for the German museum itself, another for the Mayo Clinic in Rochester, Minnesota, a third for a "Eugenics in the New Germany" tour (this statue would eventually end up in Buffalo's Museum of Science), and a fourth that ended up in Paris.

Camp's "Transparent Woman," a purported replica of a thirty-year-old, five-foot-six-inch Caucasian woman, arrived in North America in 1936 to gawking crowds and onlookers. Camp paid $20,000 for her (over a quarter of a million dollars in today's money). Made with a translucent outer shell and a skeletal frame constructed from actual human remains, the "Transparent Woman" provided viewers with a glimpse inside the female body, her artificially produced organs wired with lights so as to illuminate the figure from the inside out.

Camp immediately sent the statue on tour across the country along with a troupe of sales representatives and nurses who would provide scientific lectures, many keyed to posture and organ health.[102] Along with the touring statue, Camp instituted a "National Posture Week" to further raise awareness of the potential health and beauty ramifications of poor muscle tone. Camp's "Transparent Women" not only physically traveled from city to city but was also repeatedly reproduced in miniature, used as a logo for department store display counters.

During the 1939 Posture Week campaign, Camp's head of public relations estimated that 2,000 retail stores—including Macy's in New York City—participated in Posture Week, along with several hundred schools and 300 city health officials. To increase participation among retailers, Camp organized a window display competition, offering a cash prize for the best presentation of its products that included a public health message about the dangers of poor posture. To assist stores in their displays—and to ensure that the Camp name was in full view—the company sent promotional materials, replicas of the "Transparent Woman," posture silhouette blow-ups, and pamphlets (see figure 25). Camp public relations estimated that over 3 million educational and promotional packets were requested each year in preparation for Posture Week.

While certain contemporaries criticized the campaign as a purely money-making venture having little to do with public welfare, Camp insisted that businesses, like schools and colleges, could be "great educators, selling ideas and ideals as well as merchandise."[103] Just as the United States entered World War II, Camp established the "Samuel Higby Camp Institute for Better Posture," a Manhattan-based office (located in the Empire State Building) that collected and disseminated scientific research regarding the importance of correct posture. The Institute funded the creation of posture health films, museum exhibits, lectures, and advertising campaigns, all offered free of charge to schools, public health agencies, and the general public.[104] Leading physicians, nurses, and U.S. government agencies regularly collaborated with the Camp Institute. In 1945, a Columbia University professor of orthopedic surgery, Armin Whitman, introduced the annual Camp Posture Week on New York's WNBT, one of the first commercial television stations, located in Rockefeller Center.[105]

Camp continued to sponsor the annual Posture Week throughout World War II, despite the widespread shortages and rationing of raw materials that went into corset manufacture, namely steel and rubber. To conserve wartime resources and "woman power," the company suspended the window display contest, and instead worked with government recruiting offices responsible for enlisting nurses and soldiers, as well as with munitions plants, setting up interactive posture health demonstrations. When the AMA and the Truman administration designated 1945 as "Physical Fitness Year," Camp produced promotional literature incorporating the nationwide slogan "Get in Shape for Victory—Keep in Shape for Peace" into its posture health campaign. Another initiative from the same year—the "Try-Angles of Good Posture" depicted in figure 26—emphasized how plumb-line verticality would secure an idealized American way of life, wherein a harmonious

FIGURE 25. Two Rochester-area drug stores promote the Camp National Posture Week, girdle sales, and the promise of upright posture as an all-in-one package in the local newspaper. The top advertisement depicts, from left to right, a "stout" woman, a "slender" woman, and a man, each of whom is wearing a Camp foundational garment that demonstrably assists in upright posture. The advertisement mimics posture exam techniques that had become commonplace earlier in the century, thereby making the product to be sold appear scientifically legitimate and not simply a fashion good.

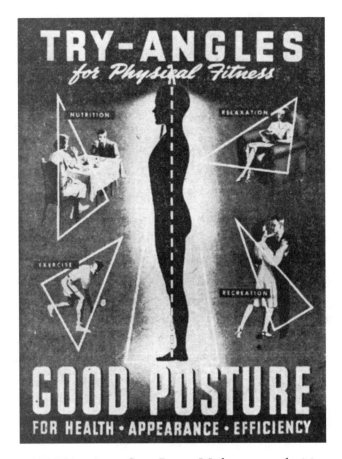

FIGURE 26.   A 1945 Camp Posture Week poster emphasizing
the foundational role that white, middle-class upright posture
plays in building a balanced, well-ordered life, and ultimately a
model of liberal democracy that is strong enough to combat
communism. The message in image and word states that
posture health makes for a more attractive individual, a hap-
pier home, and a more efficient workplace.

existence of consumerism, leisure, and workplace efficiency worked to
combat the threat of totalitarian regimes and war.

Demand for foundational garments increased during the war. Part of
the demand came from the influx of female war workers who were looking for supportive wear to help ameliorate the fatigue that came with
working on assembly lines. Combatants became important buyers as

well, specifically pilots, who demanded pressurized undergarments in order to fit comfortably into their tight-fitting anti-G suits.[106] The World War II experience normalized girdle wear for men, with one manufacturer estimating that he had sold 5 million abdominal "bracers" to men in the year 1951 alone.[107] Camp's Posture Weeks persisted well into the early 1960s, even after Samuel's death in 1944, providing stores and magazines with imagery that associated supportive foundational wear with posture health, patriotism, and heteronormative body ideals.

At first a novel arrangement of merging business interests with science, the effort to create, advertise, and market posture-enhancing body technologies took on a life of its own, a self-sustaining market that businesses soon learned could make them a profit. By the mid-twentieth century, the commodification of posture health and the creation of goods to assist consumers in their individual fitness pursuits gave way to a broader fitness industry that included an ever-growing number of stakeholders. The market for posture technologies did not create the slouching epidemic, but it did contribute to the notion that failing posture was a persistent and widespread problem that commercial goods could solve.

CHAPTER 4

Posture Queens and Fitness Regimes

THE EARLY TWENTIETH CENTURY is known as an era when many Americans became preoccupied with their physical fitness and appearance, defining their selfhood and worth more by exterior looks than by the rigor and strength of their inner, spiritual life. In certain cases, as in the muscular Christianity movement and its creation of both the YMCA and the YWCA, the physically fit body was believed to mirror a well-ordered soul. The YMCA adopted the symbol of the triangle—spirit, mind, body—indicating that an individual needed to attend to the health of all three facets of life in equal measure.

While there was a great deal of overlap between the goals of YMCA leaders and those of the American Posture League (APL), the latter focused primarily on the body as a "neuromuscular" system. Human posture, it turned out, relied on small, barely perceptible muscles that could "sense" a person's position and movement. The body's "sixth sense," as some called it, resided in the newly discovered proprioception system. The job of posture educators, then, was to foster conscious awareness of these muscles in the neck, torso, and pelvis so as to train proper posture habit formation, a task that required both physical and mental effort.

This kind of muscular training proved to be very different from the movement cures common to the Victorian era. Many nineteenth-century fitness regimes relied on equipment—barbells, Indian clubs, pulley systems, and the like—which could be found in urban, public gymnasiums, many of which required a fee-based membership. Because of these equipment needs, only the monied elite had access to—and presumably felt a need for—such exercise regimes.[1] Those involved in the twentieth-century posture crusade, by contrast, believed that in order to achieve health betterment among the greatest number of people, elaborate equipment and the inherent cost incurred by such technology created an insurmountable barrier for most Americans.

The "developmental" or "fundamental" fitness approach popular among posture scientists was comparatively simple and more economical, for it did not rely on complicated equipment. Moreover, to achieve posture fitness required attending to only a handful of muscle groups, a distinct departure from body building. In contrast to the well-known muscle-bound physical fitness idols at the time, such as Charles Atlas, Bernarr Macfadden, and Eugen Sandow, who accentuated muscular bulk throughout the body, posture advocates placed an emphasis on muscles deep within the abdominal cavity and around the spinal column, or, to use a term more commonly used today, the "core." Some of the most famous people who created popularized posture programs of the era—and still today—included Joseph Pilates, Mabel E. Todd, and Frederick Matthias Alexander. Posture work appealed to the middle and professional classes, for such training was billed as developing intelligence and self-awareness and not just crude, brute strength.

In order to sell the message of good posture to working-class Americans, posture promoters created competitions, often conscripting school-age contestants who would be judged for the straightness of their spines. The number of posture competitions that took place throughout the United States during the interwar years is difficult to accurately count, but the historical record suggests that they were widespread. Almost every school with a physical education department held such contests, beginning in kindergarten and continuing up through college. Because of this, posture contests became a shared experience among a wide range of Americans from all regions of the country.

Not all competitions occurred in schools, however. Other allied professionals with a vested interest in spinal health, such as the American

Chiropractic Association (ACA), held their own "Perfect Back" competitions, and advertised the events widely. Victors in the ACA contests regularly won scholarships to chiropractic schools and other monetary awards, often underwritten by commercial furniture or mattress companies. This outpaced school-based competitions, where winners would at best earn a top grade in their physical education class, or be featured in the local newspaper.

The rationalized posture exercise regimes that came to dominate American life emphasized the cultivation of individual discipline and surveillance, both of oneself and of others. What's more, the aim of posture consciousness combined the unique human ability of higher reasoning with primordial physicality, a combination of attributes that necessitated a kind of constant striving, even among those who were well placed to achieve this end. Sociologist Nikolas Rose argues that, in the modern era, bodily surveillance became a mandatory component of retaining one's citizenship. He writes, "The active and responsible citizen must engage in a never-ending . . . and constant process of modulation, adjustment, and improvement."[2] A vital part of the citizenship-making process, posture work and maintenance carried a lot of political weight. It was a project that required near-constant attention to the self and played a crucial role in the making of one's identity.

Posture Consciousness and Control

Shifting away from the European traditions of gymnastics and drill that many of the nineteenth-century originators of the American physical education movement promoted, the self-proclaimed "new physical educators" of the early twentieth century aimed to establish a more cohesive profession. Prior to the First World War, the overwhelming focus among leading physical educators—many of whom held medical degrees—was the building of muscles so as to achieve bodily perfection and symmetry of the human form. Dr. Dudley Allen Sargent, president of the American Association for the Advancement of Physical Education (or AAAPE, later named the American Physical Education Association), developed elaborate exercise and testing equipment at

FIGURE 27.   An early innovator in American physical education, Dr. Dudley Allen Sargent developed and manufactured his own system of exercise machines. Pictured here is Sargent's stomach pulley apparatus.

Harvard's Hemenway Gymnasium at the turn of the twentieth century. In addition to barbells, pommel horses, parallel bars, and Indian clubs—equipment necessary to train in the tradition of both Swedish exercise and German *Turnen*—Sargent installed thirty-six machines that he himself had designed and had manufactured by the Narragansett Machine Company.[3] Each machine targeted a specific region of the body in order to improve muscular development in the specified area. The most popular machines were the chest and the abdominal pulleys (see figure 27).[4]

Sargent also developed a standardized system of physical examinations that, using techniques borrowed from anthropometry, measured bodily size, strength, and symmetry. The exam consisted of fifty different tests and utilized tape measures, calipers, and spring dynameters. To keep the data organized and allow examiners to collate and compare one body with another, Sargent developed a one-page standardized chart upon which the results of the fifty tests were plotted on a single graph. Having opened his own school and separate summer camp in the 1880s for the training of physical educators—both of which enrolled more students than all other physical education training programs combined—Sargent instructed thousands of men and women in his methods.

His machines and measuring technologies never really took off in the way that Sargent had hoped. When he proposed to patent his exercise machines and tests at an annual AAAPE meeting, Sargent came under attack, especially by fellow physicians who found such a practice to be beneath professionals such as themselves. Following the wider trend of the American Medical Association, Drs. Luther Gulick (YMCA Training School in Springfield, Massachusetts), Walter Channing (Harvard), Edward Hartwell (Johns Hopkins), and Thomas Denison Wood (Columbia University) believed it necessary to distinguish the brain work of physicians from the manual laborers and tradesmen who built and sold medical devices and drugs.[5] "One trouble with us," Hartwell insisted, "is that this whole matter of physical training is in the apparatus stage," with little "insight and scientific interest in this matter" on the part of colleges and institutions of higher learning. Wood concurred, claiming that the science of physical education was "crude" and "embryonic."[6]

Hartwell, who had a doctoral degree in physiology along with his medical degree, insisted that the European systems of exercise were at once too showy and elemental. German *Turnen*, in particular, focused exclusively on muscle building, bodily form, and perfecting acrobatic performances without considering, he contended, the intricacies of muscle physiology. In order to develop a distinctly American form of physical education, Hartwell believed that the neurological side of muscular development should be emphasized. As such, he insisted on greater attention to carriage, posture, and bodily comportment than to conventional body building.

Posture exercises, then, came to be seen as an essential form not of muscular development, but of neuromuscular training, of tapping into the higher orders of the brain and central nervous system. According to Hartwell, such exercise "improved the neuro-muscular machine, facilitated the acquisition of neural pathways, led to the formation of proper

habits and enabled fathers to transmit to their progeny a veritable apti-
tude for better thoughts and actions."[7] This approach to physical education was in stark contrast to the more popularized forms of physical
culture and body building best exemplified by figures such as Eugen
Sandow and Charles Atlas, both working-class men who made a living
exhibiting their own muscle-bound bodies at sideshows and fairs, selling self-help books, magazines, and muscle developers.[8]

Materially speaking, posture training required little space or equipment. For the legions of physical educators hired by public schools and
colleges across the nation in the early twentieth century, the new, more
esoteric, and machine-free approach to physical education fit well with
the space and financial limitations of many school buildings at the time.
Posture training focused on smaller, more interior muscles, requiring
comparatively little movement, and could thus be taken up at school
desks in classrooms or in hallways, making the actual gymnasium, a
space originally built for the purpose of large-scale drill and performance,
unnecessary. What's more, posture instruction could easily be conducted at home, practiced in domestic spaces, even if they were cramped.

Jessie Bancroft, assistant director of physical education of Greater New
York City public schools, outlined the new posture-centered physical education in several books, instruction manuals intended for schoolteachers
and mothers. Bancroft told her readers to utilize her books sequen-
tially, beginning with "free-hand" gymnastics (i.e., exercises without
equipment), turning next to "light apparatus" (wands and occasionally
dumbbells), and then sports only when students had completely mas-
tered all of the physical feats in the prior books. Before students could
progress beyond "free hand" exercises, they needed to demonstrate a
"muscular sense," especially of the antagonistic muscle groups responsible
for upright posture. "The habit of incorrect posture implies," Bancroft
wrote, "a distorted muscular sense—a habitual feeling of a disproportion-
ate contraction and relaxation in opposing groups of muscles."[9] In other
words, she believed that posture training required, at base, brain work, the
willpower to overcome one's diagnosed faults.

Bancroft intended most of her physical regime to take place in classrooms, with a knowledgeable schoolteacher supervising the work.

FIGURE 28.   APL President Jessie Bancroft and her colleagues preferred exercises that did not require specific machinery or expensive gym equipment, opting for movement that could be incorporated into everyday life. Here is one example of a student undertaking posture exercises in the classroom, not in a gym.

Every exercise focused on the antagonistic actions of one muscle group. In her "feet close and open" standing exercise, the student would alternate back and forth between pointing the shoes out and back to a straight parallel position, all the while focusing on the opposite muscle groups at work around the pelvis, finding balance between the two bidirectional forces. Rather than purchasing extraneous equipment for exercise, Bancroft instructed schools and their teachers to utilize school furniture and other objects readily found in the classroom. For example, to develop trunk strength in their students, teachers could have children sit on top of their desks, securing their ankles under the bolted-in chair seat, as pictured in figure 28.[10]

Lillian Drew, physical therapist and director of physical education at Columbia University Teachers College, agreed that a "large number of mechanical appliances [are] unnecessary."[11] One of the most important technologies for Drew and others of her generation was the full-length

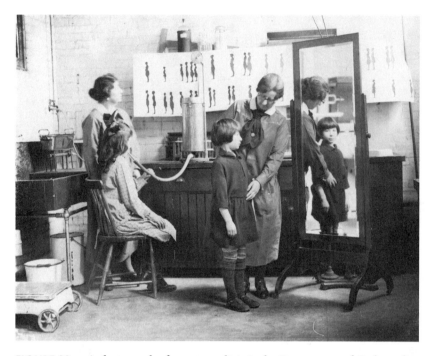

FIGURE 29.   A photograph of a posture clinic in the Department of Orthopedics at the Massachusetts General Hospital in Boston during the interwar years. Mirror work became a defining feature of posture training in the twentieth century. Also of note are the posture silhouettes hanging on the wall, depicting both good and bad form. The child on the left is undergoing a respirometer exam since posture habits were thought to have a direct impact on lung health and breathing capacity.

mirror. Mirror work helped students develop their "powers of subjective motor control," training the eyes, mind, and body all at once (see figure 29).[12] Contrary to group exercise and performance, which the European gymnastic systems mandated, Drew emphasized individualized correction.

While the techniques of posture training appeared relatively straightforward, the theories behind them were actually quite complex. Indeed, the subject of human posture captured the attention of several leading neurologists, psychologists, and philosophers in the early twentieth century. As discussed earlier, medically trained philosopher William James believed that posture training woke up an individual's deep-rooted will and intelligence. For him, posture was an ideal way for a person to attend to the oft-neglected relationship between muscular sensations and movement, between idea and action. A fellow pragmatist and centrist who sought a middle ground between materialism and idealism, John Dewey also believed that posture training could enhance embodied reflection, demonstrating a "unity of the mind and body," with each informing the other.[13]

Posture, these men maintained, was an expression and demonstration of the wholeness of a human being. According to the famed neurologist Charles Sherrington, this holism could be found in what he called "the proprioceptive system," a complex integration of reflex actions that relied simultaneously on muscular inhibition, excitation, spinal induction, summation, and synaptic events. Framed within the new neurosciences, muscles had a "sixth sense," possessing sensory organs and nervous pathways whereby muscles would send signals to the brain and vice versa. Of course, the degree to which humans had conscious, physiological control over posture was a topic of debate. The then newly discovered autonomic nervous system—the system in control of respiratory and cardiac rates, as well as swallowing, vomiting, and urination—seemed to function without consciousness or will. But posture, while reliant on the autonomic nervous system, seemed not wholly contained there either.[14]

Human posture thus became a site to explore the role of consciousness or mind. According to historian J. Wayne Lazar, there was a continuous debate among neurologists at the turn of the twentieth century about the mental versus mechanical impetus of biological processes, "about the presences of mind in various activities."[15] Reflexes, instincts, habits, and behaviors were identified by a "bewildering number of alternative and overlapping criteria . . . whether the activities were simple or complex . . . whether the activities were performed habitually, that is, with or without attention, whether the activities are voluntary or involuntary, whether the activities were intended or not, and whether the activities were accompanied by consciousness or not."[16] Certain biological functions, such as esophageal contractions, were widely accepted to be outside of conscious control, but others, such as human posture, seemed far more complex and thus necessitated intelligence by virtue of the complexity.

Consciousness was not a concern exclusive to neurologists. While Sigmund Freud made it central to psychoanalysis, W.E.B. Du Bois powerfully claimed that Black Americans suffered from "double-consciousness," a "sense of always looking at one's self through the eyes of others, of measuring one's soul by the tape of a world that looks on in amused contempt and pity."[17] The degrading two-ness that Black Americans experienced was a product, of course, of the constant oppression from living in a white-dominated, racist society. What is interesting about the effort to raise posture consciousness is that both white and Black advocates, without saying so, hoped to create the kind of double consciousness that Du Bois found inherently debilitating and degrading.

A student of Sherrington and Walter Cannon, physical educator Mabel E. Todd insisted that posture health required a "thinking body," the trademarked title of her 1937 instructional manual. Integrating engineering mechanics with anatomy and physiology, Todd claimed to provide an objective approach to human posture, in stark contrast to cultural, etiquette-based notions of proper form. Included in this objective approach was an insistence that, properly done, thinking could bring about better posture. In Todd's words, "Psychological factors and sensory appreciation are responsible for our varied postural patterns." Inspired by Sherrington's work on the proprioceptive system, Todd maintained that proper posture involved "the facility for forming adequate mental pictures and the kinesthetic sense."[18] Her technique of forming mental pictures—what we would today call mental imagery— would go on to form the basis of her trademarked brand of posture training, ideokinesis, which drew followers such as Marilyn Monroe and well-known professional dancers. Todd encouraged her students to think of the hip joint as a "wire-spoke bicycle wheel" and the pelvis as a rocker, using these visuals in her instruction (see figure 30).[19]

Throughout the interwar years, and as she gained more and more followers, Todd continued to refine her posture imagery. Evolutionary development became part of the mind-body narrative that her students

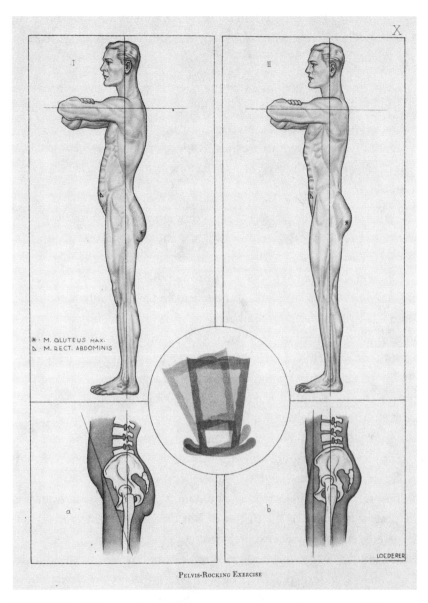

X

* · M. GLUTEUS MAX.
△ · M. RECT. ABDOMINIS

a

b

LOEDERER

PELVIS-ROCKING EXERCISE

FIGURE 30.   The use of mental imagery was common in posture training programs. Experts emphasized the necessity of brainwork in posture work so as to
set it apart from cruder forms of physical training, such as body building. This
illustration invites viewers to think of their pelvises as rockers, something that tilts
back and forth, with postural balance occurring somewhere between the two
extremes.

were to internalize. "The student thinks of an animal with a tail and what it would feel like," recounted one student under Todd's tutelage. Todd's pupil continued:

> Then the student imagines that the tail is attached to his sacrum and extends down to the floor. Next he imagines what that would feel like. With repeated practice the image may become so real that the student's body unconsciously adjusts itself to produce the feeling of the image by letting the weight of the spine drop into the sacrum and pelvis.[20]

Of course, human superiority over nonhuman animality was always assumed in such imagery work, for, as Todd and others insisted, only the former could engage and affect musculature through intellect.

John Dewey began to undertake posture training himself in 1916 after meeting Frederick Matthias Alexander at a dinner party with several Columbia University faculty members in attendance.[21] By this point, Alexander had immigrated to New York City from his native Australia, where he trained as an actor, a career cut short because of constant vocal strain. His interest in posture training and breath work grew out of his desire to gain better vocal control and health, and having developed a system of "conscious control," Alexander became a sought-after voice coach and professional reciter in New York, London, Sydney, and Melbourne.[22]

For Dewey, Alexander's posture training provided evidence that evolutionarily inherited habits could be consciously overcome if they did not serve the human race; the human organism was evolutionarily adaptive, fitted with the power of self-direction.[23] Alexander also believed that evolutionary betterment could be attained through conscious control of the body. In his book *Man's Supreme Inheritance: Conscious Guidance and Control in Relation to Human Evolution in Civilization*, Alexander wrote that modern humans had "debauched kinesthetic systems which permits defective registrations of different sensations or feeling tones."[24] Modern life led to a disassociation between the "'higher' nervous structures and functions—those which are at the basis of our conscious life—from the 'lower'—those which are involved in the execution of bodily postures and movements."[25]

Alexander and Dewey denied that human posture was a trait that evolutionarily "primitive" peoples had an inherent mastery over, and instead saw it as a complex physiological system that required "higher" order thinking, making it something that only the most evolutionarily advanced could develop properly. And yet, because upright posture was foundational to humanity, both believed that it was all the more urgent for the privileged classes to gain "primary control" over it, developing the smaller, less obvious muscles of the neck, back, trunk, and pelvis.

Alexander's contemporary, Joseph Hubertus Pilates, would come to similar conclusions even though both men ostensibly created their systems of physical fitness separately. A German immigrant to the United States in 1926, Pilates opened his first studio in New York City, training patrons in the correct balance of the "body and mind" through what he called "Contrology." "Most persons are not aware," wrote Pilates in his health manual, *Your Health*, "of the fact that, by reason of this utter lack of understanding, the human spine has been sadly neglected, for many, many generations."[26] Balance of the mind and body, he continued, formed the basis of his "science of Contrology," and like Alexander, Pilates promised that through such body discipline, the human race would improve as a whole. Perfect posture, straight spines, and mind-body balance were the "qualities of civilized man, which not only gives him superiority over the savage and the animal kingdom, but furnishes him with all the physical and mental powers."[27]

While throughout most of the interwar years Todd, Alexander, and Pilates received financial backing primarily from professional dancers and actresses, their theories and methods influenced physical educators who worked in public schools and colleges. Despite the economic downturn of the Great Depression, all three maintained their private studios while also enjoying visiting professorships and appointments in dance companies and places of higher education. At the invitation of Ted Shawn, Pilates cultivated close ties in New York with Martha Graham and George Balanchine and helped establish the Jacob's Pillow dance center in the Berkshires.

In contrast to Pilates, Todd enjoyed more of an academic career, teaching first at Columbia, then at the New School for Social Research beginning in 1931, while also maintaining her private studio. Yale University hired Bess Mensendieck, another posture fitness specialist who had studios in several European countries as well as in New York and Boston, to instruct its faculty on posture training education during the interwar years. Known for her German-inspired nude demonstrations of posture instruction, Mensendieck insisted that Yale students conduct posture training unclothed, a practice that the school adopted and continued until the late 1960s. Not until Yale admitted women did the school forgo mandating nudity.[28]

While these leading popularizers of posture training enjoyed favorable media attention and hobnobbing with the cultural elite, physical educators employed at primary and secondary schools struggled to inspire students. Writing to his friend and Smith College president William Allan Neilson, APL board member Dr. Joel Goldthwait warned that when the college set out to establish its own posture program in the 1920s, they should "expect a certain amount of protest from students who don't always know what is good for them."[29] It was one thing to train professional dancers, actors, and vocalists who depended on posture training to advance and maintain their careers; it was quite another matter when it came to the compulsory training of school-age children, teens, and young adults.

The problem of motivating students to attend to their posture became particularly acute as the culture of organized sports and games gained popularity among college students and the general public. By the 1890s, both male and female students urged more intercollegiate sports, including football, baseball, basketball, and crew, to replace the rigidity and monotony of drill-based physical education more common in the nineteenth century. While many college-level physical educators in the early twentieth century oversaw sports teams, they often required their student athletes to pass posture exams and foundational physical training first. At the University of Pennsylvania, for example, physician and physical education director R. Tait McKenzie had few positive things to say about the athlete who engaged in sports without first mastering posture training. After measuring the body of famed boxer John L. Sullivan, known as the "Strong boy of Boston," McKenzie concluded that the athlete was "not an impressive sight . . . [the boxer] was slightly knock-kneed with sloping shoulders and wide hips and his muscle form [was] not accented."[30] He derided musclemen such as Sandow as well, maintaining that such athletes were "slow and stiff" and lacked "vital strength." Such men represented great physical imbalance, having the "muscles of a giant, but the blood vessels of an ordinary man," which, according to McKenzie, put them at risk for overtaxing the heart.[31]

Posture Police

To make in-school posture training more like sports, physical educators often incorporated classroom competitions, teamwork, and play into their lessons. Educational manuals and professional journals from the time regularly advise incorporating posture plays and songs into lesson plans in order to make the acquisition of posture consciousness "fun." Similar to other public health agencies and medical specialties at the time, posture scientists turned to popular forms of entertainment and advertising to get the message of posture health across to the greatest number of people. In 1928, Boston area physicians and physical educators collaborated with the Massachusetts Institute of Technology to produce a black-and-white silent film on the importance of posture instruction, using child actors in full-body bathing suits as models of good and bad form; these scenes were then interspersed with anatomical displays of sagging skeletons and internal organs.[32]

The tradition of producing films that promoted good posture in school-age children continued for decades to come. Hardcastle Film's Social Science division and the Centron Corporation (in conjunction with the School of Public Health at the University of California) released color films in the mid-twentieth century with the message that good posture was necessary for good health, but also for social and economic success.[33] In one film titled *Posture and Personality*, the viewer watches a young, awkward girl named Adrelene realize that that her lack of popularity is due to her propensity to slouch. The film features a talking mirror—Adrelene's reflection—that taunts the real Adrelene for being a slump and urges her to make posture improvements. "Your posture is your problem," says mirror Adrelene. "What are you going to do

FIGURE 31.   High school students viewing the "Maintenance of Upright Posture" exhibit at the Museum of Natural History in New York City. The mutual interests between the evolutionary and posture sciences persisted throughout the mid-twentieth century and beyond. Displays that featured simians in bipedal positions were used to both confirm and promote vigilance against human and nonhuman likenesses.

about it?" "I'll tell you what I'm going to do," the real Adrelene yells back. "I'm going to make you a sight that I want to see because I'm going to have good posture! You just wait and see, hmmph!" The film concludes with a happy and socially popular Adrelene who, having worked on her posture exercises, can now look at a mirror reflection that is in harmony with her subjective sense of self.

Holding poster contests also grew in popularity during the interwar and World War II years. Students were encouraged to find inspiration in museum displays such as the "Maintenance of Upright Posture" exhibit held at the American Museum of Natural History in New York City, where curators encouraged onlookers to consider the human–animal divide and the evolutionary process in terms of posture health (see figure 31). Educators also prompted students to use the latest in advertising techniques, creating posters with snappy phrases, compelling

Flat - Footed and C - Curved is the Kangaroo
Please Don't Let Us Find These Failings in You

FIGURE 32.   A winning posture poster at Smith College, circa 1945. The use of nonhuman animals to promote greater vigilance over human posture was a popular tactic. Nonhuman animals, such as the kangaroo, served as model species with many posture faults, at least when anthropomorphized. At the same time, physical educators created entire exercise regimes and playground equipment for human adults and children to emulate animal physicality, such as "monkey" bars, bunny hopping, and flamingo standing, to name a few. Animal imagery continues to be used in exercise regimes around the world today, most popularly in yoga.

visuals, and humor (see figure 32). The posters judged to be of merit were then displayed in school rooms, dormitories, and lunch halls with the pupil's name attached. The poster contests allowed students who may not have had ideal physiques but possessed talents in the creative and visual arts to boost their posture grades.[34]

Assigning grades to students' posture was another way that educators motivated and, indeed, coerced compliance. For example, at Columbia University Teachers College, Drew set up a system whereby she designated a group of posture-perfect students to serve as her "Posture Policemen." The student police force—usually young college men and women majoring in physical education—were deputized to tag any student who did not stand, walk, or sit properly with a S.U.S. (Stand/Sit Up straight) badge. The posture police were also encouraged, without the knowledge of their peers, to conduct surveys of fellow classmates in lecture halls, filling out standardized "posture report cards" and turning them in to the instructor at the end of class.[35]

Although posture crusaders such as Bancroft and Drew wished to use body discipline to avoid certain forms of society-wide violence and unrest—workers' strikes and other forms of collective protest—they nevertheless encouraged a degree of interpersonal violence between students. In her instruction manual for teachers of all grade levels, Bancroft encouraged teachers to assign classroom posture averages and to then use these numbers as a way for classes to compete against one another. Such a tactic put pressure on students who—either willfully or because of a physical disability—could not make an A grade in posture. With a hint of admiration, Bancroft recounted in her instruction manual how "the boys in one class waylaid a classmate after school and pummeled him because his poor posture kept the class from one hundred per cent."[36] Students who had particularly poor posture were often placed in special "corrective" or "remedial" courses.

Disability Discrimination and Posture Fitness

According to the published record, certain parents and students appeared to appreciate posture training, while others did not. At a time when disability discrimination was legal in school admissions, employment, and immigration, enhancing one's physical form came to be seen as necessary for economic success and social acceptance. According to University of Nebraska physical educator Mabel Lee, parents wanted their children's "physical disabilities [to be] corrected . . . as far as possible."[37] One mother in particular wrote to Lee, saying, "I should like to see more attention given to detection of postural defects and work given for their elimination—particularly in the elementary school before the posture becomes too fixed."[38] Worried that her daughter with a curved spine would face a life of being sidelined—by both employers and future suitors—this mother reasoned that it would be preferable for her daughter to face a degree of ostracism earlier in her life, in the confines of school, rather than later when she believed the stakes of such stigmatization would be even higher.[39]

When the American Orthopedic Association conducted a survey of patients with scoliosis and other spinal deformities just before World War II, it found that 92 percent of people with back deformities did not suffer physical pain but nonetheless sought out medical treatment for the exclusive purpose of cosmetically remedying their curvatures.[40] Personal letters written by people seeking advice from medical doctors confirm these findings. "I'm becoming a monster to my children," wrote one patient about her back curvatures.[41] Growing up as a teenager with scoliosis in the 1930s, Joanne from Oregon told her doctor that "young people can be cruel . . . you can imagine the comments they had about me."[42] Another woman recounted how the UCLA teacher program refused to admit her after a posture screen revealed scoliosis. Ted, a coal miner from Pennsylvania, wrote that while his scoliosis "did not give him pain," or preclude him from work, the back curvature nevertheless made him "self-conscious." "I do not go swimming anymore," Ted admitted, "because of how my back looks."[43]

By contrast, for the children admitted to public schools and colleges who were fortunate enough to have avoided common childhood diseases and conditions that caused permanent disabilities, physical training could be fun and rewarding. After participating in a week-long posture drive at her high school in Goshen, Indiana, one student reported,

> The mirrors in the halls and the posters checking all the points of correct posture made a big impression. If a girl stopped to powder her nose, another pupil would read the little sign, "For Posture, Not

for Primp," and then they would both straighten up and talk about posture. I think the funny little rhymes which the art classes illustrated for us helped a lot because everyone went about repeating them.

Another classmate praised the fact that the council boys and girls were asked to grade the teachers, too. "One of the teachers told me that she tried to watch her posture when I was in her class," yet another student recounted. "I thought it was good for her to hold up her shoulders, because she usually humps over dreadfully."[44]

Boston physical educator Dorothye E. Brock asked her students to journal about their posture training. When asked what they hoped to get out of posture training, one twelve-year-old responded by saying, "I wish to correct those faults because a sunken chest might cause consumption." Answering the same question, a sixteen-year-old student wrote, "The prominent buttocks not only spoils a straight back, but it is undesirable if one wishes a good figure." Another student of a similar age wrote, "When I first started my correction it was chiefly because my family wished it. . . . Before long, though, I realized how much it would improve my appearance." This student also claimed to find "it more comfortable to stand and sit correctly."[45]

Afraid of losing student interest, Brock wanted to know whether or not her students found posture training tedious, so she asked them to comment on this in their journals. One respondent said that the work would be more interesting if she could have "five or six 'companions in misery'" to accompany her in her nightly prescribed exercise regime. The same student made clear that she found dumbbell work to be the most boring of all exercises and that she disliked having to look in the mirror to correct her posture on a daily basis.[46]

Posture Pageants

Although posture education could not be turned into a sport, it did fit quite readily within the rising popularity of beauty contests, competitions that drew crowds of onlookers and spectators. Some of the earliest

"Posture Queen" (and, in certain cases, "Posture King") contests occurred on college campuses, becoming a mainstay throughout the
middle decades of the twentieth century at the Seven Sisters colleges
and many historically Black colleges and universities. Barnard College,
for example, held its first contest in 1925, an event that continued annu-
ally until 1962.[47] Barnard junior Elizabeth Metzger was crowned the first
Posture Queen. According to a panel of judges consisting of a college
physician, the entire physical education department, and several representatives from the American Posture League, Metzger represented the
ideal body type in proportion and poise; she was five feet four and a half
inches, 120 pounds, and able to hold a perfect plumb line. The judges
described her as a "natural" beauty. According to her mother, who was
interviewed by the New York Times, Elizabeth "never used rouge or lipstick" and "she d[id] not smoke."[48]

One purpose of the college posture contests was to offer a more
"healthful" alternative to the increasing popularity of beauty contests—
such as the Miss America Pageant, first held in 1921—which heralded
scantily clad and painted women. While female beauty contests had
become ritualized across many nations and cultures long before 1921,
including May Day processions that celebrated female fertility, the
Miss America Pageant was among the first to focus solely on external
markers of beauty and sex appeal. For example, the Miss America
Pageant included a "Bathers' Revue," with contestants wearing nothing more than a bathing suit, parading their bodies before a panel of
judges and a crowd of onlookers whose collective vote accounted for
50 percent of the final score.[49] Posture experts scorned the bathing
revue, arguing that the contestants would be "exposed to grave dangers
from unscrupulous persons," and that such flaunting of the skin in a
sexualized manner would negatively influence girls and women. In the
words of one critic, the Miss America Pageant invariably provoked
"envy, malice, and vanity," ultimately "leading a girl to a career that ends
in her moral and mental destruction."[50]

The issue of women's freedom—in leisure, work, and politics—was
one of the primary reasons for the adoption of posture contests, especially at elite all-women's schools. The female leaders of the APL, many

of whom taught physical education and hygiene at the college level, worried about the moral and physical degradation of the next generation of girls. Self-proclaimed "physiological feminists" such as Clelia Mosher, Bancroft, and Drew believed that equality between the sexes would only come about if women cared for and exercised their bodies in a way that was similar to men. Doing so, they believed, would challenge the theory that women, because of menstruation, were the weaker sex and also undercut the stereotype that women, as the more emotional sex, tended to engage in artificial methods of beautification rather than a rational system of scientific exercise and improvement. As one posture expert put it, "artificial lives mean artificial faces and beauty contests are responsible for much artificiality."[51]

Agnes R. Wayman, head of the Department of Physical Education at Barnard, who crowned Metzger, used the school's posture contest as a platform to demand financial and political support from Columbia University. She also hoped that the city of New York would build pools, playgrounds, and tennis courts for girls to use. Addressing an audience of posture experts, Wayman proclaimed that the "restlessness" of the new Modern Woman was "a good thing." "She may be more radical than her mother," Wayman continued, "but this radicalism is due to greater mental and physical activity." "Without relentlessness, youth would make little progress," she concluded.[52]

But these freedoms came with limits, especially in regard to the heteronormative rules of sociability. Loose mores and heightened sexuality—represented by flappers, the debutante slouch, and the Jazz Age—would not be tolerated. Goldthwait, whose own daughter attended Smith College, urged the school's president to continue to make posture training a mandatory part of the undergraduate experience for these very reasons. "I . . . want to let you know," Goldthwait wrote, "how much appreciation I have [for] the task which rests upon your shoulders."[53] "The younger generation today," he continued in a separate letter, "needs very little assistance or encouragement in the expression of their emotions." Goldthwait urged for a "restraint of emotion" through disciplined posture exercises, tests, and competitions, which he, along with physical therapist Leah Thomas, took charge of at Smith in the early 1920s.[54]

The need to exhibit uprightness—in physical form and perceived sexual mores—was felt even more among non-white women, whom white male colonists and slaveholders had historically portrayed paradoxically as both innately hypersexualized and slothful. When Mount Holyoke crowned Fumiko Mitani, a student from Japan, the winner of its posture contest in 1926, the Black-owned *Chicago Defender* noted how only "A few years ago, [Mitani] was a barefooted, sunburnt girl in a Japanese village."[55] From her skin color to her poise, Mitani became closer to white through her Mount Holyoke education, a cause for celebration, according to the Black press. The *China Press*, otherwise known as *Ta-lu pao*, engaged in similar reportage of Miss Rose Jacobs, a Sioux Indian originally from South Dakota who won the 1930 posture contest at the University of Kansas. Picturing her in sensibly heeled shoes and a pleated skirt—mainstream fashion standards for white America—the *China Press* was quick to point out that Rose never wore lip or cheek rouge, and, most importantly, she always refrained from using "war paint."[56]

The press had much to gain from beauty contests, for reportage of such competitions sold papers. Historian Maxine Leeds Craig argues that this was particularly true in the Black press, with papers such as the *Appeal*, the *New York Age*, and the *Chicago Defender* sponsoring beauty contests decades before historically Black colleges began to hold their own posture competitions. "Throughout the 1920s," Craig writes, "fraternal groups such as the Knights of Pythias sponsored pageants, northern Black enclaves of Louisiana migrants crowned queens at Mardi Gras balls, and the Black newspapers continued to boost their sales by combining themes of racial pride with photographs of beauties."[57] In 1927, the *Oakland Western American* newspaper was responsible for setting up the Miss Golden State beauty contest, an all-Black pageant created in response to the all-white Miss America Pageant. These pageants—and their judges, who were often Black middle-class men—often rewarded light-complexioned over darker skinned women, a fact that many in the Black public resented.

Black posture crusaders saw posture contests as a way to emphasize physique over skin color, the former being ostensibly controllable through individual will and heightened consciousness. Maryrose Reeves Allen, one of the few African American graduates from the Sargent School for Physical Education, established a posture program first at Hampton University and then at Howard University, where she assumed the role of director of women's physical education in 1925. There, Allen trained scores of Black female gym teachers who would go on to take up positions in colleges and high schools across the country throughout the mid-twentieth century.[58] Upon assuming her position at Howard, Allen pledged "to mold every one of my girls so that wherever they go the world will whisper, 'I can always tell a Howard woman when I see one because she walks in such beauty.'"[59]

The men and women in charge of Black posture contests were primarily middle-class African American educators and physicians, some of whom trained alongside the white professionals from the APL. Holding their own posture contests, middle-class African Americans at once internalized and resisted white beauty norms. "Efforts by elite Blacks to assert the dignity of the race," Craig writes, "always drew upon dominant cultural codes that carried a host of conventional ways of thinking about class and gender."[60] And yet these contests also challenged the white supremacist view that Black men and women were innately ugly, depraved, and evolutionarily primitive. Black women, in particular, used physical education and posture training as a way to create and control their own self-image. In these spaces, young Black women were taught to maintain svelte, streamlined figures, disrupting the dominant white image of them as overweight, curvaceous mammies on the one hand, or promiscuous Jezebels on the other. Posture training and contests were a way for Black women, as historian Ava Purkiss writes, to "use their bodies as a site for citizenship making."[61]

Allen would have agreed. Similar to her white colleagues, Allen, who originally received her training under Sargent, saw a direct correlation between health and beauty, both of which, she believed, could be scientifically studied and taught. When Allen returned to Boston University in the 1930s to earn her master's degree, she completed a thesis that provided standardized scales to measure a woman's physique, facial proportions, and expressions. Of those measures, "proper form, proportion, symmetry, posture, and poise" were paramount.[62] Allen did not simply mimic white posture standards, for she believed that Black women had the extra task of overcoming a history of "physical and psychological degradation—a legacy of slavery, oppression, and prejudice."[63] Self-hatred, Allen observed, was particularly debilitating to the Black race, a cause for poor health, high death rates, and short life spans. Through posture, beauty, and charm instruction, Allen wished to help her students to experience their bodies as a source of pride, seeing "beauty in their dark skins, their woolly hair, white teeth, sparkling eyes [and] full mouth." Attaining proper posture, she maintained, was not simply a physical act for Black women, for it required first "an inner spirit and self-respect which makes all people feel upright and grand."[64]

Black physicians offered similar advice. Dr. A. Wilberforce Williams, a prominent African American physician and first health columnist for the *Chicago Defender*, wrote weekly columns on posture and hygiene. "The body machine," he wrote in 1921, "will function properly and correctly for many years if one is on the lookout for the first signs of sagging and immediately sets about to correct the sagging."[65] Writing health columns for the *Afro-American* during the 1930s, Dr. Algernon Jackson, director of public health at Howard University and leader of the National Negro Health Movement, noted a "racial habit" among "young colored people . . . who too often slump along, stoop shouldered and walk with a careless, lazy sort of dragging gait."[66] In another column, he wrote, "As a race, we have a high death rate of respiratory diseases, many of which I am sure can be avoided by deep breathing, upright body carriage and proper walking." "When I see the miserable slumping postures of young [Black] men and women," he continued, "I cannot help but wonder whether the physical education department of our schools, colleges and universities are doing their full duty in the important matter of race building through body building."[67]

With such well-publicized concerns coming from leading African American authorities, many Black colleges began to hold their own posture contests. The Physical Education Department at West Virginia Collegiate Institute, a Black public university first established in 1891, announced nineteen finalists in its 1927 posture contest, awarding the winners with the APL's bronze insignia (see figure 1, chapter 2).[68] Bennett College, a private school for Black women in Greensboro, North Carolina, held its annual posture contests during National Negro Health Week.[69] Although most contests and pageants remained segregated until the 1960s, there were a few exceptions. Reporting on a 1936 posture contest held at Englewood High School in Chicago, the *Chicago Defender* celebrated a few "members of the Race, who were selected as the ones having postures closely approximating perfection."[70] According to the article, only 300 out of 3,800 girls qualified to enter the competition; twelve winners were named, two of them Black.

In holding these contests and celebrating the victors, Black Americans at once internalized and resisted compulsory able-bodiedness and whiteness. Black families often celebrated posture competition wins in their local newspapers. When their daughter, Doretta, won the New York City borough-wide posture contest, the Norman family of Staten Island submitted an announcement to the *Afro-American* newspaper, heralding her for having received the "coveted gold arrow for habitually good posture."[71]

Today's associate dean at Yale's School of Medicine, Dr. James P. Comer, also remembers the importance of posture contests in his household. Comer's mother, Maggie, a Black woman who migrated north just before the Second World War, was fastidious when it came to her children's appearance, learning early the advantage of buying good clothing and shoes so as to look clean and tidy. His mother took particular pride in his sister Louise, who, as a young girl, won the school-wide posture contest at an Indiana public school. "She was the only one black," Maggie told her son. Louise ended up beating a hundred white girls. Yet such accomplishments often fueled the flames of structural racism and violence. The night after her daughter's win, Maggie, working as a caterer for a local white family, overheard a sneering guest say, "Did you see a little N—girl won the posture contest?"[72]

Although posture experts saw themselves as more scientifically qualified than commercial beauty experts on matters of health and hygiene, the two realms often overlapped in their rhetoric, practice, and modes of communication with the wider public. Indeed, both the American

Physical Education Association and the American Medical Association encouraged its members to find ways to "appeal to vanity and to pride and to physical appearance" in their students and patients. Such appeals, these professional associations maintained, were "practical and legitimate means of motivation."[73]

Women's magazines such as the *Delineator*, *Collier's*, and *Ladies' Home Journal* regularly featured physicians and physical educators involved in the posture campaign. Dr. Barbara Beattie, a 1929 graduate of Cornell Medical School, penned popular articles advising women on how to "hide their hips," improve their chin lines, and maintain a youthful appearance through posture exercises. Well-groomed and healthy women, Beattie maintained, did not "carry an ignominious projection below the small of their backs," but rather moved with "hipless grace."[74] Citing Goldthwait and other posture experts, Beattie informed readers that all such authorities "agreed that when your pelvis is in proper position, your hips . . . automatically 'fold' down and under you—where they belong."[75]

In a separate article about how the chin was a "a dead give-away of age," Beattie counseled women on neck fitness, strengthening the muscles to avoid "flabby and weak" head postures that resulted in double chins commonly seen in middle-aged men and women.[76] One of the best things that a woman could do to improve her posture and appearance, Beattie believed, was to find employment, for such a life required a woman to preserve her youthful looks, and thus her posture health. Dr. Ruth F. Wadsworth, also a Cornell Medical School graduate and author of the 1928 popular book *Charm by Choice* and health columnist for *Collier's*, insisted that "a fat abdomen is almost always a posture fault."[77]

The conflation of obesity, old age, and sagging posture—all of which became both demonized and medicalized during the interwar years—permeated health advice columns, including those authored by physicians.[78] While women bore the brunt of criticism for their so-called figure faults, the male physique came under scrutiny as well. The American Medical Association, in its public-facing publication *Hygiea*, ridiculed the "average middle-aged man," known for his "pitiful figure, his head sunk forward, chin dragging, chest collapsed, stomach protruding."[79]

To counter this trend, *Hygiea* offered advice to parents and teachers on how to motivate schoolboys to stand straight. The magazine offered entire play scripts on the theme of posture health for children to perform. In one posture play, two schoolboys were assigned the roles of "Thomas Sloven" and "Peter Posture," with the other students reciting the following lines:

> When Thomas Sloven walks to school
> He slouches and he stoops;
> He's a most ungainly looking lad
> From his head down to his boots.
> Then Peter Posture comes along,
> His face alight with joy.
> With head erect and shoulders square,
> He's a happy, whistling boy.[80]

Coeducational schools often included male and female divisions in their contests, signifying that good looks and proper form were beauty norms for both sexes. Madison Junior High School in Rochester, New York, made it a practice to name three boys and three girls as winners of its annual public school posture contest (see figure 33).[81] The all-Black Dunbar High School (known as a feeder school to Howard) in Lynchburg, Virginia, did the same, inviting local Black clergy members and YMCA representatives to serve as judges.[82] In most cases, boys and girls were assessed separately. But the winners often shared the same stage. Some schools crowned posture-right "couples," following the tradition of certain high schools' annual homecoming king and queen contests.

Promoting the assumption that straight spines signified health and attractiveness, many posture scientists contributed to the culture of disability stigma and indeed used such stigma as a means to coerce students to participate in posture contests, training, and surveillance. Many physical educators, public health officials, and physicians would tell students that if they did not have good posture, they would fail to attract the opposite sex. As one physical educator put it, "the attitude of the opposite sex is one of the strongest and most deep-seated factors."

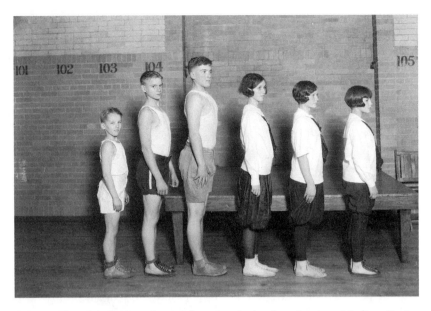

FIGURE 33.   Like many other coeducational schools at the time, Madison Junior High School in Rochester, New York, held posture "Queen" and "King" contests. Pictured here are the top three boys and top three girls, divided by age group, who won the annual competition.

"Show a lad the posture silhouette of four girls," the educator continued, "and ask him which of these he admires most or which he would choose preferably for a wife." As if betting on a horse race, the educator concluded that "Nine out of ten boys will pick the 'A' or 'B' posture."[83]

Spinal Beauties

Because school posture contests drew a significant amount of media attention, other professional and health culture organizations began to hold their own competitions. Unlike school-based competitions that mandated the involvement of all students, privately sponsored contests required voluntary participation of contestants. One such contest was the annual "Most Perfect Back" competition sponsored by the American Chiropractic Association (ACA). Beginning in 1927, the contests were held at the ACA annual convention. At first, the ACA sponsored both a male and a female division of the contest, but by 1931 it held women-only competitions, citing a lack of interest and participation among young men and boys. As one ACA member lamented: "Posture Week is intended to emphasize good carriage in both sexes, young and old. However, the women . . . are the only ones who take it very seriously." "It would admittedly be a very difficult undertaking to get a group of teen-age boys," the chiropractor continued, "or any group of males for that matter, to enter a posture contest . . . that slovenly look called the slouch is almost the exclusive property of the younger male of the species. It's a very weighty problem."[84]

The ACA's competition drew the largest interest from the southern and western regions of the country, often in places where beauty pageants had the greatest cultural presence and purchase.[85] Since Black contestants were barred from the ACA competition until the 1950s, the competitions were dominated by white, Hollywood ideals of beauty as well as the more conservative notions of the female southern "belle." Louise Goodwin, who won the contest in 1933, decided to enter the contest after her "mother happened to notice a newspaper clipping" advertising the competition. Out of 500 contestants in 1932, Goodwin won second place, but was unable to take first place because the chiropractors judging her noticed a "slight imperfection in her lumbar region," a flaw indicated by means of an X-ray, an integral part to the contest. Indeed, by the 1940s, the ACA produced life-size spinal X-rays of each contestant, having them hold their images in front of them as they lined up for the announcement of contest winners. After seeing her X-ray, Goodwin underwent a year of chiropractic adjustments and subsequently entered the following year's annual competition in Denver, where she was crowned Miss Perfect Back of America, awarded $500 in cash, and given a scholarship to attend a chiropractic college of her choosing.[86]

In later years, winners of the ACA Perfect Back won trips abroad, wining and dining with political leaders and foreign dignitaries. There was a distinct and direct linkage between perfect posture and material success. Winners of posture contests, whether from the ACA or from schools, often became darlings of the media. Newspapers and other media outlets knew that beauty contests drew readers, and posture contests were no different. Likewise, chiropractors understood that the competitions greatly benefited their profession and business, and openly discussed this fact in their trade journals.

As good posture increasingly became linked not only to health but also to personal and material success, the notion of a poor posture epidemic would become even more insinuated into the norms, practices, and beliefs of American culture and life. What was once a normal part of aging or a simple physical variation was now a bodily "defect" rife with meaning. Moreover, with internal and inter-peer surveillance being taught in the nation's schools, there was no hiding a slouch.

This structural stigmatization of individuals with curved spines and sagging heads was systematized during the interwar years, rendering such posture "failings" into a disability that could drastically affect the outcome of one's life. Of course, the stakes for maintaining an upright body were higher for those who fell outside the dominant class of healthy Anglo-Saxon men. Many women, persons with minor disabilities, and African Americans took part in posture contests and training as a form of insurance against the risk of being lumped together with "lesser" types who, because of their slouching, were presumed to be slothful, physically weak, and lacking in intelligence. Posture maintenance thus became a lifelong task, requiring constant work, vigilance, and surveillance, especially for those who lacked the protections of full citizenship.[87]

CHAPTER 5

The Geopolitics of Posture

THE SECOND WORLD WAR, much like the First, revived a sense of panic concerning the fitness of American citizens, with posture health becoming a baseline definition of individual and national strength. American bodies became crucial in the Cold War fight against communism and totalitarianism, and if there was one anatomical attribute that seemed in need of strengthening, it was the American backbone. Indeed, straightness took on a new import and meaning, becoming synonymous with heterosexuality, sobriety, and honesty. In a nation increasingly paranoid about covert infiltration by enemy regimes, the detectability of upright physical posture served as a visible indicator of normalcy, patriotism, and trustworthiness.

Just as the United States became ever more engulfed in military engagements abroad, it was undergoing significant demographic changes at home. The fractures and inconsistencies in the ideals of American democracy became clearer than ever. African Americans and women who served in the military noted the hypocrisy of fighting for freedoms that they themselves were denied on the home front. Other Americans, however, enjoyed the postwar economic boom. The 1944 passage of the Servicemen's Readjustment Act, colloquially known as the GI Bill, led to a marked increase in young men attending college and buying houses in the newly developed suburbs. With greater national wealth and significant advances in medical therapeutics—especially the development of antibiotics and blood transfusions during World War II—the threat of deadly contagious disease and infection waned. As average life expectancy increased (at least for those with reliable access to health care), concerns about chronic disease and debility grew. Epidemiologists refer to this transition in the mid-twentieth century as the epidemiological shift, when the greatest health threat at a population level went from acute, infectious (often childhood) diseases to chronic, disabling, man-made health conditions.[1]

The changes that occurred during the World War II era had a significant impact on the nature and work of the posture sciences. When it was originally formed in 1914, the American Posture League (APL) contended that there was a link between poor posture and deadly disease, especially tuberculosis; APL members offered their expertise in order to prevent, treat, and control contagious infectious diseases. But in the wake of large-scale distribution of antibiotics, the tubercle bacillus no longer presented the same kind of threat. Sensing the writing on the wall, the American Posture League disbanded in 1944.[2]

Yet posture research and surveillance continued to flourish throughout the World War II and early Cold War years, taken up, in part, by health care workers with a vested interest in new specialties such as sports medicine, gerontology, pain management, physical therapy, and biomedical engineering. The birthplace of most of these specialties was in the university setting, where researchers would conduct studies on college-age students and patients in university-affiliated clinics.[3] Hoping to receive funding through the newly expanded National Institutes of Health, many of these researchers legitimized their work by linking good posture to career success, mental acuity, and the promise of pain-free living.

Posture research and surveillance also piqued the interest of the White House. Both the Eisenhower and Kennedy administrations created presidential commissions to cajole Americans to attend to their fitness, defined, in part, by posture maintenance. Indeed, both the President's Council on Youth Fitness (Eisenhower) and the President's Council on Physical Fitness (Kennedy) grew out of Cold War fears that Americans were losing their muscular edge, falling behind other nations in posture strength and backbone, both materially and figuratively.[4]

In this context, the poor posture epidemic morphed into an endemic condition that was often trotted out in order to motivate Americans to fight the Cold War by proxy. In contrast to communist nations with their state-controlled physical education mandates and athletics, American fitness was to be undertaken voluntarily by each individual citizen. Engaging in posture work became a symbol of American patriotism and economic productivity, of citizens willing to discipline themselves for the sake of preserving the country's democratic ideals. In many ways, posture work was a logical extension of the Cold War foreign policy of containment, realized and fully embodied in each individual citizen.[5]

Back Pain Prevention

On May 16, 1961, during his first foreign trip as president of the United States, John F. Kennedy visited Ottawa, Canada, where, in front of eager crowds and camera crews, he shoveled several spadesful of dirt onto a small planting, a ceremonial tree that was to serve as a symbol of goodwill between the two nations. To onlookers, the forty-four-year-old Kennedy appeared energetic and youthful, exactly the image that his administration worked so hard to portray. Yet this rather innocuous act of diplomacy ended up haunting Kennedy, who, due to the exertion, experienced debilitating back pain for months to come.[6]

The unfortunate event marked a turning point in Kennedy's long struggle with back pain, leading him and his advisers to seek out care from one of the leading posture scientists at the time, Dr. Hans Kraus of New York City. Kraus first visited the White House on October 17, a full five months after Kennedy's injury. The doctor arrived in secret and immediately went to work evaluating the president's posture. Kennedy was no stranger to having his posture evaluated. A Harvard University undergraduate from 1936 to 1940, he would have been required to undergo several posture photography sessions during his student years. Kennedy attended Harvard when Dr. Lloyd T. Brown, head of the university posture exams, published his coauthored book, *Body Mechanics in the Study and Treatment of Disease*, research largely based on his posture photography studies. Yet Kraus's posture exam was very different from Brown's. Rather than viewing a patient's stance from afar, through a camera lens, Kraus conducted a hands-on examination, measuring the strength and flexibility of targeted muscle groups responsible for upright stance. Indeed, by the time Kraus had laid hands on Kennedy, the doctor was already popularly known for his standardized posture exam method, the Kraus-Weber test, named after himself and his co-creator Dr. Sonya Weber.[7]

Kraus first developed an interest in the posture sciences during World War II when he teamed up with Weber, a physical therapist who directed Columbia Presbyterian's Posture Clinic. A trained orthopedist, Kraus received his medical degree in 1930 from the University of Vienna, where he specialized in fracture care. Vienna was world renowned for its nonoperative approach to orthopedic injuries. The Viennese physician Adolf Lorenz, affectionately known as the "bloodless surgeon," enjoyed celebrity status in the United States, famous for having developed a nonoperative technique to correct congenital hip dislocations, a procedure that he had successfully and publicly demonstrated decades earlier in Chicago on the daughter of the wealthy meatpacking magnate Philip Armour.[8]

An avid rock climber and skier, Kraus was drawn to orthopedics because it fit with his personal athletic interests. He gravitated toward fracture care in particular because of its curative potential. Reflecting on an era in medicine when effective therapeutics lagged far behind the laboratory advances in understanding the causes of infectious disease, Kraus would tell biographer Susan E. B. Schwartz that he chose orthopedics because "in those days, there were few areas where a doctor could actually cure a patient." He explained that in medical school, "we [mostly] determined what was wrong, but then couldn't do anything to treat. . . . Orthopedics was the exception. There you could set a broken leg and make patients feel well."[9] In Vienna, Kraus became proficient in the methods of external fixation, immobilizing broken bones through casts, braces, and a set plan for gradual therapeutically oriented exercise.

When the thirty-three-year-old Kraus was forced to flee Austria (he was of Jewish descent) in 1938, he found a new home at Columbia University's well-known Fracture Clinic. But the clinic that Kraus joined would look very different by the end of World War II. Due to technological advances in blood transfusion, refinements in anesthesia, and large-scale manufacture of antibiotics during the wartime years, open operations would become more the norm rather than the exception. In the specialty of fracture care, internal fixation—whereby the surgeon rebuilds and re-aligns injured bones with metal implants—would come to replace casting and bracing as the standard of care shortly after the Second World War. Amid these changes, Kraus's Viennese style of "bloodless" orthopedic surgery became increasingly out of step with the wider trends taking place in American medicine during the middle of the twentieth century.[10]

Feeling a lack of comradery with most U.S. orthopedic surgeons, Kraus began to develop closer professional ties with physicians in the new specialty of physical medicine, also known as physiatry, a field that utilized physical agents (water, heat, electrical currents, and exercise) in treatment. As he explored this new field, Kraus made a fortuitous professional connection with Weber. Like Kraus, Weber immigrated to the United States from Austria, though she left Vienna in 1913 at the young age of eighteen, newly wed to her first husband, Hugo Eisenmenger, an Austrian engineer hired by the General Electric Company of Nela Park, Cleveland. In addition to sharing the same mother tongue, the two possessed similar worldviews, shaped by their well-heeled upbringings and connections within the Viennese medical elite. Weber was the daughter of Theodor Escherich, friend of Dr. Lorenz and chair of pediatrics at the University of Vienna, known to this day for having discovered the intestinal bacterium that shares his name, Escherichia coli. Equally important was Escherich's commitment to preventive medicine, a cause for which he raised enough donations among wealthy Viennese socialites to found an infant welfare society (Säunglingsshutz). It was in the area of preventive health that his daughter would carry on the family tradition.[11]

At the behest of Dr. Luther E. Holt, a close colleague and friend of Escherich as well as the author of the standard popular child-rearing guide of the day, *The Care and Feeding of Children: A Catechism for the Use of Mothers and Children's Nurses*, Weber took charge of the newly opened Posture Clinic at Columbia Presbyterian's Babies' Hospital in 1927, immediately after earning her doctor of science in physiotherapy in Sweden.[12] A mother of two young daughters, Weber found the Posture Clinic an ideal space for an educated, married woman with career ambitions and society connections. According to conventional wisdom, women were innately well suited to conduct work in the realm of public health and pediatrics because, as the "fairer sex," they had an innate understanding of all things relating to domesticity and child rearing. Also, the new profession of physical therapy was overwhelmingly female, a demographic similar to nursing. But while Weber held the title of physical therapist, she acted more like a physician, directing her own clinic. In her publications and newspaper interviews, she always was sure to emphasize her credentials as a "doctor" at a time well before any U.S. institutions would offer doctoral programs in physical therapy.[13]

Under Weber's lead, the Posture Clinic grew exponentially, treating as many as 10,000 patients per year throughout the 1930s, despite the Great Depression.[14] Seeing the financial benefits of Weber's work, Columbia administrators moved the Posture Clinic in the early 1940s to the university's premier teaching hospital, the Vanderbilt Clinic, and established a degree-granting program in physical therapy.

The growth of the Posture Clinic was not due to Weber alone, however. During World War II, the fields of physical medicine and physical therapy received an influx of political and financial backing from President Franklin Delano Roosevelt and his adviser, Bernard Baruch, wealthy financier and son of Dr. Simon Baruch, former chair of hydrotherapy at Columbia's College of Physicians and Surgeons.[15] In order to cope with the hundreds of thousands of war-disabled men and women returning home, Baruch in 1944 commissioned a committee of physicians to formally establish the medical specialty of physical medicine, providing over a million dollars to medical schools that agreed to develop the field. Columbia was one such school, as was New York University, where Dr. Howard Rusk, weekly medical columnist for the *New York Times*, founded its Department of Rehabilitation and Physical Medicine. By 1947, physiatry achieved specialty recognition within the American Medical Association.[16]

The medicalization of rehabilitation led to increasing physician control over areas that were once the purview of female physical therapists and other health workers who would come to be known as "ancillary" to medicine. By 1944, for example, Kraus became director of the Posture Clinic, leaving Weber, administratively speaking, second in command, a position that she and other physical therapists grudgingly assumed for the sake of securing their own image as scientifically legitimate practitioners. After its founding, physiatry adopted a highly imperialistic agenda, not only taking over the fields of physical and occupational therapy but also making key inroads into other medical specialties. Frank H. Krusen, president of the American Academy of Physical Medicine and Rehabilitation during World War II, urged his fellow practitioners to produce scientific studies on how physical therapeutics could be used in "psychiatry, dermatology, neurology, orthopedics, industrial medicine, rehabilitation and geriatrics."[17] He also supported the idea of creating programs in "prehabilitation," prescribing physical therapy in order to preserve and maintain health among the general population, a form of preventive medicine that could be used to transform an unfit civilian population into a fit military and commercial workforce.

The push to both scientifically legitimate and disseminate the virtues of physical medicine led Kraus and Weber to use the Posture Clinic as a springboard for research and collaboration. The first order of business was to standardize their own form of posture testing, one that assessed the muscular strength and pliability of the abdominals and back extensors, specific muscle groups necessary to upright standing. Their focus on muscle testing was in keeping with trends in physical therapy at the time. Aside from the war disabled, children and adults with polio constituted the vast majority of physical therapy's patient population during the first half of the twentieth century. In order to rehabilitate paralytic limbs, physical therapists developed refined methods to test individual muscles so as to create targeted therapeutic exercises for both spastic and weak muscular fibers. Kraus and Weber wished to apply a similar muscularly focused approach to their nondiseased posture patients. They veered from the conventional technique of using schematography and photography, taking instead a decidedly more hands-on approach to assessing posture.[18]

The result was a test that, rather predictably, bore both of their names. The Kraus-Weber exam consisted of six baseline measurements: three forms of graded sit-ups, two forms of graded back extensions, and one flexibility test gauging the pliability of the back and hamstrings. To make sure that the exam subject utilized the proper posture musculature, Kraus and Weber would apply resistance to the subject's legs or upper body so as to prevent the subject from using nonposture muscles to complete the tested motions. Comparatively speaking, the exam was more esoteric than posture photography, requiring tactile and anatomical knowledge that was not readily apparent to the person being tested.

The Kraus-Weber test, or K-W test for short, was not intended for clinical use, even though it was developed in that very setting on a population of disabled patients. Specifically, Kraus and Weber refined their targeted muscle assessments on adult patients who were being treated at the neighboring Low Back Clinic, housed also at Columbia Presbyterian.[19] The Low Back Clinic was one of the first of its kind, created in response to increasing concerns about the debilitating rise of back pain in America.

Since the adoption of worker's compensation laws in the early twentieth century, chronic back pain grew in prevalence, accounting for an ever-greater portion of disability claims in industry. Military physicians discovered a similar trend during World War II, and they often pointed to malingering or psychosomaticism as the cause. Meanwhile, the women who entered the workforce in droves during the war seemed to struggle just as much as their male counterparts with back injuries. In contrast to the now iconic image of Rosie the Riveter, flexing her bulging bicep, many contemporary news reports displayed women in munitions slumping and slouching from fatigue. "In terms of amount of time lost from work," a reporter for *Life* magazine would write, "the backache is one of the nation's costliest pains."[20]

Columbia hired orthopedic surgeon Dr. Barbara Stimson to take charge of the Low Back Clinic when it opened in 1946. The first woman to be certified by the American Board of Surgery, Stimson served in the British Royal Army Medical Corps during World War II, a decision she felt was forced upon her since her own country still banned female physicians from being commissioned as officers in the U.S. military. While working at the Royal Free Hospital in London, Stimson encountered "combined" clinical work—teams of specialists from psychiatry, surgery, and medicine working together—particularly in cases where patients had intractable back pain. This model of care fit neatly with Columbia's newfound commitment to physical medicine.[21]

As a diagnostician, Stimson welcomed the challenge that back pain presented. For clinicians who were called upon to treat or adjudicate a disability claim, back pain often evaded clinical measures of proof. In more clear-cut cases, X-rays would reveal a protruding disc or a compression fracture, and, at the most advanced medical centers, they would be treated through newly discovered surgical techniques of discectomy and fusion. But, as Stimson knew well from her experience in England, most patients who suffered from nontraumatic low back pain had normal diagnostic tests, from blood counts and urinalysis to X-rays. One purpose of the Low Back Clinic was to develop a refined system of differential diagnosis. To that end, Stimson recruited a cadre of specialists—a physiatrist, neurologist, psychiatrist, and rheumatologist, along with an orthopedic nurse and physical therapist—to conduct physical and laboratory workups on each patient who entered the clinic.[22]

After a couple of years of study, the Low Back Clinic found that nearly 50 percent of their patients had back pain due to poor posture, a fact that compelled Stimson to recruit Kraus and Weber's help.[23] Unlike most orthopedists, who believed that back pain was an operative matter with the main culprit being bulging discs, Stimson became ever more persuaded by arguments of inherent evolutionary weakness compounded by deleterious lifestyle. Echoing the sentiments of the generation of posture scientists before her, Stimson wrote, "The problem of the patient with low back pain has been [around] since our first great ancestor pulled himself from the four-footed to the upright position."[24] More specifically, she believed that the lumbosacral articulation—the site where the lower spine connects to the pelvis—was the weakest link, an artifact of human bipeds living in a quadruped's body. The only way to protect against such fragility, Stimson argued, was through strengthening the abdominal and back extensor muscles, hardening the postural muscular corsetry of the human body.

Drawing on their experiences in both the Posture and Low Back Clinics, and at the urging of Stimson, Kraus and Weber turned their attention to developing a diagnostic exam that could be used to detect early warning signs of potential postural back pain in the general population. The K-W test was thus a radically simplified exam compared to the more elaborate muscle assessments they conducted in the clinic. But for the purposes of speed and replicability, Kraus and Weber boiled the assessment down to six key movements and muscle groups.[25]

They first applied their test on thousands of consenting schoolchildren in New York City and Westchester County, and found that half of these youngsters failed. Sensing that European youth would likely perform better, Kraus began setting up testing sites in Austria, Italy, and Switzerland, hiring another colleague—well-heeled émigré Ruth Hirschland—to conduct the tests, all at his own expense. Publishing the initial results of the comparative study in a 1954 issue of the *New York State Journal of Medicine*, Kraus and Hirschland reported that while 50 percent of American children failed at least one portion of the test, over 90 percent of the youngsters from Italy, Switzerland, and Austria passed without much effort.[26]

"We are paying the price of progress," Kraus told *Sports Illustrated* reporter Robert Boyle in 1955.[27] And by *we*, he meant specifically white, middle-class, urban and suburban citizens of his newly adopted homeland. In his native Austria, a country war-torn and financially wrecked after World War II, Kraus witnessed an idealized way of life akin to his own childhood, where youngsters walked to school and families hiked and skied in their leisure time. American middle-class families, by contrast, traveled to and from work and school by car or bus. In their leisure hours, they watched professional sports on TV rather than engaging in physical activity. Kraus railed against the popularity of organized sports in America, seeing it as inherently undemocratic, a commercialized system that privileged a few select, highly fit students who would receive inordinate attention and resources while neglecting the weaker and less physically able children who needed physical training the most. Having fled his native home that was in the grip of a totalitarian regime bent on purging Jewish people, Kraus, along with other like-minded expats, worried about America's ability to withstand current and future threats of fascism and dictatorial rule.[28]

JFK's Personal Posture Crusade

The Kraus-Weber study touched a nerve, stoking fears concerning American weakness. Within a year's time it reached the desk of President Dwight D. Eisenhower, the five-star general who had just brought an end to the Korean War, a conflict that resulted not in victory, but in stalemate, calling into question the strength and might of the U.S. military. Based on the Kraus-Weber test's findings, Eisenhower created the President's Council on Youth Fitness in 1956, commissioning physical educators, coaches, and physicians to develop and promote exercise programs for U.S. schools and families to adopt on a daily basis.

The council operated in an advisory capacity so as to avoid any impression of state overreach. Congressional debates about universal military training and compulsory physical education reached a fever pitch after World War II. Proponents of physical fitness mandates faced powerful resistance among certain critics who likened such ideas to the Hitler Youth movement. When the Soviet Union had a particularly strong showing in its first Olympics in 1952, the debates resurfaced, but again the importance of holding up American ideals of volunteerism and private enterprise won out in the face of the enemy's system of state-controlled athletics.[29]

In place of explicit top-down control, the President's Council relied on the advertising industry and the media to convince the nation's families and schools to comply, using appeals to patriotism as the prime impetus. At the first council meeting, reporters were invited to participate in the proceedings as well as to cover the session for their respective news agencies. To bolster the effort, Eisenhower utilized the Advertising Council, a nonprofit organization originally established by the federal government during World War II in order to deliver public messages in support of the war. "Folding advertising into a governmental project," historian Rachel Moran writes, "helped develop the contrast between American and Soviet federal state-citizen relationship." Advertising preserved the veneer of "open markets, free choice, and the pursuit of individual desire."[30]

Kennedy became an even stronger proponent of the council than Eisenhower, in no small part because of the former's own physically debilitating back pain, which his administration so diligently tried to hide from public view. In a ploy to project himself as a symbol of able-bodied masculine virility, Kennedy authored "The Soft American" for *Sports Illustrated* just before assuming the office of the presidency. In addition to rehearsing the "startling" conclusions of the Kraus-Weber study, he reiterated, almost verbatim, the duo's warning about how the modern, mechanized American way of life would lead to the nation's demise. "A single look at the packed parking lot of the average high school," Kennedy wrote, "will tell us what has happened to the traditional hike to school that helped build young bodies." He continued, "The television set, the movies and the myriad conveniences and distractions of modern life all lure our young people away from the strenuous physical activity that is the basis of fitness in youth and in later life."[31] Writing for the same outlet a couple of years later, Kennedy noted how "it is paradoxical that the very economic progress, the technological advance and scientific breakthroughs which have, in part, been the result of our national vigor have also contributed to the draining of that vigor."[32]

Within the first year of the Kennedy administration, the President's Council on Physical Fitness placed thousands of newspaper, television, and radio advertisements. Several examples of the headlines for these ads include: "Let's Get the Kinks out of Our Kids!," "Could This Be Our Deadliest Disease?," and "Astronaut Glenn Discusses a Down-to-Earth Problem." Of particular note, however, was "The Silent Epidemic" advertisement. Drawing on Kraus's 1961 book *Hypokinetic Disease: Diseases Produced by Lack of Exercise*, the ad explained that "hypokinesia is spreading like a silent epidemic among our children. Increasingly large numbers of them now live such inactive lives that they can't perform simple tests requiring a minimum of strength and stamina."[33] In many ways, this ad was an extension of the earlier posture crusade, using the fear of widespread physical deficiency among the American people to instill greater bodily discipline among them. It also dovetailed with medical reportage of the newer lifestyle epidemics such as widespread obesity, cardiovascular disease, and workplace stress.

And yet conspicuously absent from these ads is any mention of posture as such. Increasingly, by the 1960s, Kraus and Weber's original

emphasis on postural health became subsumed by more generalized appeals to fitness, strength, and relief of back pain. In a way, the posture crusade that had begun at the turn of the century succeeded to such an extent
that its import seemed entirely self-evident, in no need of remark or mention. Kraus himself played a role in this evolution. While Weber shied away
from the limelight and declined to become involved with either President
Eisenhower's or Kennedy's fitness commissions, Kraus actively sought out
media attention until his death in 1996, devoting the bulk of his energies in
the 1960s and 1970s to publishing popular fitness books, using the posture
principles first established by Weber and her Posture Clinic as the baseline
for his own more muscularly oriented program.[34]

The most lasting endorsement of the benefits of postural muscle training came after Kraus's successful treatment of Kennedy. By the time
Kraus first examined Kennedy, the president had already undergone
three back surgeries. The first was a discectomy performed in 1944 after
Kennedy sustained a back injury during his World War II service in the
Naval Reserve (a posting he earned through family connections, since
he had been rejected from all other branches of the military for his poor
health). Ten years later, when he was a U.S. senator and still experiencing
chronic back pain, he underwent a lumbosacral fusion operation, with a
third, follow-up surgery in 1955 to remove the implanted hardware from
the fusion since it was causing a life-threatening staph infection.

Kennedy's entire congressional career from 1946 to 1960 was marked,
according to those closest to him, by a near constant use of crutches, at
least when he was out of the public eye. "When he came into the room
where [a] crowd was gathered, he was erect and smiling, looking as fit
and healthy as the light-heavyweight champion of the world," one
political adviser recalled. "Then after he finished his speech and answered questions from the floor and shook hands with everyone, we
would help him into the car and he would lean back on the seat and
close his eyes in pain."[35] To control the pain, Kennedy in his early presidency solicited treatment from Dr. Max Jacobson, known for his "vita-
min shots," injections laced with amphetamines in order to control
chronic pain, a treatment that, later commentators suggested, had a
"considerable and negative impact on the President's performance at

the crucial Vienna summit with Soviet Premier Nikita Khrushchev in June 1961."[36]

In many ways, Kennedy's medical history of failed surgeries and reliance on painkillers was everything Kraus warned against. Since his university days in Vienna, Kraus had built his career on opposition to high-tech medical care. He was a believer instead in daily calisthenics that would keep the postural muscles strong and supple, a regime that he prescribed for even the U.S. president. After a year and a half of strengthening exercises under Kraus's supervision, Kennedy reported that his back had improved greatly, crediting the doctor for his help. A July 1963 issue of *Time* magazine covered the president's first golf outing since the Ottawa incident nearly two years earlier. When asked about his recovery, Kennedy replied, "I . . . didn't think I was going to play golf again,"[37] a sport that the press knew was his favorite.

Against Kraus's advice, Kennedy continued to wear a back brace, a canvas corset that he had used on and off for decades. The device not only alleviated pain; it had the added benefit of making the president look taller, allowing him to sit and stand straighter than he would otherwise. Kennedy was wearing the lumbar brace on the fateful day that he was assassinated in Dallas, Texas, on November 22, 1963. Certain physician experts have conjectured that the ability of his brace to promote an upright posture may have ultimately cost him his life. Without the brace, Kennedy may have more naturally slumped forward after the first bullet hit his neck, and thus avoided the second, a direct and deadly shot to his head.[38] Instead, the brace kept him perfectly upright, making him an easier target for Lee Harvey Oswald.

Female Sexuality and Containment

Years after Kennedy's death, the council continued to have a presence in subsequent presidential administrations and remained a viable avenue for federally mandated physical fitness tests of U.S. school-aged children. It was not just the executive branch of the government that was responsible for the popularity and furtherance of these programs. Both the U.S. military and many American universities continued to invest in more traditional forms of posture testing and training carried over from the First World War.

While the standard of "good" posture continued to be universal, the reason to develop it varied considerably and in important ways, often contingent on one's social location and broader political concerns. The preparation of female military trainees during the Second World War offers one important example. Over 150,000 American women served in the Women's Army Corps (WAC) during World War II, the first time that women could enter the official ranks of the U.S. Army outside of nursing. Major Oveta Culp Hobby, director of WAC, argued that because of manpower shortages at home and abroad, women could be used for noncombatant jobs in order to "free a man for combat."[39]

Playing into the notion that women possessed certain gender-specific skills, Hobby assigned most WACs to fill positions as file clerks, typists, stenographers, motor pool drivers, and switchboard operators. Some operated filter boards for the Aircraft Warning Service for the U.S. Army Air Forces, plotting and tracing the paths of every overhead aircraft. Others worked as lab technicians for the Chemical Warfare Service. While the majority of women who served in the military during World War II were stationed stateside, some served overseas in the North African and Mediterranean theaters, where they plotted the movement of troops and coordinated the delivery of crucial supplies.[40]

Despite the proven abilities of these women, many Americans remained skeptical that a woman could do a "man's job." WAC women consistently proved their critics wrong, but as they did, they faced different prejudices. One prevalent worry was that these women would become too manly, in looks, behaviors, and sexual affections. To mitigate such concerns, Hobby created an elaborate disciplinary structure that signaled heterosexual femininity. For example, she prohibited WAC enlistees from dancing as couples in public and cautioned them against adopting mannish hairstyles. Any sign of homosexual affection was punished with decisive action. Historian Leisa D. Meyer, who has documented the lives of several WAC soldiers who were dishonorably discharged due to "masculine behavior," notes that "mannerisms and coded language were ways in which lesbians identified one another."

Many of Hobby's rules were attempts at putting an end to "butch" behaviors and dress.[41]

And yet WAC servicewomen risked being construed as prostitutes if they leaned too heavily into heterosexual femininity. Hobby upheld a version of femininity "rooted in Victorian linkage between sexual respectability and female passionlessness."[42] She adopted this ideal to quash rumors that WAC enlistees were actually call girls, there to serve the sexual needs of male U.S. soldiers, a rumor that these very men tended to spread. Resenting the enlistment of women into the Army, male soldiers had every reason to besmirch the reputations of WAC women. According to Judith A. Bellafaire, "letters home from enlisted men contained a great deal of criticism of female soldiers." "When the Office of Censorship ran a sample tabulation," Bellafaire observes, "it discovered that 84 percent of soldiers' letters mentioning WAC were unfavorable."[43]

Proper posture thus became a centralized concern of Hobby and WAC trainers, for it was a way to ensure that servicewomen could perform military femininity, an identity that would be read as strong, yet not overly manly or sexualized. Rather than being concerned simply with physical strength and endurance, WAC military trainers tended to emphasize how greater fitness led to better personal appearance. "The eyes of the Army—and of the Nation—are on you," the 1943 WAC field guide for physical fitness proclaimed. "It is of prime importance," the guide continued, "that you look well, feel well, and work well throughout your military service."[44]

Posture improvement was the first step toward achieving the twin goals of attractiveness and health. While male soldiers were encouraged to build muscular strength, made visible through the enlargement and striation of muscle tissue, female cadets were told to think more in terms of cultivating muscular tone. According to WAC officials, muscle toning exercise—especially the deep, small postural muscles—resulted in greater poise and grace, an improved silhouette without muscular bulk. Physical training and exercise, WAC trainers insisted, did not require cultivating "'new' muscles nor large muscles," but rather "*balanced* muscle control."[45]

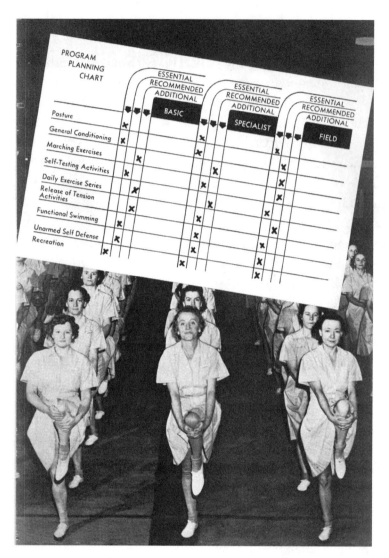

FIGURE 34.   An illustration depicting the primary importance of posture training for the women who served in World War II. Listed above all other requirements, posture exercise was understood to be the foundation of military femininity, a method of toning the body, yet not adding muscular bulk.

Female cadets were taught to think of their posture both on and off duty, constantly conducting self-surveillance using mirrors and reflective windows (see figure 34). They were forbidden to wear girdles or support garments, a rule intended to discipline the WAC enlistees and to make them distinct from civilians.[46] Or, to put it in the words of the WAC manual, "Fitness can't be faked. . . . The test of true physical fitness is hard work. A girdle alone won't hold you up throughout a tough day in the cab of a truck. Firmly toned abdominal muscles will."[47]

Posture testing and training was thus offered to WAC women as a way to negotiate the double bind of the Madonna–whore complex. Developing inner tone offered assurance that female soldiers would not morph into muscle-bound soldiers. At the same time, maintaining plumb line verticality protected women from being perceived as too flirtatious. WAC enlistees were taught how to stand, sit, and walk without protruding the hips, or swaying the back. The chest was to remain tall, but not thrust forward. The hips were to remain level, with little curvature of the lower back, lest the buttocks perk up in a sexualized way.

"Fundies" for Freedom

At Smith College, Dorothy Ainsworth continued the WAC philosophy of training for college women and eventually extended the message to a global scale in her work as founder and president of the International Association of Physical Education and Sport for Girls and Women (IAPESGW). Throughout the 1940s and 1950s, a vast majority of American universities continued to mandate posture tests and exercise. While physicians headed up these programs earlier in the century, physical educators with advanced degrees increasingly took control by mid-century.

The political landscape changed as well, and university posture programs adapted in response. While the university took responsibility for student health and discipline, it also became a site for the preservation of American democracy and the advancement of the nation's imperialist agenda. University fitness programs created war-ready bodies to fight on the front lines, but also the rear areas and at home. They were also tasked with producing elite athletes who would compete in Olympic games, earning medals that would symbolize a win over enemy states.

Much like Kraus and Weber, Ainsworth was concerned about the rise of fascism, Nazism, and communism, specifically the way that these totalitarian regimes enforced exercise upon their respective citizenries. U.S. physical education, Ainsworth contended, "provided a discipline which [is] much stronger than that of the Nazi-Fascist group because it is self-imposed."[48] From her perch at one of the most reputable women's colleges in the United States, Ainsworth aimed to use physical education as a way to cultivate American democracy and healthy individualism.

Although posture training had existed at Smith since the early twentieth century, Ainsworth promoted it as a "a new and experimental body-building class." In the context of the Cold War, when many educators and commentators worried about America's readiness in the hard sciences, mathematics, and engineering, physical education increasingly came under scrutiny, with many wondering if it was necessary to devote time and money to gym class. One way that physical educators preserved the legitimacy of their profession was to emphasize data collection, focusing ever more on the physical examination and measurement of students. To raise the stakes further, Ainsworth persuaded administrators to require that all Smith students earn at least a B– in posture in order to graduate.[49]

Under her direction, Ainsworth quadrupled the number of times a Smith student had to pose for posture pictures, making the exam a biannual event. She also modernized the exam, insisting on camera photography as the only legitimate way to record and measure student posture (see figure 35). By mid-century, most leading universities moved away from the Mosher schematograph, seeing it as an arcane measuring tool, a symbol of the older generation of posture scientists. Forgotten was Mosher's reason for developing the schematograph in the first place—namely, that tracing technology ensured and preserved student privacy, especially that of female students, better than camera photography did.

Wishing to convey an image of physical education as a hard science, Ainsworth demanded that her students pose nude for the camera during posture exams. The many ways in which she attempted to normalize

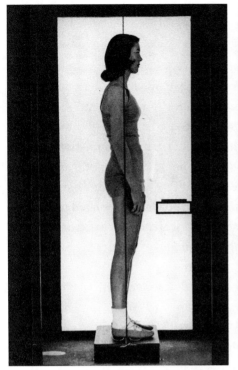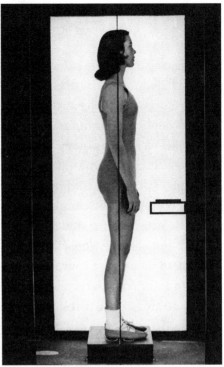

FIGURE 35. Before and after shots of a Smith student circa 1949. This set was likely used for teaching and demonstration purposes, since the photographed subject is clothed. These side-by-side photos were used to train students to compare faulty and ideal posture, and also to monitor progress over time.

student nudity can be seen in her personal papers. In a memo from Ainsworth to her faculty instructing them on how to conduct classes in posture and biomechanics, she asked all instructors to take class attendance while the students were in the shower. She urged a near constant gaze on Smith students:

Please take [attendance] rolls in shower and explain bells; i.e. they should be on the floor and in costume at five after the hour, then the bell will ring about twenty of the hour when they . . . go to the dressing rooms, put on a sheet and shower shoes and go . . . into the showers. Another bell will ring when they enter the showers, where they will remain until a second bell rings and then they will go out. Be sure all classes are dismissed when the bell rings.[50]

Facing a shortage of faculty and staff who could undertake the labor of conducting biannual exams, Ainsworth developed a master's degree program in physical education for matriculating Smith students so that they could help examine their peers and take course work in the basic sciences and kinesiology. Ainsworth taught her graduate students how to conduct anthropometric exams, orthopedic exams, and various forms of posture recording that were used at other colleges and elementary schools.[51] The comprehensiveness of the data accumulated under Ainsworth's tenure attracted students and researchers from across the globe, pupils who would then bring the techniques of posture measuring back to their home countries.[52]

In addition to data collection, Ainsworth developed a unique system of posture training that she called the "Fundamentals of Movement," or, as students liked to call it, "Fundies." Ainsworth described the class as one that focused on "the more simple and primitive forms of movement, i.e. walking, running . . . and folk dancing, but always with an emphasis on erect carriage."[53] Her inspiration came from Niels Bukh, a Danish gymnast who had developed a system known as *primitiv gymnastik* during the interwar years.[54]

Bukh's *primitiv gymnastik* was a romantic, vitalist-*völkisch* ideal of a return to nature to counterbalance the physical and mental weakness brought about by industrialization, cosmopolitanism, and elitism. At his large school in the rural town of Ollerup, where he primarily trained the sons and daughters of farmers, Bukh revived what he believed to be the fundamentals of Ling's gymnastic—namely, the "faults of [human] carriage." In his handbook *Fundamental Gymnastics*, Bukh wrote that he offered a "thorough working and toning of the whole body" in order to address "all the stiffness due to habitual postures . . . which gives the heavy and clumsy appearance to [people]."[55]

There was little overt homage paid to Bukh in Smith's "Fundies" classes, especially after World War II. By the mid-1930s, Bukh pledged his allegiance to the German Nazi Party, angering his fellow Danes, especially after the Nazi invasion of Denmark in 1940. Bukh's rather public display of homosexual relationships, however, would make a lasting alliance with the Nazis impossible, leaving him, by the 1940s, to become persona non grata. Nonetheless, his system of exercise that promised greater muscular tone and grace fit perfectly with the perceived needs of American educated women during the Cold War.[56]

With its emphasis on movement essentials, the Fundies class was intended to address the needs of all students, no matter their physical build or athletic capabilities. Ainsworth portrayed this individualized attention as necessary to the upholding of democratic ideals. Similar to Kraus and Weber, she believed that the popular focus on American sports and competitions worked to the detriment of the health of all those who exhibited little athletic prowess. Before the passage of Title IX, this was particularly the case for women who were marginalized in college and professional sports. Rather than fight the inherent double standard that precluded women from engaging in organized sport, Ainsworth in many ways enabled it, restricting Smith students from engaging in intercollegiate sport, which she believed fed a masculine sense of competitiveness.[57]

To Ainsworth, posture training through the Fundies would preserve the traditional gender binary and, by extension, protect against the threat of communism and its promotion of unsexed womanhood. The image of the husky Soviet woman, who worked as a ditch digger or mechanic, performing heavy labor, often served as a foil for the ideal Smith woman. The prevalent trope of "Soviet women as graceless, shapeless, and sexless," historian Robert Griswold writes, "functioned to discredit Communist women and, more important, Communism itself."[58] And yet to remain competitive with the Soviet Union, the United States needed to make use of women's minds and physical labor.

In 1954 when Ainsworth founded and became president of the IAPESGW, she delivered the gospel of proper posture to developing nations with the goal of staving off the spread of communism. She was rather explicit about the potential of this kind of soft neocolonial rule. "The social significance of physical education," she wrote in 1950, "is appreciated today more than the early days. . . . It is brought home to us because of the varying [political] ideologies and the use of physical education as a means of indoctrination in these ideologies."[59] And yet, for all the talk of democracy and self-determination, the IAPESGW selected only European and North American leaders to direct the effort, with the expectation that non-white participants would comply and appreciate the "goodwill" effort of women such as Ainsworth.[60]

Straight Men

According to historian Jonathan Katz, the first documented use of the word "straight" to signify heterosexual appeared in 1941.[61] To exude straightness, men wishing to present as heterosexual would don crisp, tailored suits. The classic gray flannel suit of the 1950s was "severe and subdued, with a streamlined, military stiffness that was . . . confined, controlled, columnar . . . nothing swings loose, nothing is tufted."[62] It was a suit, as fashion historians have commented, that brought a sense of security and conformity. It was also a suit that promoted upright posture, and thus the appearance of sexual conformity.[63]

Just as "straight" became the idiom to describe heterosexuality, "bent" became shorthand for homosexuality. During the early Cold War years and in the hands of red-baiting Senator Joseph McCarthy, homosexuality became bound up with communism, both being seen as subversive and lurking threats that needed to be rooted out.[64] What made the so-called problem particularly pernicious, according to vigilantes like McCarthy, was that individuals could hide their political beliefs and sexual lives. In an attempt to police such subversives, the U.S. government turned to the medical and forensic sciences to provide material proof.

The science of biotyping—reading a person's outward body to determine inner thoughts and character—enjoyed a renaissance in this context. The practice of physiognomy dates back to Hippocrates, with a notable resurgence seen among eighteenth- and nineteenth-century racial scientists such as Johann Kaspar Lavater, Francis Galton, and Cesare Lombroso. The paranoia that defined the Cold War and the Lavender Scare created the perfect political conditions for these techniques to gain popularity once again.

William H. Sheldon led the way, and he did so by relying on the extant archival record of posture photos held at private and federal institutions across the United States. Sheldon was not particularly interested in posture per se. Rather, he used posture photographs to develop a system that measured the shape, muscular tone, and fatty tissue of a body. Taken together, these numbers, he claimed, offered a window into an individual's psychological make-up. Later in his career, Sheldon would invent his own specialized photo technique that required a total 360-degree view of a human body: front, back, and side. Posture photos, by contrast, were often only taken in silhouette.

Sheldon concluded there were three types of male physiques, all existing along a continuum. At the extreme poles—the greatest deviations from the presumed norm—were "endomorphs" (soft roundness) and "ectomorphs" (linear fragile). The "normal" or mid-range types were known as "mesomorphs," who were characterized as muscular, square, and firm.[65]

The most troublesome type, Sheldon maintained, were ectomorphs, whom he believed were men who had hidden homosexual tendencies. According to his somatotyping system, these men were brainy, yet worried and harried. And while this type presumably "loved to be alone," he was also an enigma, for his "sexual drive . . . was very high." Endomorphs, who he portrayed as "fatties," did not pose as much of a threat, for they tended to spend their "evenings enthroned in [their] easy chairs absorbing chocolates and television." The ideal type of man was the muscular mesomorph who, "free of doubts, ma[de] decisions quickly . . . [was] a born executive."[66] Sheldon was a biological determinist. He believed that a person's somatotype was genetically determined and thus be unchanging. During his lifetime, Sheldon was criticized for both his substandard scientific practices and his blatant racism, and would ultimately be discredited by the time of his death in 1977.[67]

Without an archive of posture photos, it is possible that he would not have been able to develop his typology. His interest in biometry was first piqued when he was a medical student at the University of Chicago. There he viewed thousands of posture photos and began to dabble in it himself. During the Second World War, when Sheldon was stationed at the School for Aviation Medicine in Texas, he mined draftee posture photos, and tested a beta version of his somatotyping system on pilots, claiming that his method of inquiry served as an important indicator of each man's success in flying. After the war, he joined Columbia University's Constitution Clinic, where he eventually perfected his signature system of tripolar photography, which he used on nearly 4,000 psychiatric patients at a nearby hospital.[68]

Sheldon's personal papers, particularly the bulk of his team's correspondence, offer a sobering reminder of the pervasiveness of posture photography, examination, and training in America at the time. He hired several assistants who scoured the country for nude posture photos. His staff wrote to prisons, physical education departments, psychiatric units, and state medical schools asking for negatives of photos, which, if acquired, would be reproduced and analyzed using Sheldon's system of biotyping.[69] A good portion of these photos still exist in his personal papers, which are held at the Smithsonian's National Anthropological Archives, but the photos themselves are restricted, a topic taken up in greater detail in the last chapter of this book. Although it is impossible to get an accurate count of the number of posture photos Sheldon had in his possessions, we know that at least some 46,000 pictures were used in the 1954 publication of his *Atlas of Men*. And this is likely just the tip of the iceberg.[70]

Black Strength and Posture

By World War II, many Black university students had become highly skeptical of anthropometry and physical anthropology, having been historically degraded by white scientists wielding calipers and rulers. Recalling his days as student body president at the Hampton Institute during the interwar years, St. Claire Drake contended that the "race-conscious Hampton student generation" would never have permitted physical measures like that of Sheldon's. When asked about the work of white physical anthropologist Melville Herskovits, who taught at Howard University during the interwar years, Drake responded, "I suspect that many of us would have resented a white man at Howard using his post to measure the students' heads, to check under the arm ... we would have been hostile toward the Howard students who cooperated too."[71]

But posture exams, it seems, did not carry the same level of distrust. When the Hampton Institute hired its first Black physical educator, Charles H. Williams, to take over the program in 1931, his first order of business was to mandate a yearly posture examination of each student. This was but one step in Williams's campaign to revamp Hampton, moving it away from its origins as a vocational training ground for rural Blacks and Native Americans into a four-year college with a strong liberal arts offering. In his appeal to invest in camera photography for physical education examinations, Williams promised to "definitely attack the posture problem of the Hampton students." At stake was not only the health, efficiency, and respectability of Hampton students, but also the institution's reputation since, as Williams claimed, posture was of "great scientific interest" at most other institutions of higher education.

In Williams's mind, devotion to posture training was more than just optics and raising the institution's reputation. He understood postural strength to be inherent to developing race pride. A former Hampton star athlete himself and cofounder of the Colored Intercollegiate Athletic Association, Williams was a firm believer in the importance of athletics and its ability to foster racial dignity, challenging white stereotypes of Black physicality. Certain Black activists and educators worried that competitive sports would threaten race relations; some believed that white violence would escalate racist sentiments among white Americans if Black athletes demonstrated too much physical prowess.

Williams walked a fine line between placating whites and developing a racial consciousness among his students through dance and posture training. Having developed strong connections with the famed choreography and dancer Ted Shawn, Williams founded the Hampton Institute Creative Dance Group. Williams believed that Blacks had a "native capacity to dance," and an inborn sense of rhythm. Fascinated with the life and customs of native Africans, he choreographed pieces that drew upon traditional African dance as well as older plantation dances such as the cakewalk. According to dance historian Katrina Hazzard-Donald, "the older plantation cakewalk functioned as an expression of derision." The cakewalk's elevated, erect postures and high kicks, Hazzard-Donald argues, "ridiculed elite whites," who assumed Blacks did not have the bodies or discipline for such poise. The dance juxtaposed stiff upright-ness against postures and gestures used in a number of dances from the Senegambian region of West Africa, including the sabar and the lamba.[72]

The male and female Hampton dancers, most of whom were physical education majors, demonstrated controlled athletic movements in their performances, focusing much of their training on the postural muscles and developing posture consciousness. Williams hired Char-lotte E. Moton, a Radcliffe graduate and daughter of Robert Russa Moton, principal of the Tuskegee Institute, to assist in the posture training. Underscoring upright movements that emphasized power and grace helped Williams and Moton choreograph dances that would preserve racial dignity, cautious not to confirm stereotypes of Black people as hypersexualized or savages.[73]

During World War II and the years immediately thereafter, the newly formed "Hale America" campaign adopted Hampton's posture program. A predecessor to JFK's President's Council on Physical Fitness, Hale America specifically targeted African American physical education, with the U.S. government appropriating 10 million dollars for the effort. John B. Kelly, Philadelphian and Olympic sculler who headed up Hale America, tapped two well-known Black leaders to take over the African American division of the movement. The 1936 Olympic gold medalist Jesse Owens served as the national coordinator of Negro activities, and Hampton's Charlotte Moton filled the role as director of women's ac-tivities. The effort targeted men, women, boys, and girls, a "civilian army, tough in mind and body." "Only by keeping every civilian physi-cally fit," one representative said, "can this nation keep the tanks moving, the planes flying and the ships floating. America must not grow soft—at any cost!"[74] Moton instituted the posture training programs based on her experience at Hampton, which she argued were ideal for "mass par-ticipation" among all citizens, young and old, male or female.[75]

Beautiful White Women

For American women living during the Cold War, posture work became a crucial part of beautification. By the mid-twentieth century, beauty management became credited courses in places of higher education, taught by women holding PhDs in physical education. Certain colleges billed their posture training as a specialized type of beauty training, a consumer good that promised to enhance a co-ed's beauty, much like cosmetics and fashionable clothing. When informing the Smith entering class of 1958 of the school's robust posture program, Ainsworth told the students that it was "the sort of practice and instruction for which a model pays a handsome price."[76]

To beautify was not a simple appeal to vanity. As Ainsworth put it, poor posture was a "handicap [that] may be so serious as to hinder [a student's] progress in social or business and professional life after college."[77] According to historian Kathy Peiss, this line of thinking was pervasive at the time. The attractive woman, she writes, "bespoke the American way of life," a symbol of "a free society worth defending."[78] The cosmetics industry, for example, linked women's grooming and beautifying to the preservation of national morale, workplace productivity (among both men and women), and democratic order. Even with the nationwide rationing of certain chemicals and raw materials during World War II, women's makeup was deemed indispensable. After the war, when the cosmetics industry was booming, one 1948 report claimed that 80–90 percent of American women used lipstick.[79] Expressing ambivalence about the demands of the beauty ideal, one woman in 1950 wrote, "economic pressure—I have to earn a living—forces me to buy and use the darn stuff."[80]

At Case Western Reserve University in Cleveland, physical educator Florence C. Whipple developed a "School for Models" program in her department. The explicit goal was to "make the health program more attractive" to incoming and prospective students. According to Whipple, the program appealed to an increasing number of students with each passing year after the war. Posture training was at the heart of this program. Whipple found that "a forward head and tense shoulders seemed to be faults common to practically all of the girls, in varying degrees." Since a "'proud' head position was essential for a good 'model' carriage," Whipple contended, she developed specific posture exercises to ensure such a stance.[81]

Whipple hoped to capture the same attention to her program as the "Norma Look-Alike" contests did among Cleveland-area women in 1945. Norma, and her mate Normman, stood in the Cleveland Health Museum, statues purchased in 1945 with great fanfare. Carved of white alabaster, Norma represented the statistical average of 15,000 "native white" Americans who measured themselves for the Bureau of Home Economics in 1940 for the purpose of establishing a system of standardized sizing for the ready-made clothing industry.

Shortly after the statues arrived in Cleveland, the local *Plain Dealer* held a competition to find the "living embodiment of Norma." The newspaper ran a daily announcement of the competition throughout September 1945, printing a detailed anthropometric chart that the contestants submitted after having entered their own personal measurements. Approximately 4,000 women entered the contest, reflecting the degree to which bodily self-surveillance was becoming routine and rewarded.[82] Etiquette guides and women's magazines frequently encouraged women to intimately scrutinize their bodies with exacting measures. In her *Glamour Book* guide, the well-known fashion designer and milliner Lilly Daché advised her readers to "stand naked in front of a mirror," a form of self-appraisal that she admitted required bravery since the unadulterated truth was "going to be a shock." Women needed to look long and hard at all the "bulges in the wrong place," the ghastly sag in the abdomen, the rolls around the midriff. To remain vigilant against "secretarial slump" and the "dowager's hump," women needed to develop a habit of checking their own figure faults, adopting habits and exercises to remedy the problems.[83]

Black Beauty and Female Respectability

Black physical educators at historically Black universities and colleges adopted a similar kind of instruction for their female students, but with an awareness of the higher stakes involved in the politics of respectability for women of color. In many ways, proper posture, much like cleanliness, thrift, and sexual purity, served as a "behavioral entrance fee . . . to the right of full citizenship."[84] In Evelyn Brooks Higginbotham's rendering of the Progressive-era politics of respectability among Black women, the performance of proper etiquette served two main functions: the uplift of African Americans and the amelioration of white fear and suspicion of Black Americans.[85]

Maryrose Reeves Allen, head of the Department of Physical Education for Women at Howard University from 1925 to 1967, taught her students to think of beauty as the linchpin of female respectability. As discussed in the previous chapter, Allen conducted a quantitative analysis of female beauty for her master's thesis, measuring the posture of over a hundred women of various ethnicities and races and having a panel of judges (both Black and white art teachers, photographers, and physical educators) evaluate the faces of her subjects, determining which attributes and proportions were necessary to female beauty. One of her goals was to create "scientific measures of beauty" that could then be taught in college so that every co-ed could be "her own model." Allen felt that Hollywood and women's magazines had made a "racket" of the beauty ideal, offering only a "surface, empty kind of beauty which tends to increase surface living and thinking." Like many health and education professionals before her, Allen appealed to objectivity and the rhetoric of science in order to create an area of expertise that could not be simply accessed through mass media or folk knowledge.[86]

A member of the Black educated elite herself, Allen did not, however, simply emulate white beauty standards. Self-hatred, Allen observed, was particularly debilitating to the Black race, a cause for poor health, high death rates, and short life spans. Her efforts to scientize beauty was a way to resist a larger trend in America, where, as she observed, "most races are trying to lose their identity and all become the white race."[87] At Howard, she offered courses in "Body Aesthetics"—a required course for nonmajors—and "Body Sculpture through Movement," and held a Beauty Clinic, a space that Allen often referred to as "the laboratory" where students engaged in a thorough "self-analysis" of their own posture, dress, makeup, hair, and social graces.[88]

Allen tied physical education directly to race politics in America, referring repeatedly in her lectures to how beauty maintenance was essential to combating the legacies of slavery and life under Jim Crow violence and segregation. "All races are in need of [beauty] training," Allen began her Foundations of Physical Education lecture, "but because of our handicap of slavery, we have the economic system which affects us drastically, prejudice, lack of education, plus the lack of educational opportunities, and the problem of fitting into society in general."[89] While later civil rights activists and Black pride advocates would pin the so-called "problem of fitting in" on centuries of white supremacy, refusing to adhere to the socially constructed assumptions regarding race hierarchies, Allen's stance was at once radical and conservative for her time, articulating an early iteration of "Black Is Beautiful," yet also adhering to conventional gender and social norms.[90]

Regarding posture training, Allen drew upon the work of white experts. Her lectures regularly featured the writings of Bess Mensendieck, Mildred L. Albert, and Bonnie Prudden, the latter a collaborator and colleague of Kraus.[91] And yet Allen found her own meaning in the promise of proper posture for Black women. For example, in her class notes from her Boston University years, Allen placed a special emphasis on the lesson that "upright carriage tends to abolish fear," offsetting these words with bold print and underlinings. When it came to teaching her own students how to maintain proper posture, she emphasized the need to harmonize both inner and outer beauty—that posture was an external measure of beauty that was impossible to attain unless a woman also attended to the health and well-being of her internal organs and mind. Allen believed posture to be one of the purest expressions of beauty, for it could not be hidden behind any artifice. She made her students conscious of the power of posture, teaching them that while beauty encouraged "vanity and deceitfulness, posture expresses the emotions of the soul . . . posture tells the truth."[92]

Although Allen made important gains in dismantling the assumptions of white supremacy inherent in the American beauty ideal, she was fairly conventional when it came to gender roles, using posture training as a way to prop up the importance of marriage and having children. The overall goal of beauty education, Allen maintained, was for Black women to "perform [their] most noble function—motherhood."[93] For homework, she assigned her students to read articles written by experts of etiquette. One in particular, titled "Slouch, Poor Curves Rob You of Sex Appeal," discussed the reality and importance of the male gaze, especially when it came to a woman's figure, urging female readers to "concentrate on the MAN!" in their own self-presentation.[94] While Howard men engaged in sports and athletics, Howard women, under Allen's leadership, participated in Beauty Bazaars, posture pageants, and May festivals, all believed to be more fitting (i.e., less aggressive) forms of competition for the female sex.

To Allen, posture and poise went hand in hand. In the Howard "Charm Clinic," Allen emphasized to her students that good posture would enhance one's grace, an important attribute for "putting others at ease."[95] In these classes, students were taught that etiquette, beautifying, and maintaining physical health were all essential to racial uplift (see figure 36). Allen's classes largely mirrored the kind of instruction that women would gain at for-profit, Black-owned charm schools and modeling agencies, industries that came into their own in the early post–World War II years. Ophelia DeVore, a pioneer in the field of modeling for African American women, opened both a modeling agency and a charm school in the late 1940s, hoping to change the way that Black women were depicted in popular American culture.[96] Pushing back against the assumption that only white—or those who can pass as white—is beautiful, DeVore instilled in both her models and the ordinary Black women who took her charm classes that a woman's attractiveness was not dependent on skin color.

But this effort, historian Malia McAndrew rightly points out, "was not without its own set of dilemmas and contradictions."[97] In her charm school, which attracted primarily working-class Black women, DeVore taught the essentials of grooming and posture, upholding the conventional view that a woman's social and economic worth was

FIGURE 36.   A posture training session taking place at the Palmer Memorial Institute in North Carolina, a private charm school founded by Black entrepreneur Charlotte Hawkins Brown. Part of the instruction included having Black teenagers practice head-carrying with books.

fundamentally tied to her physical appearance. Writing about the history of etiquette guides targeting Black readers, Katharine Capshaw Smith notes that such advice often veered toward performing whiteface, advising actions that are racially infused, and "scripting behavior in 'proper' dress, manners, speech, worship, leisure, entertainment, and relationships. 'Proper' is sometimes code for 'white.'"[98] Nonetheless, DeVore and other Black beauty culturists can be found time and again holding out the promise that with the right grooming and posture training, economically disenfranchised Blacks could secure a better future for themselves. *Afro-American* beauty columnist Natalie Scurlock advised her readers that at no cost, Black women could greatly enhance their charm if only they attended to their posture. "There is nothing that adds more to your appearance," she implored, "than good posture and erect carriage."[99]

Yet while the Black press and colleges like Howard offered classes in beautifying, defining success as heterosexual dating, marriage, and personal happiness, many female students began to press for more permissive rules concerning dating and heterosexual sociability. Allen found herself caught in the middle of this paradox many times in her years as Howard's physical education director. In one telling example, when she caught Howard student Lucy Hueston kissing her classmate and star basketball player Louis T. Coates in the bleachers, Allen quickly informed the dean of women of the impropriety and reportedly said that "she was going to do all she could to see that [Hueston] was sent home." Hueston was suspended from Howard, while Coates was merely put on probation. The student body protested, took over the floor of the gymnasium during a basketball game against Virginia Union, and halted play until the University Disciplinary Committee agreed to reconsider the case. Due to the student protests, Hueston was permitted to return to campus after only a ten-day suspension.[100] While Allen upheld the double standards commonly held at the time—a view that would increasingly become out of step with the beliefs of the student body—her program had staying power, for her emphasis on posture and beauty fit with Howard's mission to be a training ground for what W.E.B. Du Bois called the "talented tenth" of educated Black leaders. In this regard, Howard was no different from most universities at this time, when, as historian Susan K. Cahn points out, "heterosexual relations informed one—if not *the*—essential element of college life," contributing to a kind of marriage mania that permeated both the student body and the administration.[101]

With poor posture serving as a sign and signal for everything from sexual deviancy and racial degradation to unemployability and chronic disease, it is little wonder that the posture epidemic became an endemic condition woven into the fabric of American culture in the mid-twentieth century. Posture examinations became a way for government officials, employers, educators, and medical scientists to evaluate not only overall health but also moral character and capabilities at the individual and population levels. A powerful tool that addressed the twin concerns of national fitness and character during the Cold War, posture evaluation and training piqued the interest of several U.S. presidents, leading to the institution of various President's Councils that promoted such work in order to combat national "softness."

In this context, Americans, both young and old, were taught to believe that they could contribute meaningfully to the containment of communism and the advancement of free-market capitalism and democracy through their individual body work. Posture control was thus touted as a means to mitigate a whole host of threats—from that of sexual deviancy and racial instability to ongoing warfare and the specter of totalitarianism. In short, developing a strong and erect backbone became an expression of American patriotism.

CHAPTER 6

The Perils of Posture Perfection

RETURNING TO CAMPUS in the fall of 1950, Cornell students were greeted with a scandal of the flesh. According to the *Cornell Daily Sun*, posture silhouette photographs of all freshman and sophomore women mysteriously disappeared from the Sage Gym Office the previous May, just before graduation.[1] This was not the first report of missing photos, at Cornell or elsewhere. At Vassar, an all-women's college, students worried about the security of their posture photos at every annual Yale mixer. Word on campus was that the men at Yale would "break into the [gym] . . . to steal the pictures and use them to arrange blind dates."[2]

While the 1950 incident would bring posture photography into question at Cornell, it strengthened the resolve of other university programs. Responding to the Cornell incident and to a medical report that claimed that compulsory posture programs were "useless," a Yale professor of physical education and head swimming coach, Robert J. H. Kiphuth, claimed that there was "a pile of scholarly material a foot high" supporting the value of posture training.[3] Kiphuth also reminded his colleagues of the widely reported Kraus-Weber studies that showed American physical fitness levels lagging far behind those of Europeans.[4]

Rather than being swayed by the denunciation of posture photography and training, Kiphuth insisted that, if anything, the program at Yale and elsewhere needed to be strengthened. So as to steel the program against criticism from other scientific experts and the student body,

Kiphuth thought that the personal relationship between the teachers and the students needed to be strengthened. The primary goal, in his mind, was to "make the students more sympathetic with our aims." Speaking on behalf of his colleagues, Kiphuth opined: "We believe in the program, the University believes in it, and now we must lead the student body to accept it."[5]

Others agreed. Ellen Kelly, director of the Penn State University posture clinic, where she examined local elementary schoolchildren along with college students, insisted that "there is no reason why the taking of posture photographs should not become an accepted procedure in any school system." In particular, Kelly believed that posture photography was the single best way to motivate students, the surest "method of securing active interest and enthusiasm . . . toward posture improvement." She warned her colleagues, though, to steer clear of the press and to avoid over-involved parents. "If publicity is avoided while the procedure is being initiated," she counseled, "easy acceptance should follow."[6]

While professionals continued to debate the worth and scientific legitimacy of posture training, students who attended colleges and universities after World War II began to resist the mandatory physical exam, claiming it was an invasion of privacy. At the same time, disability rights activists fought against the wider policy of school-based physical education mandates, required courses that were inherently ableist and exclusionary. What experts such as Kiphuth and Kelly sensed—and what motivated their felt need to double down on the virtues of posture training—was student indifference and hostility; postwar students were no longer the docile subjects to which posture experts had become accustomed earlier in the century.

This chapter explores the posture sciences from a bottom-up perspective, with an attempt to capture the thoughts and actions of those who underwent posture assessments and instruction in the early to mid–Cold War years in America. These personal accounts include an array of voices from white and non-white men and women. Reactions varied; some favored posture training, while others, citing feelings of great discomfort and, in certain cases, violation, opposed it. Posture exams provoked particular distress among students who had nonnormative bodies, those with scoliosis and other spinal irregularities who, through no fault of their own, could not meet the standards of mainstream posture, beauty, and health. Posture educators did little to ameliorate the stigma, and instead placed these students into "remedial" classes, segregated from "normal" students.

Due in part to student resistance, nearly all U.S. universities and colleges by the early 1970s shut down their posture programs. Yet unlike the more storied accounts of the 1960s counterculture movement, there is no one event or explicit protest that brought this practice to an end. Instead, the demise of university posture programs happened gradually, influenced by larger transformations in higher education beginning in the 1950s. The gradual dismantling of *in loco parentis*, for example, afforded students far more autonomy and the right to refuse intrusive and dehumanizing physical examinations. Additionally, the rise of coeducation and efforts to desegregate higher education led to greater diversity among the student body, a context in which the homogeneity of posture norms seemed anathema.

For the most part, the closing of posture laboratories and clinics on university campuses occurred without remark. Many of the scientists who conducted posture research turned to more specialized inquiries, gravitating toward the fields of sports medicine, ergonomics, and rehabilitation, while also adapting to the new world of patient rights and legal safeguards that protected research subjects. Meanwhile, the new, post-1970 generation of college students, free from physical education mandates, expressed dismay and disbelief that the generation of students before them had had to endure such indignities.

The end of mandatory university posture exams led to a significant decline in reportage of a poor posture epidemic—the numbers and visual data upon which the epidemic came into existence ceased to exist. As a result, the incidence rate of poor posture appeared to fall. What is important to note about this dénouement is that it did not occur because of a new cure or because of increased compliance with posture fitness initiatives. Rather, it occurred because students literally refused to stand for such examinations, bringing an end to the data collection that scientists needed in order to make the poor posture epidemic legible and a statistical reality.

Posture Photography and In Loco Parentis

When asked about the mandatory posture exams during her time as a student at Pembroke during the World War II years, Marcella Hance summed up her participation and that of her classmates by saying, "It never occurred to us to flout authority." "The Physical Education department . . . told us to do these things," she continued, "and we did it."[7] Before the 1970s, both private and public universities enjoyed great leverage over student life. In loco parentis, Latin for "in place of the parent," granted university officials the right to heavily regulate student conduct and character—including speech, association, and movement—without concern for student autonomy. One of the first legal cases pertaining to in loco parentis occurred in the late nineteenth century when a University of Illinois student disputed the school's mandatory chapel attendance. Ultimately, the Illinois state Supreme Court ruled in favor of the administration, stating that a student "necessarily surrenders very many of his individual rights. How his time shall be occupied, what his habits shall be; his general deportment . . . his hours of study and recreations . . . he must yield obedience to those who, for the time being, are his masters."[8] Many colleges required nightly curfews and daily room inspections. They also demanded "right" character and physical discipline. At the historically Black Hampton Institute, for example, students were expelled for bad work habits and "weakness of character" for much of the first half of the twentieth century.[9]

Female students, referred to as co-eds, were far more regulated and surveilled than their male peers. When women first entered institutions of higher learning in the late nineteenth and early twentieth centuries, the male educated elite worried that women, due to their presumed inherent biological frailties, would suffer from a whole host of infirmities if they devoted too much time to book learning. In an effort to assuage these concerns, universities measured and tracked the health of their co-eds religiously.[10] College physical education and health departments churned out study after study, pointing to improved physical measures—posture as well as student height, weight, and lung capacity—as proof that women actually got healthier at college, not sicker.[11]

Before the Second World War, many co-eds seemed rather proud of their physical education achievements. About her posture exam and training in particular, Pauline Ames Plimpton, a 1922 graduate of Smith, said that she felt "very lucky" to be made aware of her posture faults. She specifically appreciated "Dr. Goldthwait's exercises and personal attention." "It was one of those things, like Freshman English," she recounted, that continued to be "appreciated afterwards," long after her student days. Similarly, Mary Evans Boname (Smith '27) insisted: "the pictures *were* good for me, not so all important in themselves but definitely increasing our receptivity to the program of physical education which put muscle-building in perspective." Most memorable, she claims, were Goldthwait's lessons in which he "showed . . . X-rays of before and after that made clear where inner organs were sagging and how susceptible to multiple complaints the offender would be."[12]

One purpose of college physicals and quantification, especially among female students, was not only to encourage self-surveillance but also to instill the importance of good posture in future progeny. Boname would go on to raise a son with the same posture principles she had learned during her student days. When her son underwent his own posture exam at Yale in the 1950s, the examiner asked, "How come you stand so straight?" Boname's son replied, "[because] my mother went to Smith." According to Boname's telling, she "lived with [a] definition of *healthy* posture ringing in my ears . . . head up, chin in, chest high, tummy in, hips rolled under." And she regularly disciplined her children according to these dictates.[13]

By the mid-twentieth century, when college physical examinations had become the norm, students became less convinced about the worth and necessity of such requirements. In 1950, a young Sylvia Plath penned a letter home, reporting, "My height is an even 5'9"; my weight 137; my posture, good; although when my posture picture was taken, I took such pains to get my ears and heels in a straight line that I forgot to tilt up straight." "You have good alignment," the examiner told Plath, "but you are in constant danger of falling on your face."[14] Contrary to earlier generations, Plath conveyed a sense of nervousness and embarrassment about the exam. And unlike Plimpton, who claimed to have experienced an ever-greater appreciation for her posture training after graduation, Plath appears to have been scarred by the experience. In her 1963 novel *The Bell Jar*, for example, she used the exam as a literary device to convey a sense of dread. When the main protagonist, Esther Greenwood, finds herself facing the first intimate encounter with her boyfriend Buddy, Esther confesses that "undressing in front of Buddy suddenly appealed to me about as much as having my Posture Picture taken at college, where you have to stand naked in front of a camera, knowing all the time that a picture of you stark naked, both full view and side view, is going into the gym files."[15]

Several key changes regarding both college posture exams and campus culture writ large help explain the changing attitudes toward posture evaluations among co-eds during the mid-twentieth century. One significant shift was that as camera technology became more advanced and inexpensive, increasing numbers of physicians and physical educators began to adopt it for posture exams rather than using alternative means. Many lower-prestige medical specialists embraced clinical photography, seeing it as a tool that would project scientific objectivity and progressive modernity in an era of rapid innovation in medical technologies.

Of course, debates about the use of photography for posture exams had been happening since the early twentieth century. Clelia Mosher invented the schematograph precisely because she found camera photography to be a threat to student privacy.[16] Misgivings about the dangers of photography were quite common in Mosher's time. During the late nineteenth century, when photography became instantaneous and automated enough for amateur use, commercial outlets capitalized on amateur photography, using snapshots of everyday people—who had no knowledge of having been photographed—in print advertisements. Outraged, legal scholars Louis Brandeis and Samuel Warren argued for the "Right to Privacy" in an 1890 issue of the *Harvard Law Review*. According to Brandeis and Warren, photography and the reproduction of images in advertisements and newspapers "invaded the sacred precincts of private and domestic life," overstepping the "obvious bounds of propriety and decency."[17] Their essay, historian Sarah Igo argues, "signaled an important rethinking of privacy's reach," extending a property-based understanding of privacy to include a person's emotional and psychic sense of self. Having a picture taken without consent, and making that photograph public, Brandeis and Warren argued, could create "mental pain and distress, far greater than could be inflicted by mere bodily injury."[18]

Women stood at the forefront of some of the earliest privacy lawsuits against amateur photographers. In the 1900 *Roberson v. Rochester Folding Box Co.* case, the seventeen-year-old plaintiff, Abigail Roberson, sued Franklin Mills Flour Company for having used her portrait on product packaging without her consent, leading to one of the first legal implementations of privacy rights in the country. Twenty years later the highly accomplished Harvard physician Alice Hamilton found herself in a similar situation, discovering her face and name on the cover of the Angier Chemical Company's cod liver oil advertisement without her prior consent. Writing about the incident in a professional journal, Hamilton expressed her dismay, feeling it was "an unwarrantable invasion of privacy."[19]

Eventually, by the 1930s, medical photography came under legal scrutiny. In response, the Eastman Kodak Company began manufacturing specialized "clinical cameras" to promote the standardization of photo-taking in the clinic, making the practice appear to be more like X-ray technology than popular picture taking. At the same time, medical photography became ever more professionalized, with its own training manuals and established rules of best practice, many of which revolved around how to preserve patient privacy. One of the first rules was to avoid photographing the patient's full body and instead focus only on the anatomical area of concern. When, as in the case of posture photography, the object of study demanded a full body photograph, the camera technician was expected to mask the subject's eyes, either during the camera shoot or afterward during the developing process.[20]

Professional medical photographers advised extra caution when photographing female patients in the clinic, knowing they were the most likely to press charges. Male photographers were advised to have a female chaperone in the room when taking pictures of female patients. In the end, as historian Heidi Knoblauch demonstrates, more and more clinics ended up hiring female photographers in order to further insulate themselves against potential litigation.[21]

Posture clinics followed suit. But as Wayne University would learn by the late 1930s, female camera operators and chaperones did not placate all co-eds. In a highly publicized case brought before the Detroit Board of Education, three female Wayne University (now Wayne State University) students claimed that the mandatory posture photographs were "outrageous" and demanded that the board open an investigation. When interviewed by the *Chicago Daily Tribune*, Wayne University health director Dr. Irvin W. Sander claimed that the students' health privacy had been fully protected: there was a nurse chaperone in the photography room and the students wore masks. Depicting these complainants as outliers, Sander told the *Washington Post* that out of the 1,350 photos he had taken, no other women had lodged a formal complaint about being photographed.[22] The president of Wayne University sided with Sander, concluding, "I can't see a thing to criticize about it. In fact, I believe [the posture program] to be a step forward in student health."[23]

The demand for nudity upset Wayne University co-eds the most. By the mid-twentieth century, many professionals in the fields of physical education and physical medicine increasingly insisted that their subjects pose nude, maintaining that anatomical landmarks be in full view, unshrouded by clothing. Insisting on nudity served as a professional distancing technique, making the clothed examiner entirely distinct from and in command of the unclothed subject. When Wellesley College first adopted camera photography, it required its students to stand fully unclothed so that examiners could affix light aluminum pointers along the women's spines and sternum, a technique used to better demarcate spinal curvature in the resultant photograph (see figure 37). This procedure, known as the "Wellesley Method," was touted by physical educators as an advancement in scientific precision when it came to posture measurements.[24] According to the creators of the method, the Wellesley Method eliminated discrepancies between examiners, measured posture "faults and merits impartially," and prevented the examiner

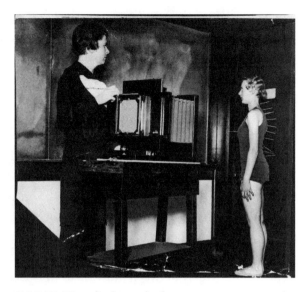

FIGURE 37.   A physical education instructor at the
University of Minnesota demonstrates the "Wellesley
Method" of posture photography in 1934. The tell-tale
sign of the Wellesley method is the use of aluminum
pointers running up and down the examined subject's
spine. This image was published in the *Minneapolis Tribune*, which explains why the student is wearing a leotard
rather than posing naked, as was demanded according to
the rules of best practice.

from basing the entire posture grade "on a single pronounced defect."
The authors of the original study distinguished their work from that of
older methods (such as Mosher's schematography), claiming that posture grades up to that point had been "subjective" and "unintentionally
biased."[25]

The projection of scientific objectivity was particularly important for
posture scientists on university campuses where college athletic departments grew exponentially after World War II, threatening the very existence of physical education. Additionally, some of the major tenets of
posture science were coming under scrutiny not only from the wider
medical community (especially in the wake of the pharmacological
revolution), but also from factions within physical education and

orthopedic medicine. Kenneth D. Miller, professor of physical education at Florida State, concluded a 1951 survey of posture research contending that there was a "bewildering lack of unanimity" among experts in the field, with no agreement about what constituted good posture or an agreed-upon method to measure it. He also noted growing doubt within the field about the ability of physical training to actually bring about meaningful change in body posture.[26]

Miller's study built on several studies conducted by physiologically oriented physicians and orthopedists. The most extensive study, headed up by University of Pennsylvania cardiologist Louis B. Laplace and orthopedist Jesse T. Nicholson, found that posture correction did not uniformly bring about physiological benefits. Focusing on cardiac and pulmonary functioning, the authors found that "the results of correcting faulty posture differ widely between individuals irrespective of the grade of the defect." In testing circulatory efficiency, they discovered that a corrected posture moderately improved blood flow in 50 percent of their subjects, but that in over 15 percent a "certain degree of postural slump" was far more advantageous than an erect posture.[27]

Dr. James Frederick Rogers, chief of the Division of Health and Physical Education in the U.S. Office of Education, also railed against the gospel of good posture. "Posture grading," he insisted, "has been done essentially for aesthetic reasons." "It is manifestly unfair to seriously compare children as regards posture," he continued. "It would be just as unfair as to classify them as beautiful, plain, homely and ugly. We cannot standardize the human body."[28] Rogers insisted that the human spine was elastic, constructed to bend in all directions, and therefore should not be trained to maintain one singular erect position. He also disputed any correlation or causality between posture, physical health, and mental acuity. "To say that posture has any effect upon immunity," he quipped, "is just a little absurd." He further pointed to the fact that examiners would inevitably find "very delicate and stupid and immoral subjects at the A end of posture distribution and very vigorous, bright, and honorable children at the D side."[29] A physiologist by training, he believed that good food and rest were more important to health than posture training. His research led him, similar to Laplace and

Nicholson, to question the assumption that erect posture improved respiration. In one study, he found that individuals who had let their arms hang loose by their sides demonstrated better lung functioning than those who adopted a stiff, erect posture.[30]

The growing criticism of the posture sciences during the mid-century years had the effect of making certain university educators more committed than ever to demonstrating the scientific worth of their programs. In schools such as Wellesley, where students could earn PhDs in physical education, dissertations devoted to posture exams abounded. In a 1943 survey of the field, two university physiologists claimed that the sheer volume of work devoted to the scientific establishment of posture programs in the U.S. "would require a monograph of large proportions."[31] Following this trend, many universities modernized their posture laboratories and clinics, replacing the old technologies of schematography and modesty drapery with camera technology and enforced nudity.

For the most part, white male students went along with the edict to shed their clothes. A 1948 Harvard graduate and veteran of the Second World War, Irving W. Knight adopted a fairly indifferent attitude toward the exam. "Nudity was a normal part of the college experience," Knight reflected, recounting how at Harvard (and many other universities), men were expected to swim in races in the nude since it was believed to be more hygienic. From his years serving in the armed forces, Knight was accustomed to stripping naked for repeated physical exams. In many ways, the military normalized both male nudity and posture exams. Recalling his time as a World War II veteran, D. W. Schulenberg writes: "We were just in from Japan, mostly draftees, with two years in the service [and] a frontal [posture] picture was taken as part of the discharge procedure."[32] Knight himself said that "none of us gave [the posture exam] a second thought. We had been living lives without privacy for several years in the armed services."[33]

Yet in taking white men as the norm, examiners gave relatively little thought to how more marginalized populations—women, minorities, or those with disabilities—might experience the exam differently than those who, by virtue of being part of the majority, enjoyed far greater autonomy and security in their daily lives. In a 1953 letter to her fiancé stationed overseas in the Korean War, Smith college senior Bonnie Sharpe wrote, "We had senior posture pictures the other day and the ordeal was positively traumatic!" "They're always taken in the nude," Bonnie continues, "and the idea of 'dropping the sheet' and posing is abhorrent."[34]

Sharpe was not alone in her personal protest. In 1948, several Smith freshmen told the *Hartford Courant* that the exam was "'grim!'" "First," the students explained, "they make you wrap yourself up in a sheet, or maybe it's a shower curtain, and you feel just like a Roman senator, and then they make you unwrap and they take your picture in your birthday suit."[35]

Graduates from Pembroke College voiced similar feelings of degradation. Teresa Mellone ('39) described it as "such an embarrassing experience." Ann M. C. Anderson ('59) depicted it as "nerve-wracking." Finally, Elissa L. B. Arons ('66) recounted more of a collective unease about being "stripped to the waist," saying, "We all had concerns that it was really—there was something uncomfortable about this whole thing."[36] In an attempt to explain why more women did not refuse to stand for these photos, Ulle Viiroja Holt ('66) blamed it on the fact that college women did not have the words to express feelings of violation or sexual harassment. Holt also pointed out that there was no academic process for women to safely protest and report abuses. "We were powerless," said Holt. "We didn't even have the language to talk to each other except, 'Ew.'"[37]

University examiners scrutinized not only posture, but also the shape, size, and weight of students. Gloria Elizabeth Del Papa (Pembroke, '46), an Italian American student who described herself as "overweight," recalls the "great influence" the director of physical education, Bessie Rudd, had on her. "I can remember Miss Rudd throwing this thing [the silhouette photograph] . . . on the desk and saying, 'You realize that this would make two of any other girl at Pembroke College.' She was disgusted. She felt I was terrible [*sic*] overweight. Well, I was big, but I really wasn't an obese monstrosity."[38] Speaking in an interview more than forty years after the fact, Del Papa said she felt generally unharmed by Rudd's harsh words because she was preoccupied with her chosen extracurricular activities. Yet the way she was treated by Rudd and the institution after her posture photo was taken indicates that her extracurricular time may have suffered in order to make room for mandated weight loss efforts. Rudd "put me on diets constantly and sent me to the school psychiatrist—we used to have a school psychiatrist/psychologist—to see why I was overeating and all that. Because I never lost any weight. I just kept my merry way. So, there was a concerted effort to really make me lose weight."

Minority students experienced a double bind, for they felt that they could not risk articulating any displeasure with the exam, even though some of them abhorred it. Expressing the need to fit into Pembroke in the late 1960s when there were only six Black students on campus, African American student Bernicestine McLeod Bailey (Pembroke, '68) says about her posture photo: "I went along because I didn't know what else to do." "I guess I thought of it," she continues, "as a crazy white thing." The number of indignities that Black women experienced on predominantly Anglo-Saxon white campuses such as Pembroke likely made posing for nude posture photos just one more item in a long list of transgressions. Facing everything from outwardly racist white roommates to fraternity-sponsored minstrel shows, the Black students at Pembroke in the late 1960s largely remained silent, describing themselves as timid and shy, working hard to pretend that race was not a factor in their daily lives.[39] Women such as Del Papa and Bailey likely felt powerless to refuse or question posture examinations, for as non-white women their voices were already marginalized and their bodies pathologized for not adhering to Eurocentric beauty standards.

Co-Eds Take Revenge

For many co-eds, the practice of posture photography amplified feelings of sexual anxiety in an era already rife with concerns about sexual deviancy and the place of the nude female body, particularly in the growing industry of popular pictorial magazines. Since the early twentieth century, popular magazines and advertisers utilized seminude images of women to sell products such as soap, corsets, and perfumes. Print media used the slang term "cheesecake" to describe the practice of covering the female body enough—but not too much—to make the mass-produced images of seminude women at once alluring and respectable. By the mid-twentieth century, magazines such as *Life, Esquire, National Geographic, Ebony*, and eventually *Playboy* and its African American counterpart *Duke*, pushed the notion of acceptability, exposing more and more flesh in the hopes of boosting circulation and ultimately creating what became known as "borderline material," erotic imagery that, according to historian Joanne Meyerowitz, "stretched the gap between respectable cheesecake and illicit pornography." The rise of the pinup girl, Meyerowitz argues, "moved images of women's bodies from the margins of obscenity to the center of mainstream popular culture."[40]

Leading up to the 1946 U.S. Supreme Court case of *Hannegan v. Esquire*, wherein the U.S. Post Office wished to revoke the second-class mailing privilege of *Esquire* magazine due to the perceived obscenity of Alberto Vargas's pinup girls, women provided testimony in both opposition and support. Certain child welfare activists and social workers testified in favor of the magazine and its use of cheesecake, seeing the semi-clad woman as a progressive step toward shedding the repressive Victorian past. These women found the Vargas girls a fitting tribute to female beauty, which, they believed, would help the cause of women's liberation. Only the truly abnormal onlooker would derive sexual pleasure from such images, they concluded. By contrast, the feminist activist Anna Kelton Wiley argued that such images degraded women and threatened the fight for equality by minimizing women, seeing them as little more than mere sexual objects.[41]

Similar disputes erupted when *Ebony* began to mass produce semi-nude images of Black women during the 1940s, but the arguments took place against the backdrop of entrenched racism and the centuries-long history of Black enslaved women being sexually exploited, having no right to privacy of their bodies. Some Black women, taking a stance similar to that of Howard University's Maryrose Reeves Allen, believed that cheesecake helped to promote the idea of Black beauty, a form of resistance against the notion that only white women could be

considered attractive. Others, however, worried that these images perpetuated racialized stereotypes of the animalistic, hypersexualized
Black woman, and wished to place the blame on the immorality of white
popular culture.[42]

Posture photography went far beyond the limits of cheesecake, or
even borderline material. Even *Playboy* and *Duke*, both of which challenged the limits of borderline material by including images of bare
female breasts in the immediate postwar years, still covered the female genitals.[43] It is thus little surprise that college-age women who
were forced to pose entirely naked for these photographs felt exposed
and violated.

Nude posture photos *may* have been acceptable, at least for a time, if
it were not for the inadequacy of university safeguards, measures that
would have protected the photographs, negatives, and identities of the
students standing in the frame. Posture photos were stolen, either in
reality or according to rumor, and this made college co-eds worry about
their own social status if their photos were to fall into the possession of
a male student. Exactly what, after all, was the difference between a
Playboy or *Duke* centerfold and a university nude posture photo, with
a co-ed's name attached to it, freely circulating among college fraternity
houses? To university co-eds, there was a world of difference.

Reading against the grain, one can see co-eds taking up subtle and
covert forms of resistance in the published and archival records. In creative ways, co-eds chipped away at the veneer of objective professional-
ism that physical educators and physicians liked to project. They did so
by refusing to accept that the photos were, to use the words of Dr. Sander
from Wayne University, "as impersonal as a clinical examination."[44]
They insisted on the sexual nature of the photos, not in the way that
male students did, but in a way that would bring the entire pursuit into
question.

Take, for example, a cartoon in a 1950 issue of the *Vassar Chronicle*,
which depicts—through parody—the nude posture photograph as a
prop in heterosexual courting (see figure 38). In the image, the female
Vassar student pulls away from Arnold's attempt at physical intimacy, a
model of masculine aggressiveness being countered by feminine

But Arnold, I hardly feel I know you well enough
to exchange posture pictures.

FIGURE 38.   Female co-eds often engaged in humor as a form of subversion regarding the practice of nude posture photography.

restraint. Her demure gaze, upright posture, crossed ankles, tidy pin-curled hair, and starched blouse all indicate that she is, as marriage and dating guides recommended, in control of this physical encounter. Her response—"But Arnold, I hardly feel I know you well enough to exchange posture pictures"—while humorous, also points to the premium placed on female modesty. Co-eds needed simultaneously to flaunt and protect their breasts and genitals from male onlookers. For the unmarried woman looking to attract a potential spouse, form-fitting sweaters and skirts signaled a proper degree of flirtatiousness. But if the co-ed moved in the direction of exposing too much skin, of baring her breasts and genitals entirely, then she risked sullying her reputation and social standing. The line between harlot and appropriately attractive was very thin, making the concealment of certain fleshy parts of the female body a high-stakes matter. In this context, a misplaced posture photo, whether in reality or rumor, could tip the reputational balance for a woman who ultimately had no control over the situation.

To fight back, certain co-eds engaged in speculation about the true intent of the professionals behind the camera, claiming that examiners had prurient aims. "I think the feeling was that the gym teachers were probably homosexual," Arons said of her posture photo experience at Pembroke. "It was just a *type* of person who became a Phys. Ed. teacher in college," Arons continued, "and it was just *known*, it was just sort of— you felt it."[45] Living at a time of rampant homophobia, when same-sex relations were illegal and pathologized, these kinds of accusations could end a career, and students likely knew this. Such claims—whether based in rumor or reality—had the potential of subverting the conventional power differential between examiner and examinee. By shifting the concern of the exam encounter from that of compulsory able-bodiedness to that of compulsory heterosexuality, Arons was able to transform herself into a kind of normative authority, casting doubt onto the character and virtue of her examiner, turning her teacher into a devious, nonnormative, sexually predatory authority.[46]

The most surprising form of resistance that I encountered in the archive, however, relates to the opening story of this chapter. When news of the stolen co-ed posture photos broke on Cornell's campus in the fall of 1950, one unidentified source, speaking to the school's newspaper, claimed that, in an act of resistance, several female students did it.[47] It was not, as many presumed, a plan hatched and executed by male students.

It is worth pausing to consider this scenario. If true, it would mean that several co-eds strategically exploited the sexual double standard that governed campus life at the time (and the larger U.S. society, for that matter) for their own gain. Putting aside who exactly stole the posture photos, one thing is certain: the university administration responded by bringing an end to nude posture photography on Cornell's campus the following year.

Stolen posture photos—an act almost always assumed to be carried out by male undergraduates—proved to be an administrative and public relations headache for university officials at Cornell and elsewhere.

For starters, it was a sign that the university was failing to uphold in loco parentis, both in terms of student protection and in maintaining a semblance of orderly operations. While the 1960s counterculture movement is often held up as the exemplar of campus unrest during the Cold War, the seeds of discord were already being sown in the early 1950s.

One example is the widely popular springtime tradition of campus panty raids. Most panty raids were undertaken by large groups of boisterous male students who circled female dorms, demanding and/or seizing lingerie. On certain campuses, the annual spring ritual sometimes turned into riots, requiring police enforcement to calm the scene.[48] Although later dismissed by even male student activists of the 1960s as mere "pranks," the raids created chaos across campus, challenging the university rules set in place to discipline all sexual encounters, while at the same time defying conventional courtship rituals that assumed young people were to be sexually restrained until marriage. While not overtly political, the riots challenged the mandates of a conformist society, a so-called Silent Generation no longer content to remain silent.[49]

In the interest of appearing in control of the situation, university officials downplayed the raids. Even the famed sexologist and Indiana University faculty member Alfred Kinsey saw the raids as little more than youthful play, a harmless spring frolic. In many ways, officials responded with the all-too-familiar adage of "boys will be boys," with the administration refusing to enforce an end to the raids.

But to those on the ground, it was anything but harmless. In the spring of 1953, Princeton students took to the streets chanting "We want girls! We want sex! We want panties!"[50] A year earlier, 850 Cornell men marched through the center of the campus, bugles blaring, shouting "We want sex! Hey! Hey! Hey! All the way!" Campus police reported damage to electrical fixtures, railings, drainpipes, and doorways. The police did not intervene until the mob attempted to enter a co-ed dorm, which was quickly locked down, and thereafter, according to certain reports, "the raid fizzled."[51] At Howard University, student Larry D. Coleman recalled that approximately 1,000 men were involved in the spring ritual during his freshman year. "They swept upon the Quad and

Bethune Hall like a horde of locusts," Coleman reported. "Fortunately, none of the women were hurt, although some of the men suffered injuries as a result of leaps from third-story windows, assorted abrasions, and contusions."[52]

It was often on the heels of these panty raids that reports of stolen posture photos would surface. "Everybody lived in terror," one female co-ed explained, "that the guys were gonna get a hold of the posture pictures."[53] Reports of male students raiding university gyms at all-women schools abounded during the 1950s. "Those posture pictures, I want to tell you," another co-ed reminisced, "were a source of great concern as to what was going to happen to those things or if people could break into the gym and steal [them]; it was regularly threatened that they were actually out on the campus with our names on them."[54] Fraternities made the pilfering and circulation of female posture photographs part of their hazing rituals. At Dartmouth, fraternity pledges admitted that they were frequently "sent out in search of posture pictures, particularly those rumored to be held in a specific secure location at Wellesley."[55]

On one level, the raids and stolen posture photos perpetuated the notion that men were biologically wired to be sexual aggressors while women played hard to get. Historian Erika Milam describes how this dichotomy, first articulated by Darwin, resurfaced in the years immediately after World War II, when evolutionary biologists and primatologists "bestialized [men] because of their association with aggressive, warlike behavior, whereas women were exempted from such degenerate stereotypes."[56] In this context, women were thought to be "coy," possessing a flirtatious but passive demeanor, requiring some element of courtship.

Certain co-eds found that they could use the biologically assigned attribute of coyness to their own constructive ends. Ruth A. Atwell, director of the physical education department for women at George Washington University, voiced frustration with female students who objected to having their posture photos taken, describing these women as being "coy." Atwell claimed that those of "her charges who objected to the procedure" did so out of a "false modesty." In other words, Atwell accused her students of adopting a kind of behavioral conservatism only to serve their own ends.[57] When interviewed about the resisters at Wayne University, Harvard's head of posture training, Norman Fradd, replied, "I have to laugh at the charges brought at Wayne. . . . It certainly is funny," he continued, "when the thing that conservative, puritanical New England has accepted has drawn censure in the more radical mid-West."[58]

While Fradd may have found such refusals "funny," the co-eds found nothing humorous about the matter. His reaction could also be characterized as hypocritical since, by the mid-twentieth century, many universities and colleges expanded their definitions and surveillance of student decorum. On most campuses, the list of "parietals"—the rules that governed student life—grew during the World War II and postwar years, as did the student handbooks that detailed disciplinary repercussions for even the most minor infractions.

With more first-generation students matriculating into institutions of higher education than ever before, postwar universities felt greater pressure than ever to socialize their charges, inculcating them with the mores and behaviors of the professional middle class. Historian Beth Bailey points out that such cultivation pervaded Cold War universities that saw it as their mission to foster democratic personalities. In order to address the needs of a new era, universities and colleges began to move away from an emphasis on academic offerings to instead focus on student life outside of the classroom, creating new administrative positions, such as deans of students and counselors.[59]

One of the most pressing matters, they believed, was controlling sexual behavior. The middle-class culture of respectability demanded that young adults refrain from premarital sex. Instead of attempting to control the male sex drive, which seemed like a fool's errand, universities cracked down on co-eds, mandating that they live on campus, where all of their comings and goings could be tracked. In essence, parietals became a set of rules almost exclusively targeted at controlling and governing female students.[60]

The theft of posture photos was but one way in which the system of parietals—of preserving the modesty and dignity of co-eds—seemed to be failing. While not tantamount to sexual intercourse, stolen photos could ruin a co-ed's reputation, especially if the pictures were made available for popular consumption. The nude female body became a surprisingly acceptable form of male entertainment once media companies began to produce "girlie" magazines. The traffic in nude photos created a context in which women felt unprecedented pressure to protect themselves against male sexual aggression in person and now, by extension, in print.[61]

One could argue, then, that those who refused to comply with the posture photo mandate adhered more closely to the rules of respectability than the posture scientists did. In an era when etiquette and dating manuals advised women to think of courtship in terms of an investment, with men as buyers and themselves as "items," women needed to preserve their modesty at all costs, especially since female nudity was a commodity readily purchased at any newsstand, or, in the case of posture photos, freely accessed if the school did not adequately secure the storage of the images.[62] In the end, the women who resisted posture photography exposed the hypocrisy of a university system that, bent on preserving sexual double standards and the cultivation of respectable women, failed so miserably in doing so.

Disability Activism

Cornell's decision to put an end to nude posture photography in the wake of the 1950s scandal was fairly bold given what its peer institutions were doing. Despite legal threats and scandals, many universities continued to conduct posture exams throughout the 1950s and 1960s. Cornell did not, however, close its posture program entirely. On the contrary, the school maintained its requirement of mandatory posture training courses until the late 1960s.

Posture advocates at Cornell and elsewhere insisted that physical training was necessary for students to be employable, presentable, and physically fit. In the era before the disability rights movement, these scientists claimed that they were assisting students who exhibited poor posture, without questioning the core problem of structural discrimination. The job of the posture scientist was to discipline the bodies of the students so that these young adults would fit societal demands, not to have society fit them.

But what about those who had bodies that did not and could not conform to established posture norms? Since most institutions of higher education barred students with severe disabilities, university examiners and researchers did not actually encounter true human variation. By and large, if scientists wished to research posture standards among severely disabled people, they would need to gain access to hospitals, nursing homes, and prisons.[63] Since physical homogeneity tended to reign on college campuses, researchers ended up magnifying the few physical differences that existed between nondisabled students, making what would have been mild physical aberrations in the general population into disabilities that needed to be managed and controlled. By mid-century, several university physical education departments developed a new categorization system, dividing students into "normal" and "subnormal" according to their posture exam results (the category of "abnormal" was reserved for those with severe disabilities).

An early advocate of "remedial" posture coursework, George Thomas Stafford, a professor of physical education at the University of Illinois from 1923 to 1962, developed a system whereby he separated students into two groups: "regulars" and "correctives." Believing that the latter needed more individualized attention, he required students with posture deviations to enroll in small-group, intensive exercise classes. Since "correctives" automatically earned a failing grade based on their posture photos, class attendance and participation became the one and only way that these students could pass the university's physical education requirement.

Remedial exercises often took the same form as regular ones, but for "correctives," educators like Stafford emphasized the need to rehearse the stages of normal child development. As such, instructors would engage in neuromuscular re-education techniques, having young adult students work on muscular tone while lying down or on all fours, reserving bipedal standing exercises for only advanced students. In Stafford's first year of providing remedial work, over 500 University of

Illinois students were placed in the correctives class, taking specialized training five times per week.[64]

Scoliosis turned out to be one of the most common postural irregularities found among college students. A condition with a long history dating back to ancient Egypt, spinal curvatures, in and of themselves, were not particularly worrisome. But the scientific and cultural meaning attached to scoliosis changed over time, especially in the middle third of the twentieth century. In the early years after the advent of the germ theory, physicians sought to isolate the cause of spinal curvatures in the hopes of standardizing treatment, which, at the time, was highly individualized. The medical establishment discovered that a whole host of diseases caused scoliosis, including everything from rickets and cerebral palsy to tuberculosis and—increasingly from the 1930s to 1950s—infantile paralysis, or polio. Given the murky set of etiological factors that governed spinal curvatures, several physicians decided to instead establish a universal system of measuring spinal curvatures to better document disease progression as well as treatment successes and failures.[65]

Dr. John Robert Cobb developed one of the more popular systems in the 1940s. Appling a geometrical analysis to X-rayed spines, Cobb determined a more precise and numerical way to measure curvature degrees. The method, known as the "Cobb angle," is still used today. Using Cobb's technique, the American Orthopedic Association deemed curvatures greater than 45 degrees to be "severe" and any curve less than 20 degrees to be "mild," assigning the qualifier of "moderate" to any that fell in between. For the most part, orthopedic surgeons used these categories to dictate and standardize treatment. Invasive surgery (fusions and steel rods) was recommended in severe cases, bracing in moderate cases, and discretionary exercises and close monitoring in mild cases. All treatments shared, relatively speaking, the same goal: to halt the progression of the curvature. Some bodies responded to medical and surgical interventions, others did not. Thus, while the adoption of Cobb's angle brought about uniformity in terms of standard of care, the unpredictability and mystery of the condition remained.[66]

At the same time, polio became one of the leading causes of scoliosis in the 1940s, especially as cases of infantile paralysis became more severely debilitating. Bulbo-spinal polio left those who contracted it unable to breathe or swallow without artificial support; it also caused significant leg and trunk paralysis as well as severe scoliosis, with some curvatures measuring 100 degrees or more. Curvatures like these—along with the accompanying muscular imbalance from trunk paralysis—often defied medical intervention. Undeterred, medical practitioners met aggressive polio spinal curvatures with equally aggressive therapeutic interventions. As medical historians have demonstrated, in an era when President Franklin Delano Roosevelt himself required mobility assistance due to adult-onset polio, the national campaign of disease prevention and the insistence on curing disabilities related to polio defined much of American politics and culture of the first half of the twentieth century.[67]

Yet even as the incidence of scoliosis due to paralytic polio, rickets, and tuberculosis declined precipitously after the mid-century discoveries of antibiotics, dietary supplements, and the Salk vaccine, the same degree of medical and cultural vigilance continued to be applied to scoliosis cases, even to those categorized as having "idiopathic scoliosis." Although the cause and natural history of idiopathic scoliosis remained unknown, medical professionals, educators, and family members tended to assume that spinal curvatures, no matter the cause, meant a life of grave debility. In other words, idiopathic scoliosis became a disability by association.

Universities generally denied entry to applicants with moderate to severe scoliosis. Laura Bond, a woman with scoliosis and a single mother of two young sons, was denied admission to UCLA's teacher training program in the 1950s. The university was not so much concerned about how she would juggle parenting and college work, but instead refused her entry into the program because of her "physical disorder."[68] Mildred Scott, a polio survivor with scoliosis and a leading voice of the American Federation for the Physically Handicapped, was also barred from getting her teaching degree because of her disability.[69] Students with mild curvatures, on the other hand, had more success in gaining entry to college. Yet, regardless of their feelings of well-being and ability, these students would frequently be categorized as "subnormal," disabled relative to the rest of the "normal" student body.

While Stafford saw himself as an advocate for disabled students, many of his students felt stigmatized by virtue of being segregated. One University of Illinois student who underwent physical training for scoliosis and kidney disease reported, "It was very embarrassing . . . to be viewed by others as 'one lungs [sic], limps or what not,' because of our defects."[70] The student further complained about the constant "critical gaze" of fellow students and instructors, the latter of whom possessed an unwavering belief that "this sort of training was beneficial."[71]

In all likelihood, college students with scoliosis would have already experienced the "critical gaze" before ever stepping foot on campus.[72] Written accounts of those who lived with spinal curvatures during the interwar and World War II years reveal that many first discovered their scoliosis not in a clinical setting, but rather in more personal and intimate settings wherein a loved one or friend would first detect the curvature. In other words, healthy young boys and girls first learned of their potential unwellness from the penetrating gaze of family and friends.[73]

Outside of the home, children with scoliosis would often be mocked by their peers. Common jeers would include "Hey, hump back," or "You'd make a good Quasimodo."[74] If the family sought advice from a physician, medical treatments would frequently result in more unwanted attention. Many doctors prescribed heavy, cumbersome braces to children and teens, orthopedic devices that could not be hidden under clothes and required nearly twenty-four hours of constant wear. In more severe cases, certain teens and their families opted for invasive surgeries, requiring the person with scoliosis to miss school altogether. About her experience with scoliosis surgery as a teen in the 1950s, Iris Halberstam-Mickel says she felt traumatized having missed "a lot of socialization during the ages of 12 to 14," and that this had a lasting "emotional and psychological impact."[75]

"I think in this country we are visually oriented," wrote one woman, who went by the name Martha, in her published account of her scoliosis experience. "[We have] preconceived notions about what normal men and women, boys and girls should look like," she lamented. "When people deviate from that too much," she continued, "there is trouble."[76] Whether they refer to it as "the gaze" or staring, many scoliosis survivors felt shame and discomfort from unwanted visual attention. "It drives me crazy when people stare," said another woman of her time growing up with scoliosis in the 1950s.[77] Another contemporary, writing under the name of "Frank," made a similar observation, attributing his inability to find a spouse or to seek certain kinds of employment and promotions to the "unattractive" appearance of his back. "In my estimation," Frank commented, scoliosis "made me introverted and an outcast."[78]

Remedial posture classes often made the lived experience of stigma and conspicuousness worse. In the clinical setting of the university gym, where posture photos and exercises were taken and assessed, unwanted stares turned into the clinical gaze. While experts such as Stafford found fault in his professional peers who created a "fetish of symmetrical development" and sought "standardized results in a quite unstandardized universe," the goals of posture training did little but reify such fetishes, which permeated not only the posture sciences, but also the wider culture.[79]

Students with mild to moderate scoliosis rarely protested college posture programs outright. Many adhered to the demands of university physical education to avoid being failed out of school. Some embraced professional physical training, hoping it would teach them how to correct or disguise their curvatures as they moved into the workplace. Before the 1973 passage of the Rehabilitation Act, the first civil rights law to protect against anti-disability discrimination, those with spinal curvatures could be legally refused employment. In her interviews with fellow "scolies," Iris Halberstam-Mickel includes the story of "Marlene," a college graduate who admitted to feeling "as normal as anyone else, except when it [came] to the job market." "I applied for many executive secretarial positions," she recounts, "only to be turned down after the interview." The potential employers "are most impressed with my resume," Marlene continues, "but once they meet me, my qualifications seem to dwindle before their eyes." In addition to her frustrations in the job market, Marlene also felt that "scoliosis is a real handicap when you are looking for a husband."[80]

Based on oral interviews, it appears that there were a handful of students with idiopathic scoliosis who, in small measure, started to protest school-based posture mandates beginning in the mid-1960s and early 1970s. One female student who entered Cornell University as a freshman in 1964 recalls how when physical educators were sure that, through proper remedial posture exercise classes, they could "cure" her scoliosis, she did not believe them. In a move toward disability pride, she and her fellow posture "defectives" called themselves the "Drastic Spastics," refusing to accept the programmatic goal of trying to achieve body normality.[81]

Physicians began to observe a similar kind of resistance in the clinic. Concerned about what they dubbed "noncompliance," orthopedic surgeons turned to psychologists to help explain what they saw as the pressing issue of brace wear refusal. Although Dr. John H. Moe dismissed "problem" patients by insisting that the "emotional obstacles to brace treatment ha[d] been greatly overemphasized," one study from the mid-1970s found that at least 20 percent of teens and young adults who were prescribed a brace refused to wear it. In the words of the researchers, these adolescents and young adults were generally the types who "are not awed by the physician and are not overly impressed by what he says. Scolding, threatening and loud exhortation to increased [sic] cooperation," they observed, "are unlikely to be effective."[82]

Ultimately, individuals with severe disabilities, those who had been historically marginalized at institutions of higher education, helped the most in bringing an end to mandatory physical education requirements, including posture exams. Exposing the inherent ableism and overt discrimination of mandatory fitness courses and evaluations, Edward Roberts fought many battles to gain admission to the University of California at Berkeley in 1962. A post-polio, respirator-dependent quadriplegic who required a wheelchair for mobility, Roberts is remembered for founding the "Rolling Quads" and creating the Independent Living Movement, making Berkeley's campus more accessible for other student wheelchair users such as himself. Yet he never would have achieved these feats if he had not first made an appeal to the Burlingame, California, public high school to excuse him from completing the state-required physical education in order to graduate.[83]

Reflecting on the ableism that permeated all levels of public and private education at the time, fellow disability activist Judith Heumann remarked, "nobody knew anybody who was teaching in a wheel-chair."[84] A post-polio quadriplegic and wheelchair user as well, Heumann gained entry to Long Island University, but when she graduated in 1969 and passed the New York teaching certificate examination, she was refused employment on the grounds that she was not able to walk. In 1970, Heumann won one of the earliest disability-based employment discrimination lawsuits against the New York City school system.[85]

The disability rights movement, along with the burgeoning patient rights movement of the 1970s—especially its explicit concern for informed consent in scientific research—would go far in making practices such as school-based nude posture photography impossible to carry out. If the institution did not realize it, then the newly emboldened students did.[86] Soon after these early wins, alumnae who had undergone the posture exam began airing public complaints. In a tone that was in keeping with the various civil rights movements, one Smith alumna said of the experience: "I felt like a piece of meat," comparing the posture photo to having to "pose for a girlie magazine" or engaging in "making skin flicks."[87] If the personal was political, so too was the decision of whether or not to stand for a posture photo.

An Unremarkable End

How and when each university decided to close its respective posture program is murky at best, as are the precise reasons for each institution deciding to do so. Writing for the *Yale Alumni Magazine*, Andrew Letendre ('58) claims that "the photography was discontinued in the late '60s after word got out that pictures of naked Vassar women had been stolen and were circulating outside the academic world."[88] Yet in a 1977 interview, a Vassar professor of art history, Christine Havelock, claims that it ended in 1969 because the female students "didn't like it. It was an incipient sign of the women's movement, the women on campus were no longer docile. They objected to the removal of clothing, that was a violation of privacy."[89]

Havelock's reference to the women's movement is surely a part of the reason for the demise of posture photography. After all, only a year earlier in 1968, over one hundred women from the New York Radical Women (NYRW) mounted a highly publicized protest at the annual Miss America contest held in Atlantic City, New Jersey. That same night, African American activists staged a Miss Black America contest to protest the exclusion of Black women throughout the pageant's history. Whereas the Miss Black America contest targeted segregation and not the competition itself, the NYRW wished to bring down all such beauty contests, believing that they perpetuated racism and sexism. A pamphlet from the event read: "The Pageant contestants epitomize the roles we are all forced to play as women. The parade down the runway blares the metaphor of the 4-H Club County Fair, where the nervous animals are judged for teeth, fleece, etc., and where the best 'Specimen' gets the blue ribbon."[90] Nude posture photos, as we have seen, evoked similar feelings.

Perhaps most importantly, with the retreat of in loco parentis, it became illegal to require nude posture photography. The century-long customary and legal practice of in loco parentis, with all of the rights and responsibilities of legal guardianship, eroded as students fought against university rules aimed at disciplining and controlling their daily lives. Many university officials largely acceded to student demand, for administrators were "relieved of their responsibilities for the physical well-being of their students, a change that protected them from an increasingly litigious society."[91] With the passage of the Twenty-Sixth Amendment to the U.S. Constitution in 1971, granting those eighteen years of age or older the right to vote, along with other key court rulings, places of higher education could no longer appeal to in loco parentis. Students of the 1970s enjoyed newfound freedom of movement, dress, and political activism. They also, as historian Craig Forrest points out, could refuse to participate in the once-mandatory requirements such as ROTC and physical education classes and exams.[92]

Not everyone celebrated the end of mandatory physical education exams and the posture training programs on college campuses. Havelock, for one, lamented the loss, saying that good posture "was essential to the educated woman," "equivalent [to] a man flexing his muscles on the football field." "Good posture was a sign of beauty," she continued, adding that "one cannot be beautiful if the spine is curved. It will inter-fere with the ability to walk properly, and very likely the ability to marry properly."[93] For every woman protesting the Miss American Pageant, there were just as many others who were devotees of public personali-ties such as Helen Gurley Brown, editor-in-chief of *Cosmopolitan* maga-zine from 1965 to 1997, who insisted that "good posture is one of the most important thing[s] anybody can do to look better."[94]

Indeed, the ending of posture programs on university campuses did very little to challenge the outsized importance of personal appearance. In the postwar service sector economy, one's value became even more dependent on appearance and beauty, especially as companies demanded public-facing employees to possess pleasing personalities and good looks. "The more one's livelihood depends on appearance and personality," the scholar Jenny Smith Maguire writes, "the more the individual has a stake in cultivating the body as an enterprise."[95] These new pressures appear to explain why after Vassar became coeducational in 1969—the year posture training and examination were no longer mandated—men at the school urged for a continuation of the school's posture fitness offerings. "I felt wonderful after taking the course," Vassar student Jim Selfie reported in 1977, having undergone two years of posture training. "I got a steady stream of compliments." "My body is as important to me as my mind," Selfie continued. "One doesn't have to be a jock to feel good."[96] Having no reason to fear violations of privacy or a threat to his reputation as his female counterparts had, Selfie and others like him embraced the school's posture training with zeal.[97]

Although the language of a poor posture epidemic would drastically subside by the early 1970s, the beliefs that led to the initial articulation of the epidemic remained intact, especially in the popular imagination. In place of school-based posture exams, individuals were expected to engage in self-surveillance, policing their own alignment. Media outlets encouraged and taught readers how to engage in self-policing, often couching the effort in terms of self-help and empowerment. For ex-ample, in 1975 Alma Pryor, a Black fashion model, told *Chicago Defender* readers that "beauty in a woman is the sum total of her looks, the way she moves and her general posture." Pryor advised readers to "contract the buttock muscles slightly, and hold the stomach in." When walking, she continued, "the hips should be level, not swinging from side to side." A simple way to self-regulate, Pryor suggested, was by "glancing at your reflection in shop windows."[98] Twenty years later, the historically Black newspaper the *Philadelphia Tribune* ran an article with essentially the same message. "Do a posture alignment check every 20 minutes throughout your day," the paper stated, for "good posture can keep you feeling and looking young, healthy and confident no matter how old you are."[99]

The white press offered similar advice. "At the risk of sounding like your mother," wrote Jane Brody, health columnist for the *New York Times* in 1993, "stop slouching and stand up straight." "Poor posture can make you look older and fatter than you are," she continued. "It can also set you up for all kinds of chronic and recurring aches and pains." Contradicting what some of the earliest medical critics of the posture sciences had demonstrated, Brody told her readers that "when you slump, your diaphragm collapses, there is less room for your lungs to expand, and the resulting shallow breathing means there is less oxygen available to nourish your body."[100]

Despite the closure of university posture laboratories and clinics, posture panic held purchase in the popular imagination. While the epidemic may have ended, the felt need to remain vigilant against slouching persisted, and would eventually become woven into the burgeoning fitness industry of the late twentieth century, a time when commercial gyms offered Jane Fonda–inspired aerobics, secularized hatha yoga, and mainstreamed Pilates, exercise systems that each, in their own ways, promoted the improvement of posture. Ironically, the shuttering of posture laboratories helped keep the culture of posture panic alive well into the late twentieth century. Though posture photography came under attack, the theories that informed the establishment of the practice did not. Since the study of human posture through large photographic data sets ceased to be a legitimate form of inquiry, influential social commentators such as Brody could continue to emphasize the importance of good posture in disease prevention without risking damning criticism.

CHAPTER 7

The Posture
Photo Scandal

IN MARCH OF 1995, archivist Ben Primer boarded an Amtrak train bound for Washington, DC, against his better judgment. Days earlier, Princeton University officials ordered him to travel to the Smithsonian's National Anthropological Archives, where he was to oversee the destruction of thousands of nude posture photographs of the school's alumni taken during the 1950s. Outfitted with safety earmuffs, Primer, along with archivist Nancy MacKechnie of Vassar's Rare Books, fed photograph after photograph into a SEM Model 22 disintegrator. Within a matter of minutes, decades of recorded history had ceased to exist.[1]

Primer and MacKechnie found themselves in a professionally compromised position largely because of a *New York Times Magazine* exposé that had run less than two months earlier. Journalist Ron Rosenbaum claimed to have discovered a cache of mid-century Ivy League nude photographs in the William Sheldon papers held at the Smithsonian. The reaction of the Ivy League university officials to the existence of these photographs was swift and unquestioning. Most university spokespersons responded with words of scandalized disgust and outrage, assuring the public and implicated alumni that all records would be destroyed. Journalists like Rosenbaum applauded and urged the destruction of these photographs, characterizing the practice as a "bizarre" pseudoscience, or worse, a nefarious eugenic scheme in which America's ruling class became unwitting guinea pigs. Feminists such as

Naomi Wolf also wanted the photographs completely erased from the historical record, arguing that they debased women.[2]

What few journalists or university officials cared about was why and when the practice of measuring posture came about and to what ends. Instead, much of the conversation focused on how Sheldon, by then a deceased constitutional psychologist, utilized the photos for his own purposes in the mid-twentieth century, creating a body typing system purported to predict a person's psychological makeup based on his or her physique. Nearly twenty years dead by the time of Rosenbaum's exposé, Sheldon served as the perfect scapegoat for dismissing an ostensibly absurd practice. An anti-Semite who endorsed neo-eugenic ideas of biological determinism, Sheldon had already fallen into disrepute among a majority of scholars and scientists.[3]

Sheldon's morphology was indeed spurious and racially bigoted. But the story behind the existence of nude posture photos involves a larger and more complicated history in which Sheldon plays only a very minor role. Similarly, to dismiss the posture photos as being uniquely injurious to women, as Wolf did, is also sorely lacking in historical context. And yet, because of these rather simple—yet effectively sensationalistic—attributions, a significant portion of the historical record pertaining to the practice was destroyed.

This chapter reopens the nude posture photo scandal of the 1990s in an attempt to understand why a practice considered necessary and scientific just decades earlier provoked such a strong reaction. Those involved in the destruction of the historical record—from journalists and social commentators to university officials—paid surprisingly little attention to the merits of the scientific basis for such physical evaluations, or reflected on the fact that they were living in a nation that spent millions of dollars on pharmaceuticals and fitness goods that purported to improve posture health. Outrage among elites who wished to preserve their privacy also dominated the discussion, leaving little space for a more thorough and measured investigation into whether human posture had any indicative value concerning health in the first place. Refusing to dig into the longer history of the posture sciences and their basic tenets, those who fueled the scandal of the 1990s actually perpetuated the well-established linkages between human stance, health, and personal worth. The very records that made the slouching epidemic a reality had been destroyed, but the assumptions upon which the posture sciences rested were left intact.

Early Photo Destruction

When Rosenbaum set out to write his essay, he was already well acquainted with school-based posture exams, despite his tone of shock and surprise. Upon his arrival at Yale University in 1964, Rosenbaum, along with all other entering freshmen, would visit the Payne Whitney Gym to have their posture photographs taken, as had been custom since the World War I years. In preparation for the exam, Yale officials demanded that Rosenbaum and his classmates strip nude. Recounting the moments after being told to take his clothes off, Rosenbaum writes, "And then—and this is the part I still have trouble believing—they attached metal pins to my spine. There was no actual piercing of skin, only dignity." The "metal pins" to which Rosenberg refers were actually lightweight aluminum pointers that examiners would affix up and down a subject's spine, a method first developed and adopted by Wellesley physical educators in the early twentieth century, as illustrated and discussed in the previous chapter (see figure 1, chapter 6). Seeing archived photos of students with these pins, he confesses, was an "embarrassing reminder of a particular practice, which masqueraded as science and now looked like a kind of kinky voodoo ritual."[4]

The rhetoric of spectacle that Rosenbaum deployed in his exposé helped to distance himself and his generation—the ones who would go on to mount the counterculture movement—from posture photography, making it seem like a unique and strange practice, rather than the routinized, largely accepted tradition for much of the twentieth century. Rosenbaum positioned himself as a journalist with a shocking discovery about a secret, nefarious past. Worse still, he insinuated that material proof of the practice still existed, with nude posture photographs of past college goers out there for anyone to see.

In reality, by the time Rosenbaum's exposé appeared in print, certain key colleges and universities had already disposed of their posture photos decades earlier. For institutions of higher education, the photos proved to be an inconvenience in at least two respects. First, since student privacy was protected by law after the 1974 passage of the Family Educational Rights and Privacy Act, posture photos (even though they were taken before the law's passage) had the potential of creating a significant legal challenge for many universities. Second, storage of such documents required ample space, money, and personnel, a kind of luxury that many colleges did not have, particularly those like historically Black colleges and smaller, lesser-known schools.

Oftentimes, the destruction of a school's posture photos occurred in an ad hoc fashion and largely off the record. Arlene Gorton, the athletic director at Brown's Pembroke program during the 1960s, burned the photos a few years after the school discontinued its yearly posture exam mandate.[5] At Yale, a janitor purportedly found the school's nude posture photos in a locked closet sometime in the early 1970s. When he reported their existence, the athletic director had them shredded and burned with ostensibly little or no consultation with the school's archivists.[6]

At Smith, "Darth Vader" destroyed the college's cache of posture photos. Custodian Mike King, who, according to his colleagues, bore a striking resemblance to the *Star Wars* villain when he donned his welding helmet to operate the incinerator, burned the photographs in November 1984 by order of Smith's president and distinguished scholar of women's history, Jill Ker Conway.[7] Contrary to Yale's rather ad hoc and impulsive method of disposal, Smith officials deliberated the fate of their photos for several years, and along the way maintained a detailed record of the decision-making process.

Ironically, the first sign of trouble regarding the photos at Smith came about in the late 1970s when a PhD student and alumna asked to use them for her dissertation research. After completing her studies at Smith in 1976, Gretchen Dieck became a doctoral student in Yale's School of Public Health, where she built a dissertation project around Smith's archive of posture photos. Proposing a historical cohort study, she hoped to determine whether young women who had detectable spinal deviations in their late teens and early twenties had an increased likelihood of back pain and other musculoskeletal disabilities later in life. Dieck had attended Smith shortly after the ritual of annual posture exams had come to an end. However, her dissertation adviser, Jennifer Kelsey (Smith, '64), had personal experience being photographed and alerted Dieck to the research potential that the collection presented. In 1979, Dieck wrote a letter to Smith President Jill Ker Conway requesting access to the photos, hoping to test the assumption that posture scientists had been making for nearly a century—namely, that postural deviations led to a future of disability and ill health.[8]

Had Dieck proposed to conduct her study over a decade earlier, she likely would not have needed institutional permission. But her request came in the wake of journalistic and scholarly reports demonstrating that the U.S. Public Health Service had enrolled unsuspecting, impoverished African American sharecroppers in a forty-year longitudinal study of untreated syphilis, withholding effective treatment from these men who had been diagnosed with the sexually transmitted disease, but never informed of their condition. The public exposure of the infamous Tuskegee Syphilis Study in the early 1970s led to increased federal regulations on human experimentation, broadening the scope and process of institutional review boards (IRBs) in order to ensure patient safety and self-determination in any setting that received government funding to conduct research. The resultant National Research Act of 1974 and the Belmont Report of 1978 prioritized patient autonomy, restricting exploitative relationships between researcher and subject, and upholding the principle of "do no harm." The sea change in research protocol, subject protection, and growing federal funding for university-based research projects led nearly every university to set up their own internal IRBs.[9]

When President Conway received Dieck's letter, she immediately sent a memo to Dr. Mary Laprade, director of Smith's Science Center and chair of the school's Committee on the Use of Human Subjects in Research. The tone of Conway's memo indicates that, in the beginning, Smith officials saw no red flags with Dieck's request. In a straightforward manner, Conway asked Laprade to arrange a visit for Dieck to view the photos, which were kept in the basement of College Hall, along with other unprocessed boxes of archival material belonging to the Physical Education Department.[10]

For the first part of the study, Dieck wished to use the latest technology—a computer digitizer—in order to measure the spinal deviations in the student photographs, determining the incidence of spinal curvatures and misalignments in each class cohort and also across all of the alumnae. While conducting an initial review of the decades upon decades of archived posture photos, Dieck found that the images taken during the late 1950s would work best for her project since it was during this time that students were pictured not only in profile but also from the back, providing a posterior view of the students' spines. In the end, she narrowed her sights to three class cohorts: 1957, 1958, and 1959.[11]

From the outset, Dieck assured Laprade, Conway, and other college administrators that student anonymity would be preserved—that she would replace the name of each alumna with a numerical code. To offer more protection and to make it possible for her to conduct her research at Yale, where the computer technology existed, Dieck made slides of the photographs for transport and analysis, a process that removed any identifying names from the actual photos. Additionally, since she was interested only in the pictures showing the students' backs and spines, she would not be viewing faces or be able to detect any other identifying features. In her laboratory at Yale, Dieck stored the slides and other documentation in two locked cabinets, one for the slides of the photos and a separate one for the master list of names. Only a few authorized individuals had access to the keys. Writing about her experimental design for the academic journal *IRB: Ethics and Human Research*, Dieck contended that she and her adviser had "determined that no legal statutes were being violated by using the posture pictures, either in the state in which the college was located (Massachusetts) or in the state of Connecticut."[12]

In virtually every way, Dieck had met the research standards of the time. She received IRB approval from both Yale Medical School and Smith College. She got the blessing of Smith's President Conway, and she also received permission to transport the slides to Yale. She had determined that for the first part of the study she did not need alumnae consent to measure the pictures, since identity had been fully protected; university officials at both Yale and Smith agreed.[13]

Problems arose, however, as soon as Smith alumnae learned of Dieck's project. Dieck needed alumnae cooperation in order to conduct the second—and arguably most vital—part of her study. To determine the causality (or lack thereof) of postural deviations and later-life health problems, she needed a way to communicate with the alumnae, most of whom were around forty years of age, having graduated from Smith approximately twenty years earlier. Dieck developed a questionnaire to assess the incidence and degree of back pain among each class cohort, comparing the original posture photos to the individual subjective assessments of back health. To recruit as many participants as possible, Dieck turned to the Student Alumnae Association (SAA) for help, asking the class presidents to advertise and endorse the study. The act of completing and returning the questionnaire, Dieck concluded, would be taken as informed consent on the part of the alumna-participant.[14]

When SAA Executive Director Trudy Stella and President Nancy Lange first heard of Dieck's research, they voiced caution; from the start, they were worried that alumnae might take offense at having their photos accessed. Learning of the first part of the study, Lange wrote to Conway, "I think most alumnae would be pleased to have the posture pictures used to further medical knowledge, however, in view of the sensitivity of the subject, I think we have to be very careful about the protocol." "We could be criticized," Lange continued, "for letting an unauthorized person look at the pictures without permission."[15]

The study nevertheless proceeded. But when the SAA received the request to assist in the second part of the study, Dieck met resistance. "When contacted," Dieck explained, "all three class presidents had initially agreed to write a cover letter for their class." "Sometime later," she continued, "one class president became upset because the pictures still existed."[16] While not named in Dieck's publications, the president in question was Adelaide Ray Robb, a 1959 alumna. In her 1981 spring newsletter to the class, she alerted her peers: "Gretchen S. Dieck . . . will be asking each of you for permission to use your posture picture in a study at Yale University." With pointed language, Robb wrote, "*Read her letter carefully and follow your own inclinations.* I have nothing more to say on this matter as I find I cannot react."[17]

But Robb did react. Before bringing the newsletter to a close, she confessed, "Your president thinks that [Smith should] destroy our posture pictures, (why even the IRS keeps records only seven years)."[18] That the posture photos still existed twenty years after the class of 1959 had graduated shocked many alumnae. Some blamed university officials for being vague all along, and others claimed that university officials engaged in outright deceit concerning the fate of the records. Around the time of Dieck's study, Marcella Hance (Pembroke '44) told one interviewer, "somebody assured us that those posture pictures had been destroyed; I hope so. They were terrible." She insisted that at the time of being photographed, "it was not a feeling that it was going to be on permanent record."[19] Alumnae such as Robb and Hance harbored deep distrust of their respective institutions, a feeling not assuaged by the passage of time. Painting a scenario in which Smith College agreed to destroy the photos, Robb quipped: "Who would believe them anyway? They always have prevaricated about those pictures and now they want to send them down the thruway to Yale!"[20]

Fearing for the viability of her dissertation, Dieck abandoned the first part of the study, the part that, according to university officials at both Yale and Smith, complied with the legal and research ethics mandates of the time. "Even though we had permission from both IRBs to look at all of the pictures," Dieck and Kelsey wrote, "we decided that it was not in our best interest to do so."[21] Dieck herself admitted that the problems she faced with her research had nothing to do with design or legalities, but rather with public relations.

In an attempt to salvage the project, Dieck persuaded Conway to send an additional cover letter to the classes of 1957 and 1958 expressing the administration's enthusiastic support for the project. Dieck also found a champion in alumna Katherine O'Bryan Canaday, president of the class of 1957. Canaday assured her classmates that the study was "a unique opportunity for us to contribute (painlessly) to medical research." Yet her letter also pointed to discomfort with the provenance of the photographic records, voicing concerns about potential misuse and violation of privacy. Resurrecting the lore—and in some cases the reality—that university men had access to and sexualized interest in the photos, Canaday's letter begins:

> Remember: "Posture pictures used as bait,
> Bound to get you an Amherst date"?

Addressing a question that seemed to be on every alumna's mind, Canaday continued, "Were the pictures destroyed after our graduation, or had they really been stolen by an Amherst fraternity?" "Neither is correct," she informed her classmates. "Posture pictures have survived and are now on their way to Yale."[22]

It is hard to believe that Canaday's letter offered much assurance to her classmates, especially for those who still felt traumatized by having to pose nude for school officials and feeling as if they had no control over the permanent record. Nonetheless, Dieck was able to get a 50.5 percent response rate from the class of 1957 and a 43.5 percent response rate from the class of 1958. Likely due to Robb's open criticism of Smith College's "prevarication," the class of 1959 only provided Dieck with a 29.1 percent response rate.[23]

Despite such public relations difficulties, Dieck's dissertation produced rather remarkable results. While her study confirmed decades of research that demonstrated a relatively high incidence of postural aberrations at the population level (70 percent, according to her results), her findings, in stark contrast to earlier studies, suggested that such detectable bodily asymmetries held little or no predictive value.[24] Speaking to the more specialized diagnostic category of scoliosis, Dieck found that such curvatures played a "relatively unimportant role in the development of spinal pain in the adult years," nor did such asymmetries appear to increase the risk of one developing herniated discs or arthritis later in life.[25]

This conclusion contradicted decades of posture-related research and, most boldly, challenged the general trend among American orthopedic surgeons, who, by the early 1970s, adopted a highly interventionist stance regarding scoliosis care, prescribing heavy metal bracing and, in certain cases, implanting steel rods along the spinal columns of young patients who exhibited curvatures greater than 20 degrees.[26] While by the early 1980s a few researchers had come to similar conclusions as Dieck, the majority opinion in medical and public health circles was to aggressively treat scoliosis.[27] This was an era when many states mandated scoliosis screenings in elementary and middle schools. Recounting her middle-class experience of growing up in Newton, Massachusetts, during the 1970s and 1980s, comedian Amy Poehler writes that she feared two things the most: nuclear war and scoliosis. "My generation was obsessed with scoliosis," Poehler recounts. She continues, "Judy Blume dedicated an entire novel to it. At least once a month we would line up in the gym, lift our shirts, and bend over, while some creepy old doctor ran his finger up and down our spines."[28]

Dieck offered sound physiologically and biomechanically based rationales for her findings. First, she explained, most postural asymmetries were likely not severe enough to exert forces that would result in pathology (such as annual tears) or a degeneration of the facet joints (such as arthritis). And even if scoliosis led to some degeneration, she concluded, it may not necessarily lead to pain. Finally, she found that in many cases the body adaptively compensated for asymmetric loads in a pain-free way. Aggressive treatment that aimed to bring about symmetry might actually create new pain symptoms, she suggested, causing more pathology than if the postural asymmetries were left untreated, permitting the natural progressive adaptive responses. With this study, Dieck joined a small and unpopular contingent of researchers who questioned the need for posture-based screenings as well as casting doubt on surgical preventive treatments entirely.[29]

Given the near universality of posture examinations in higher education, the military, and certain medical institutions, one could imagine how medical scientists and other researchers would want to preserve the photos in order to conduct further in-depth studies, either to challenge or to confirm Dieck's results. Population-wide data such as these were a rarity, especially in postwar orthopedic surgery studies. After all, other medical researchers at the time of Dieck's study had already begun large-scale cohort studies to great public enthusiasm and federal funding. Both the 1948 Framingham Heart Study and the 1976 Nurses' Health Study served as gold standards for epidemiological research.

But no one came to the defense of the potential scientific worth of the posture photographs and their preservation. Indeed, quite the opposite happened. Knowing that Dieck continued to get requests from alumnae to have their photos destroyed, Laprade felt mounting pressure to come up with a solution to the problem. If the college did not do something, it risked jeopardizing alumnae—and thus potential donor—relations. When she entertained the idea of mailing the photos back to each individual alumna, thereby granting the women (instead of the university) ownership of the pictures, a colleague from another university warned Laprade that when another institution tried to do this, it "got in trouble" because the "pictures were classified as pornographic material."[30] When she proposed to have individual alumnae personally pick up their photographs, President Conway found this option to be too expensive and managerially unviable.

Having vetted all other options, the only choice left was to destroy the records. Before she made the decision, Laprade reached out to Dr. O. Donald Chrisman, a clinical professor of orthopedics at Yale Medical School and co-owner of a private practice, Hampshire Orthopedics, in Northampton, asking him to appraise the scientific worth of the records. Given the professional trends in orthopedics at the time, Chrisman rather unsurprisingly told Laprade in no uncertain terms: "I do not think that any further scientific studies on that subject would be additive." Refusing to address Dieck's findings and essentially offering Laprade permission to destroy the photographs out of hand, Chrisman wrote, "[I] really think that the time has passed for use of that material."[31]

In all likelihood, Chrisman's assessment came as a relief to Smith officials, since his opinion squared with the demands of the angry alumnae. Indeed, alumnae demands to destroy the photos grew in the early 1980s, especially as word got out that Smith still had the pictures. By 1983 the SAA, now in the thick of the controversy, reversed its original stance of wishing to participate for the betterment of medical science. In a letter to President Conway, Judith Marksbury, SAA president and Lange's successor, wrote, "Trudy Stella doesn't see any constructive way to deal with the pictures. . . . She said that they belong to the College," the letter reads, assigning clear ownership to Smith. "Thus she doesn't feel she should really have any say as to what is done with them. . . . However, from a personal point of view," Marksbury emphasized, "[Trudy] *thinks they should be destroyed, and probably should have been destroyed after each class graduated.*" Marksbury concurred: "My opinion is that they should be destroyed."[32]

Although Laprade had requested that Conway resolve the issue of the posture photos since Dieck had completed her dissertation and graduated with her PhD in May of 1982, Conway refused to give an answer until the SAA made plain its unequivocal view that the pictures should be destroyed. Finally, in October 1983, Conway wrote to Laprade: "Now is the time to have these pictures destroyed . . . and please let me know when and how they were destroyed."[33] Over a year later, King fed photo after photo into an incinerator housed in the Clark Science Center. As an early Christmas present to Conway, Laprade sent the president a photograph of King, aka Darth Vader, at the incinerator, providing visual proof of the destruction. The Darth Vader photo is one of the few remaining pictures left in the Smith archives that speaks to both the existence and subsequent annihilation of the College's decades-long posture photo program.[34]

The Feminist Sex Wars

The destruction of posture photos at Smith and elsewhere did not end the controversies surrounding them—they would live on in people's minds and in the lore of college life, representing a time before the height of the counterculture movement. Indeed, throughout the last two decades of the twentieth century, the photos served as exemplars of a past that several progressively minded commentators wished to render morally reprehensible. The photos became, in other words, potent symbols in the writing and rewriting of American history for the educated elite in the final decades of the twentieth century.

The same year that Smith burned its photos, comedian Dick Cavett (Yale '58) took the stage at Yale University as the 1984 commencement address speaker and breathed new life into the posture pictures. The campus was significantly different than in Cavett's days; most notably in 1969, Yale College matriculated its first class of women. By 1984, women made up approximately 40 percent of the graduating class. Despite the rather radical shift in student demographics, Cavett reverted to stories from his days as a "brother" of the then-sitting Yale president; both men belonged to an all-male secret society during their undergraduate days. In one of Cavett's comedic bits, he referred to the posture photos, a practice that, it is important to note, none of the graduates sitting in the audience would have experienced. Setting up the joke, Cavett recalled that when he attended the university, he and his fellow Yalies would sneak off to New Haven's red-light district, where they would peer at nude posture photographs of Vassar women. His punchline: "The photos had no buyers."[35]

Sitting in the audience was a young Naomi Wolf, who would later describe the day as the "graduation from hell." "I'll never forget that moment," Wolf writes in her 1991 bestseller *The Beauty Myth.* "There we were, silent in our black gowns, our tassels, our brand new shoes. . . . We dared not break the silence with hisses or boos," she continues. "Consciously or not, Mr. Cavett was using the beauty myth aspect of a [male] backlash: when women come too close to masculine power, someone will draw critical attention to their bodies."[36]

Wolf and her generation enjoyed many of the achievements brought about by second-wave feminists, from abortion rights and greater educational access to the highest employment rate in history. Yet the success seemed to come at a cost, with a perceptible cultural backlash against feminism and female empowerment. Wolf, in particular, identified the continued insistence on female beauty as one of the most potent political weapons against women's advancement. "The ideology of beauty," Wolf maintained, "is the last one remaining of the old feminine ideologies that still has the power to control those women whom second wave feminism would have otherwise made relatively uncontrollable: it has grown stronger to take over the work of social coercion that myths about motherhood, domesticity, chastity, and passivity no longer can manage." "It is seeking right now to undo psychologically and covertly," Wolf contended, "all the good things that feminism did for women materially and overtly."

Wolf was thus not necessarily upset by Cavett's insinuation that the posture photos were being traded and viewed as pornographic material, which is noteworthy. Throughout the late 1970s and early 1980s, certain feminists were highly distressed with the unparalleled growth of the porn industry. Only a year before Cavett's Yale address, feminists Andrea Dworkin and Catharine A. MacKinnon developed a model anti-pornography ordinance that was proposed—but ultimately vetoed—in the city of Minneapolis, Minnesota. Taking a zero-tolerance stance on pornographic materials, Dworkin and MacKinnon believed that through pictures and words, pornography was a sexually explicit subordination of women, harming women by conditioning men to view them as inferior, as sex objects to be used. They argued, in short, that pornography was a violation of women's civil rights.[37]

Dworkin and MacKinnon's tactic of deploying legal strategies—instead of protests, consumer boycotts, and other avenues of resistance—created significant fissures in the feminist movement. Before Dworkin and MacKinnon, second-wave feminists who opposed pornography focused most of their efforts on depictions and acts of sexual violence, hoping to stem the growing rates of rape and domestic violence. But in 1979, with the formation of Women Against Pornography (WAP) in New York City's Times Square, the protests grew to include a rally against all sexual imagery in public spaces, including an effort to get magazine displays of *Playboy*, *Penthouse*, and *Hustler* to be placed on top shelves in bodegas, newsstands, and candy stores. An area known for its hustlers, peep shows, and pornographic movies, Times Square was one of the most visible places on the international stage for WAP to take to the streets. Their chant—"Two, four, six, eight, pornography is woman hate!"—quickly became international news.[38]

To other feminists, WAP treaded dangerously close to the New Right, a group that defined pornography as everything from sodomy to same-sex relationships and thus wished to institute state-mandated censorship of nearly everything outside of the heterosexual nuclear family unit. Phyllis Schlafly, for example, endorsed Dworkin and MacKinnon's anti-pornography ordinance. As a result, the feminist movement became splintered into what would become known as pro-sex and anti-pornography feminism. By most accounts the "sex wars" between these two groups brought an end to second-wave feminism.[39]

Wolf, representing the newer, third-wave generation of feminists, generally sided with pro-sex feminists. Cavett's joke was an underhanded way to take aim at feminists such as Dworkin and MacKinnon, implying that they were prudish in their attempts to ban pornography and that pictorial representations of nude women would always be available to the male gaze, despite attempts to restrict it. Wolf did not necessarily disagree. What Wolf found deleterious was "beauty pornography"—whereby standards of female beauty, including naked beauty, were in the hands and eyes of the patriarchy, reified not only in the commercial pornography industry run by men, but also in the cosmetics industries run by men, in the magazines run by men, in the plastic surgery industry run by men. For women to take control, Wolf believed that there needed to be a "radical rapprochement with nakedness." Advocating more opportunities for women to see a variety of bodies, Wolf maintained that the fastest way to demystify the beauty myth was to "promote retreats, festivals, excursions, that include—whether in swimming or sunning or Turkish baths or random relaxation—communal nakedness." "Men's groups," she pointed out, "from fraternities to athletic clubs, understand the value, the cohesiveness and the esteem for one's own gender generated by such moments."[40]

What Wolf found pernicious, in short, was Cavett's punch line. Saying there were "no buyers" perpetuated the beauty myth wherein Yale men had the power to deem the pictured women worthless based wholly on their looks, both in the past and then again in 1984 in Cavett's retelling of the story. Other Yale alumni told equally injurious and scandalous stories. C. Davis Fogg (Yale '59), who attended Yale at the same time as Cavett, claimed to have stolen Vassar photos to create cards that could be traded and sold. Similar to Mark Zuckerberg's original Harvard-based Facebook, Fogg and his friends would rank the nude posture photos, giving them each a grade: "'A' for too hot to imagine, 'B' for ravishing, 'C' for nice." With the intention of selling the cards to men at Harvard, Princeton, and Yale, Fogg assigned a price to each photo card based on the degree of "hotness," with values ranging from $75 to $150. According to Fogg's autobiographical-turned-fictionalized short story, "the female jocks got the most attention, and the medieval history majors the least."[41]

Most troubling to Wolf was not necessarily the vile way in which men such as Cavett and Fogg graded and objectified women, but that most American middle-class women accepted and internalized such rankings to the point that they were competing against one another, with nearly everyone trying to avoid looking like the loathsome "Ugly Feminist." While the men in Wolf's graduating class got the "traditional grand tour of Europe," many of her female peers were asking for nose jobs and breast implants. The beauty myth, she lamented, was "marshaled to neutralize the achievement of women, especially in rites of passage." Casting herself as a latter-day Betty Friedan, Wolf made explicit the connection between *The Beauty Myth* and *The Feminine Mystique*, writing that the "inexhaustible but ephemeral beauty work took over from inexhaustible but ephemeral housework."[42]

If the photos had retained their original purpose (that is, if they were merely scientific objects solely concerned with the study of human posture), Cavett would not have had a laugh line, nor would Wolf's graduation have been ruined. But the photos—although destroyed by the time of Cavett's speech—endured beyond their material obliteration to become powerful cultural touchstones, which men like Cavett could trot out to get a laugh, to poke fun at feminists, and to make young women who had not even had a posture photo taken squirm. While they did not live in fear of their photos being discovered, women such as Wolf could still be degraded by the stories and jokes referring to stolen posture photos, since it was a simple yet effective way for men like Cavett to belittle women across multiple generations.

The Science Wars

Unlike Wolf, Rosenbaum's objective in exposing the history of the posture photography practice was not about righting the wrongs of gender inequality. Instead, Rosenbaum set his sights on casting doubt on the scientific merit of the very posture programs that made nude photography of students possible. Skepticism in the United States about the scientific enterprise was pronounced in the 1990s. In 1994, Richard Herrnstein's and Charles Murray's *The Bell Curve: Intelligence and Class Structure in American Life* set off a heated and highly publicized debate about the validity of the biological and psychological sciences, especially when deployed with what seemed to be an obvious political agenda. Based on IQ measures, Herrnstein and Murray maintained that there was a deterministic relationship between class, race, intelligence, and heredity, and that because of this, Black Americans were on average less intelligent than whites. To left-leaning academics such as Stephen Jay Gould, *The Bell Curve* was little more than "a manifesto of conservative ideology." Accusing Herrnstein and Murray of shoddy statistical analysis and weak correlation coefficients, Gould maintained that the data did not live up to the rigors of science, and that the book contained "no compelling data to support its anachronistic social Darwinism."[43]

The bell curve debate was just one of many instances that seemed to point to a rising intellectual interest among nonscientists in the growing visibility of science in the wider American culture. At the time, science studies scholar Dorothy Nelkin contended that heightened public interest in science was a direct result of the Cold War coming to an end. The "marriage," as she put it, between the U.S. government and the American public had suffered to the point that the latter no longer trusted the former. During the Cold War, the two abided by a "social contract . . . built on the premise that scientific information was a public resource, and what was good for science was good for the state."[44] The erosion of public trust had already begun by the 1960s with the antiwar and environmental movements. It then grew with the animal rights movement and revelations of unethical uses of human subjects in research projects past and present. The scientific enterprise during the Cold War years enjoyed a wealth of federal spending and extraordinary optimism about its potential. As a result of a growing national deficit and federal cutbacks in defense spending, researchers in the 1990s faced overwhelming cynicism concerning the moral authority of science.[45]

Concerned with the growing public distrust, certain scientists engaged in the "science wars," attempting to discredit all "outsiders," including social scientists, the media, and academics from the humanities who voiced criticism. Other scientists, however, joined ranks with nonscientists, voicing real concerns about the wider public's view and literacy in scientific matters. If debates such as the one surrounding *The Bell Curve* became the norm, how would the general public know whom to believe? What if, in seeing the extent of disagreement among scientists, the general public became convinced that scientists did not know what they were talking about at all?[46]

It was in this climate of extreme distrust that Rosenbaum penned his exposé of the posture photos. "What," he asked, "could have possessed so many elite institutions of higher education to turn their student bodies over to the practitioners of what now seems so dubious a science project?"[47] Regrettably, Rosenbaum never actually tried to answer the very question he posed (nor did he ever challenge his own assumptions about the obvious dubiousness of it all), and instead made himself a character in his own story, a detective following a trail that would eventually lead him to actual copies of the posture photos, which he liked to remind his readers were "NUDE, NUDE, NUDE."

And these were not just any nude pictures. Posture photography, he wrote, "was a long established custom at most Ivy League and Seven Sisters schools. George Bush, George Pataki, Brandon Tartikoff and Bob Woodward were required to do it at Yale. At Vassar, Meryl Streep . . . at Wellesley, Hillary Rodham and Diane Sawyer. All of them—whole generations of the cultural elite—were asked to pose." Speaking about the possibility of still finding nude posture photos in the archives, Rosenbaum directly addressed the presumed *New York Times* readership, warning: "If you attended Yale, Mount Holyoke, Vassar, Smith, or Princeton—to name a few of the schools involved—from the 1940s through the 1960s, there's a chance that yours may [still exist]."[48]

Rosenbaum wrote this even though early in his investigation, he learned that most of the photographs at the Ivies and Seven Sisters had been destroyed—in other words, most of the pictures of the named famous people no longer existed. Yale's photos were destroyed by a janitor in the 1970s, Harvard's and Brown's only a few years later, and Smith's in 1984. Learning this, he could have focused his efforts on discovering

the fate of the far greater number of photos taken at *non-elite* institutions—
state schools, prisons, historically Black colleges, the military—places
that he only mentions in passing. Or he could have pointed his readers to
the extensive published record of the twentieth century that included
reprints of the posture photographs, available to anyone with a public library card to see. Or he could have decided to change course entirely to
investigate the history of the posture sciences as an intellectual and cultural pursuit.

Instead, he fastened onto constitutional psychologist William H. Sheldon, who had already been discredited by the wider scientific community during the 1970s. For a brief time, at least given the decades-long
practice of posture photography, Sheldon did work with university posture scientists in order to advance his own somatotyping research. Beginning in the late 1940s and early 1950s, Sheldon collaborated with several
universities where he took his own photos—and in some cases used already recorded posture photos—to perfect his system of body typing.
In 1954 he published the *Atlas of Men*, a book heavily illustrated with
nude photos of Harvard men. With plans to create an accompanying
Atlas of Women, Sheldon had already established relationships with many
of the Seven Sisters and other non-elite and state schools to gather his
photographic data. His assistant, Barbara Honeyman Heath, for example, worked at Smith College in 1950 to make records for Sheldon's use.

As discussed earlier, Sheldon's project and photographic method
differed from the longer tradition of posture measures in significant
ways. First, Sheldon was not interested in human posture, in spinal
deviations, or assessing the symmetry or lack thereof of other bony
markers. Rather, he was concerned with the constitutions of bodies—
muscular or lanky, thin or stocky—and he used these physical markers
to determine a person's inherited and unchangeable personality and
mental makeup. His aim was not to remedy or prevent physically based
illness or disability, but rather to categorize bodies which he believed to
be predictive of human psychology. During his time at Harvard just before World War II, he worked with psychophysicist S. S. Stevens, who
built Sheldon a somatotype calculating machine, which relied on a very
different method of quantitative assessment than the plumb line analysis

used by most posture scientists. When it came to posture, a student would be given a grade ranging from an A to a D. When it came to Sheldon's somatotyping, a student would receive a set of three numbers. Sheldon calculated a person's type by measuring the presence and degree of three components—endomorphy (soft roundness), mesomorphy (squareness, firm muscularity), and ectomorphy (linearity, fragility). He would then quantify these characteristics, placing them on a seven-point scale, with the ultimate well-balanced physique receiving a 4-4-4.[49]

To arrive at these numbers, Sheldon required three different photographic angles of a person's unclothed body: front (anterior), side (lateral), and back (posterior). Most posture scientists, however, were primarily interested in the lateral view, and only occasionally took anterior and posterior photos. By the 1950s, certain posture scientists would come to adopt the Sheldon three-angle approach to photo examinations, but the purpose of such pictures still remained the same, that is, to assess bony markers and bodily asymmetries.

Another significant difference between Sheldon and the work of most posture scientists revolves around informed consent. By and large, posture photography on university campuses was not conducted with explicit consent of the students; because of in loco parentis, examiners believed that if a student matriculated into the university, agreement to be involved was implied.[50] Sheldon, by contrast, appears to have developed a process of informed consent, at least in certain cases. In 1950, when Smith College allowed Sheldon—and more specifically his assistant Heath (Smith '32)—to photograph its student body, the women received an information sheet prior to their participation. The consent form assured students that the pictures were "medical photographs . . . taken for research purposes only." Furthermore, they were assured that their privacy would be protected, with "the features of the face and other identifying marks . . . blocked out so that the individual identity is concealed." Students were also told that the somatotyping pictures would be "taken, in lieu of, or in addition to, the conventional posture picture."[51] Such written assurances were not made during the annual posture examinations.

The most striking difference between Sheldon and posture photography, however, is the degree to which racism and eugenics informed each of the respective enterprises. While many posture scientists were heavily influenced by theories and assumptions of race science, they did not speak in terms of limiting births among non-whites or placing restrictions on reproduction. Moreover, posture practitioners—physical educators, physicians, and physical therapists—by and large believed that poor posture could be improved through training. A strict determinist, Sheldon believed that body constitution was inherited and could not be changed or manipulated in any way. In his 1949 *Varieties of Delinquent Youth*, Sheldon lamented the rampant reproduction of the world's unfit and urged radical social and political measures that favored Anglo-Saxons, who he believed to be biologically superior.

Ostensibly unaware of these differences, Rosenbaum conflated Sheldon and the much longer history of posture photography, seeing the two scientific traditions as one and the same. Although Rosenbaum himself likely stood for a posture picture and not a Sheldon photograph, he assumed that everyone's photo fell into the latter category, and thus deemed the entire posture science enterprise a "bizarre pseudoscience" that could be linked to Nazism and America's sordid history of sterilization. "Entire generations of America's ruling class," Rosenbaum declared, "had been unwitting guinea pigs in a vast eugenic experiment run by scientists with a master-race hidden agenda." From his own telling, Rosenbaum felt as if he had uncovered a heretofore unknown chapter in the history of human experimentation in the United States. Only this time, the victims were not the underprivileged Black men of the Tuskegee experiment; instead, the unwitting targets in Rosenbaum's piece were "the most overprivileged people in the world," people who had "[earned] the one status distinction it seemed they'd forever be denied."[52]

These overprivileged "victims," of course, would not be denied justice. Almost as soon as Rosenbaum's piece hit the newsstands, angry alumni flooded the phone lines of the named institutions, demanding action. Over the next couple of months, university lawyers and public relations departments sprang into action, putting pressure on university archivists to scour their holdings for any extant photos, to corroborate the truth of the Rosenbaum piece (which did not always hold up), and to ultimately provide their respective institutions legal cover in a situation that looked increasingly dire. The resultant story is one in which, because of a sensationalistic media report, archivists were forced to prioritize the needs of the institution over the informational needs of the general public.

Archivists on the Front Line

When archivist Margery N. Sly arrived at her Smith College office on January 16, 1995—the Monday morning after Rosenbaum's piece appeared in the Sunday *New York Times*—the phones were already ringing off the hook, with colleagues and concerned alumnae on the other end of the line. The first professionally trained archivist to head the Sophia Smith Collection, Sly had spent her early career in Northampton during the late 1980s and early 1990s creating a professionally maintained and housed university archive, with searchable and accessible source material for researchers to use. Before her arrival, the university records were not centrally located or housed—the posture photos that Dieck accessed, for example, were in the basement of College Hall, an administrative building rather than a dedicated archival space. The decision to burn the photos at Smith in 1984 came about not at the hands of an archivist, but at the behest of the college president.[53]

Before Rosenbaum's exposé, Sly was aware of the posture photo tradition because it was detailed in the course catalogues, which she referenced often to respond to questions pertaining to the school's history. And she also knew that the photos at Smith, at least, had been destroyed. But she knew little beyond this—at least not until the *New York Times* set off a scandal that required her to know.[54]

Under the pressure of angry alumnae threatening to bring legal action against the university and a college administration anxious about damage control, Sly immersed herself in the archival system she helped to create, looking for detailed information about the fate of the posture photographs. Within a matter of days, she tracked down the pertinent records and created a two-page information sheet that Smith officials used when speaking with enraged alumni. She also collaborated with the public relations office, writing an article for *NewsSmith*, assuring alumni that their posture photos no longer existed, that all pictures had been incinerated in 1984.[55]

But many of the alumnae, she recalls, "did not believe us—after all, the article had been published in the *New York Times* so it must be accurate." Sly and her colleagues wrote to the *Times* asking the paper to publish an erratum to Rosenbaum's piece, but to no avail. No matter how many times Sly spoke to college administrators, no matter how many times she published articles offering material proof that the posture photos had been burned, many of the aggrieved alumnae persisted. "Some alumnae believed us," Sly recalled, but "some continued to believe Rosenbaum."[56]

Sly's professional peers faced similar pressures and frustrations in the wake of the *New York Times* exposé. At Princeton, Primer, who had at least some forewarning of Rosenbaum's piece since a *New York Times* fact-checker had called him weeks earlier, was nevertheless unprepared for the way the university administration would react. Within twenty-four hours of the exposé's publication, Princeton's public relations office and legal counsel demanded more information, asking Primer to comb the archives looking for the existence of any photos. Primer found that, compared to other institutions such as Harvard and Yale, Princeton had a fairly limited posture program, that the photos were only taken sporadically, and that some were linked to Sheldon's project. But as far as he could tell, the photos no longer existed in the university archives. "After significant digging," he wrote in an email to the college communications office two days after the Rosenbaum piece hit the newsstands, "there is nothing in the President's office on this . . . or in Health and Athletics." Signing off the email, he wrote, "I do not plan to pursue this further."[57]

Universities that remained unnamed in Rosenbaum's article also reached out to their archivists, taking every measure to control a potential public relations and legal disaster. Amy S. Doherty, the archivist at Syracuse University, searched health service records, yearbooks, and correspondence. "We have 35,000 photos in our archives," she told one reporter, "but not those."[58] The University of Wisconsin–Madison, which had a robust and federally funded posture science program under the helm of Dr. Frances Hellebrandt, promptly investigated its own records too. To the administration's relief, J. Frank Cook, director of UW-Madison's archives, was able to quickly assure university officials that the photos no longer existed—he had taken care of them in the mid-1970s when a concerned researcher had come across a stack of posture photos in her department files. "Everyone saw this as a ticking time bomb," Cook recalled. He thus proceeded to arrange with state officials for a "confidential destruction" of the records. When asked why he did not preserve the records, Cook replied that "the chance of . . . embarrassing the university made this a totally untenable holding. I could see no justification for it," he concluded.[59]

Within only a couple of weeks after Rosenbaum's article, most university archivists had confirmed that if their institution had ever had posture pictures, most had already been disposed of. Some photos still existed, but these were often of students clothed, wearing leotards, swim suits, or other tight-fitting clothing.[60] Some universities, it turned out, did not demand nudity for posture photography.

What Rosenbaum had discovered was a cache of photographs in only one archival holding: the William Sheldon papers at the Smithsonian's National Anthropological Archives. But since Rosenbaum referred to the Sheldon photos as posture photos, many alumni and university administrators understandably conflated the two. University archivists tried to bring clarity to the situation by first demonstrating that their respective school's posture photos no longer existed (if they even had in the first place) and by explaining to their institutional employers and alumni that Sheldon's photos were not only distinct from posture photos but also that they existed in a place outside of their purview, for the somatotyping photos lived in and belonged to the Smithsonian.

Having cleared up the confusion created by Rosenbaum's piece, archivists such as Sly and Primer rightly believed that the posture photo scandal was behind them. But within only a couple of weeks after the exposé, the Smithsonian announced that, after fulfilling a request from Yale university officials, it had wiped clean the school's history of Sheldon's somatotyping, shredding a hundred pounds of photos and negatives.[61] When asked about the purging, Donald J. Ortner, director of the

Smithsonian's Natural History Museum, said that his team would "probably destroy all of the photos if the universities asked." From where Ortner stood, "any historic or scientific value of the pictures 'would be minimal.'"[62]

Determining the worth of a collection is a vital part of an archivist's job. And in many cases, it is anything but straightforward. Archivists must balance the modern deluge of data with the material reality of limited resources and space. Assessing the value of a collection also requires a commitment to preserving the past, while making predictions about whether and how future generations will use the holdings. On top of that, they must work in concert with many stakeholders and within an institution that has its own, often troubled, history.

Ortner was not an archivist; he was a physical anthropologist. Moreover, he was a physical anthropologist working for a federal institution with a checkered past, a history that was regularly featured in the news. The National Anthropological Archives (NAA), where the Sheldon papers and posture photographs are still held today, is a repository largely built on the nineteenth- and twentieth-century colonial practices of pilfering and extracting Indigenous cultural artifacts and biological data. During the late 1980s and early 1990s, several tribal nations succeeded in holding the Smithsonian accountable for its past. With the passage of the 1989 National Museum of the American Indian Act and the 1990 Native American Graves Protection and Repatriation Act, the Smithsonian could no longer claim rightful ownership over many of its holdings, especially the human remains and relics that men such as Aleš Hrdlička had taken from native peoples earlier in the century. In 1991, for example, the Larson Bay tribe of Alaska, after years of politically pressuring, reclaimed from the Smithsonian 756 sets of human skeletal remains for reburial, establishing an important precedent in the future of repatriations to come.[63]

When faced with calls from elite universities and their alumni who claimed ownership over the posture photos in the Sheldon papers, Ortner likely saw the pictures in a light similar to the demands made by Native Americans. But of course, the issue of provenance in these two cases is significantly different. For starters, the power differential between Smithsonian officials and the two respective groups are not the same. Whereas Native Americans had to fight over the course of a century, taking significant legal recourse in order to successfully prove ownership of their ancestral remains, the predominantly white elite only needed to make a couple of angry phone calls for the Smithsonian to act with immediacy, granting ownership without question.

Additionally, Ortner was not bound by the same professional standards of stewardship that define the work of archivists. But even physical anthropologists were mystified by his reflexive urge to destroy the posture photo records. Ann M. Palkovich, professor of physical anthropology at the Krasnow Institute for Advanced Study at George Mason University, was stunned by the speed and certitude of the Smithsonian's actions. To her, the Sheldon photos had obvious scientific value. "There are many data archives of metrics collected from a large number of individuals," she wrote, "that have subsequently been productively datamined for other purposes—for example, the Fels Longitudinal Study of Skeletal and Dental Biology, and the Cross-National Longitudinal Research on Human Development and Criminal Behavior are just two that come to mind."[64] To Palkovich, it was power and politics that shaped Ortner's appraisal of the holdings, not scientific worth.

Professional archivists, Primer and Sly found the destruction reprehensible, not only because of how the power differentials inherent to most archives seemed to once again be working in favor of the white elite, but also because it undermined the very purpose of their profession: namely, the preservation of information for the general public, for potential research projects in the present—and if that was not possible, the future. A simple and less destructive solution to protecting privacy with the Sheldon papers would have been for the Smithsonian to restrict access to all the photographs, not allowing researchers to view the photos until decades later, when the subjects who were photographed were dead. Placing restrictions such as this is a common practice that archivists use to protect individual privacy while at the same time preserving records for future use.[65]

But when Primer, Sly, and others in the archival profession made this kind of argument, few outside of their professional circle were moved by it. "University counsel and administrators," Sly recalls, "were totally freaked out by public relations." For her part, Sly remembers feeling "really upset that people weren't understanding the value of this material and also that they weren't believing us." Recounting the frustrations of having her professional opinion entirely overruled by university counsel, Sly remembers being dismissed on two counts: "They weren't believing us about the [university posture] photos having been destroyed and they weren't believing us when we argued that the Smithsonian should keep the [Sheldon photos] because they have research value."[66]

In the end, university administrators found the appeal and ease of record destruction too enticing. And they had the purportedly happy outcome from Yale to serve as an example. Speaking on behalf of Yale administrators once they learned that the Sheldon photos of Yalies had been destroyed, Gary Fryer exclaimed, "We are delighted that the privacy of the individuals in those photographs will be forever protected."[67]

Taking their cue from Yale, Princeton officials ordered Primer in March 1995 to visit the Smithsonian to oversee the incineration of its institution's pictures, despite Primer's protestations that such an act was an affront to the ethics of his profession. After his trip to Washington, Primer told the *Princeton Alumni Weekly* that he took "no pleasure in having done this deed."[68] "One always wonders," he went on, "what some future research might have been able to learn from these photographs that would be unrelated to the original purpose for which they were taken."[69] (See figure 39.)

Sly's marching orders would be passed down months later in August of that year, well after Rosenbaum's exposé appeared in print. In the intervening months, she had hoped that the Smith officials would forget about the Sheldon photos. But they didn't, and like Primer and many other archivists before her, she too traveled to DC, where Smithsonian officials took her to a "massive shredding facility," where she would erase a part of Smith's history. To this day, Sly feels regret and deep frustration at having to do "damage control" in this way for her employer. "I really didn't want to be party to records destruction," she told me with a tone

FIGURE 39. A plastic bag of the shredded remains of posture photos that Princeton archivist Ben Primer preserved from his trip to the Smithsonian Institution.

of remorse still detectable in her voice, even though it had been nearly thirty years since shredding the records.[70]

The Cost of Destruction

Given the amount of attention paid to the posture photos during the 1980s and 1990s, it is noteworthy that few, if any, discussions addressed the basic question of what defined good human posture or why it mattered enough to have entire archival holdings devoted to it in the first place. Instead, Rosenbaum deemed the inquiry that lay behind the posture photos a pseudoscience, even as his colleague, Jane Brody, writing for the Health section of the same newspaper, told readers that "perpetually allowing your body to stand or sit in misaligned positions can exact an unpleasant toll on your physical well-being and possibly your social stature." Studies such as Dieck's did not make headlines, for her research challenged the notion that poor posture begot poor health, a scientific truth that had been taken for granted for decades.

A fundamental disjunct in thinking concerning the posture sciences and health was mainstream in the late twentieth century, and it became so acceptable that no one seemed to see the contradiction. Aside from political motivations, university presidents, archivists, trained anthropologists, and highly educated alumni could deem archived posture pictures of little scientific worth partly because it seemed self-evident that poor posture led to poor health. There had been, after all, entire industries built on this presumed truth—ergonomic chairs and offices, newly developed pharmaceuticals to prevent osteoporosis, medical and surgical devices to straighten spines, a multi-million-dollar fitness and beauty industry. Very few seemed to question the connection between posture and health; the crusaders who created the slouching epidemic nearly a century prior were still making an impact by the twentieth century's end.

While the preservation of the Sheldon photos may have had little bearing on the future of the posture sciences, the destruction of them made sure of it. The same can be said for the university posture photos, most of which ceased to exist by the 1970s and early 1980s. But the destruction of posture photos, it is important to keep in mind, had little to do with scientific curiosity or the relative merit (or lack thereof) of the posture sciences. Instead, most photos were destroyed in the name of protecting the privacy of the educated elite. But even when student privacy had been ensured by researchers, as in the case of Dieck's study, alumni and schools would not stand for it. Scientific investigation, in this case, was overridden by elites who had the power to demand destruction of the record in order to guarantee that their bodies—or at least photographic representations of their bodies—would never be used again.

There are many lessons to be learned from this chapter in history. Most striking, of course, is that privacy is often more of a privilege than a democratic right. Thousands of nude posture photos still exist in the Sheldon papers to this day, part of a 122-box collection of restricted materials. The photos that were incinerated in 1995 are only a fraction of the photos contained in the Sheldon's papers. According to the inventory list, photos of identifiable subjects from places such as the Oregon State Prison still exist, as well as those of patients from the Elgin and New York State Hospitals.[71] In other words, photographs of historically more vulnerable populations remain in the archive, while the images of white elites have been actively erased.

Historians often speak of archival silences, absences, and erasures, pointing to the ways in which certain populations of people—often powerful, well-resourced elites—can be easily found in the permanent record, whereas the disenfranchised are often invisible, by virtue of their not having the cultural or economic capital to ensure the future preservation of their life's work or experiences. But as the history of the posture photo scandal demonstrates, the elite also have the power to create archival silences of themselves when it suits them, while more vulnerable populations generally do not. If and when the restrictions on the Sheldon papers are lifted, researchers will see a misrepresentation of the past, with fewer white bodies than had originally existed.

Which nude bodies from the past are available for public viewing and which are not is an issue that transcends the Sheldon papers. But a distressing trend seems to be happening whereby the white elite are protected from such indignities, while the less powerful are not. Take, for example, the Zealy daguerreotypes held at Harvard's Peabody Museum, where researchers can view seminude images of enslaved men and women, such as that of Renty and Delia Taylor, with identities and nudity paired, known to any onlooker.[72] Or consider the materials preserved at the National Archives at Atlanta of the Tuskegee experiment, an archive open and available for the public to see. In these instances, concerns of privacy never arose because the subjects in the photos were never thought to have a right to it in the first place.

This is not to say that all of the photos still contained in the Sheldon papers should be automatically destroyed. Rather, multiple concerns and a wide range of stakeholders should be taken into consideration when evaluating and balancing the worth of the holdings.[73] In regard to the posture sciences, the total destruction of the Sheldon holdings would potentially mean that studies such as Dieck's will never happen again. In the absence of these kinds of data with which a researcher can study the natural history of postural deviations, the basic tenets upon which the initial poor posture epidemic was built will continue to be an almost uncontested truth. As a result, the notion that poor posture has predictive value, spelling a future of ill health and disability, persists, informing the outlook of leading health journalists, of advertisers for fitness devices and pharmaceuticals, and of how many people to this day view their own well-being and self-image.

iPosture

LARGE-SCALE formalized posture evaluations are largely a thing of the past in the United States. Without such population-level biometrical data, reports announcing a poor posture epidemic have dwindled as well. One could rightly assume, then, that the epidemic has come to an end.

To arrive at this conclusion is reasonable. Yet it is important to carefully consider the process and the degree to which the epidemic has come to an end. It did not, for instance, subside because public health officials succeeded in getting Americans to permanently correct themselves so that slouching no longer occurred. The promise of the American Posture League and its slouching eradication program never, in the strict sense, succeeded. Instead, the condition of poor posture became endemic to U.S. society, subsumed by what professionals and lay citizens perceived to be more pressing health matters, such as rising rates of obesity, low back pain, and cardiac disease.

As poor posture became more of an endemic rather than an epidemic problem, the assumption that it served as a valid indicator of current and future health became so entrenched as to remain unquestioned. Along the way, concerns about widespread failings in human posture have served to fuel other public health concerns. The most notable has been the dramatic rise in low back pain since World War II. Epidemiologists have found that approximately 568.4 million cases of disabling low back pain exist worldwide, with the highest prevalence seen in the United States, with Denmark and Switzerland following close behind.[1] In the

United States, spending on low back pain exceeds that of hundreds of other health conditions (including diabetes), with an estimated $134.5 billion dollars devoted to the condition in 2016.[2]

As spending on low back pain management has increased, so too have investments in posture health, for the two are believed to be inextricably linked. While clinical studies support the use of posture-focused exercise and targeted physical therapy in order to ameliorate already existing back pain, there is a surprising dearth of evidence demonstrating that these same treatment modalities are successful in *preventing* back pain among people who are asymptomatic. The latter is often assumed to be self-evidently true, in no small part because of the enduring success of earlier posture crusaders.

To insist that back pain is an inevitability for people who may slouch, slump, or have detectable deviations but are otherwise pain-free is a type of therapeutic reasoning that essentially makes the risk of disease or disability acquisition a disease state itself. The advent of risky medicine, as medical historian Robert Aronowitz calls it, is a much broader and more complicated phenomenon characteristic of the early to mid-twentieth century.[3] The idea that poor posture posed a serious health threat became entrenched in American public and popular health culture as early as the turn of the twentieth century. And because of its ties to the evolutionary sciences, the number of people who were presumably in jeopardy of becoming unwell because of poor posture seemed to include nearly everyone. Thus, despite myriad changes in American culture, politics, health professions, and the evolutionary sciences that have taken place over the past hundred years, the belief that human posture is an indicator and a manipulable variable in health projections is surprisingly tenacious.

Similar to a century ago, today's evolutionary biologists are at the forefront of the posture sciences, presenting health care providers with what appears to be incontrovertible proof that back pain primarily stems from flaws, or a maladaptation, in human evolution. Known on Harvard's campus as the "barefoot professor," a nickname he earned for his habit of running long distances without footwear, Daniel Lieberman is a leading expert in the evolution of the human body and exercise.

Lieberman readily admits that the tendency of human beings to laze about is intrinsic to our evolutionary makeup. "Exercise," he writes, "is a truly odd sort of medicine," an "abnormal behavior from an evolutionary perspective."[4] According to Lieberman, the earliest human beings adopted a bipedal stance in order to save energy. About 4 million years ago, when climatic change made African forests more seasonal and variable, early hominids adopted bipedalism in order to travel longer distances in search of food sources. The ability of these early hominids to conserve energy when it came to food acquisition gave them a reproductive advantage.[5] In other words, according to the theory of natural selection, human beings are biologically inclined to find food using the least amount of energy doing so, and thus exercising for the sake of improving one's physical fitness is entirely artificial.

Unlike certain first-generation paleoanthropologists who argued that human beings were never meant to be bipedal in the first place, Lieberman follows in the footsteps of those who have long tended to blame civilization—or, in his case, industrialization—for chronic health problems such as back pain, obesity, hypertension, and osteoporosis, conditions overwhelmingly found in North America, Europe, and Australia. Adopting rhetoric very similar to the early twentieth-century physician and anthropologist Sir Arthur Keith, Lieberman contends that, "from the body's perspective, many developed nations have recently made *too much progress*. For the first time in human history," he continues, "a larger number of people face excesses rather than shortages of food. Two out of three Americans are overweight or obese."[6] Obesity, according to Lieberman, is not the only concern. "Depending on where you live and what you do," he warns, "your chances of getting low back pain are between 60–90 percent."[7]

Why, one may wonder, would so many chronic health conditions be passed down from generation to generation? Wouldn't natural selection—the driving force of Darwinian biological evolution—make sure that the physically weak would fail to get their genes into the gene pool? Lieberman is not a pure biological determinist. Instead, he points to cultural evolution as the culprit. To him, cultural evolution is "an even more powerful and rapid force" than natural selection, a force that alters genes through changing environments. The "culturally inheritable" conditions of poor posture, back pain, and obesity belong to a group of diseases Lieberman calls "mismatch diseases," maladies that arise due to novel environmental conditions for which the human body is poorly adapted. Low back pain, one of the costliest of the so-called mismatch diseases, is an inefficiency that Lieberman believes evolutionary medicine and targeted therapeutics can mitigate.

In order to solve this (evolutionarily speaking) new problem of industrialized peoples, Lieberman has turned to sub-Saharan Africa in order to study the habits and physical activity of subsistence farmers and tribes that he categorizes as hunter-gatherers. Once again, "primitive" peoples are understood to be model organisms in order to solve the problem of "civilized" people. Ostensibly unconcerned with the problematic history of physical anthropology and its perpetuation of race science, Lieberman upholds the inhabitants of the Kenyan village of Pemja as having "natural" postures and spinal curvatures. By his telling, the people living in Pemja do not suffer back pain, for they spend most of their days walking—fetching water, hunting, and traveling—as opposed to sitting for long periods of time in chairs, working at computers, and staring at screens. The Pemja people, in other words, are exemplars of back health.[8]

Attracted to the logic of evolutionary biology, certain therapeutic body workers and self-designated ethnophysiologists have engaged in similar work, finding indigenous populations who exhibit "primal posture" for the sake of curing those in industrialized nations. One of the most prominent North American adherents to this approach is Esther Gokhale. Raised in India by European parents and later educated at Harvard and Princeton in biochemistry, Gokhale today is known as the "posture guru" of Silicon Valley, where she treats corporate heads of Google, Facebook, and other prominent online personalities, such as conservative journalist Matt Drudge.[9] Gokhale developed an interest in human posture at a young age. With a tendency to exoticize the other, Gokhale recalls of her childhood in India, "I remember listening to my Dutch mother marvel at how gracefully our Indian maid went about her duties and how easily the laborers in the street carried their burdens."[10]

Later in life, during her travels to Burkina Faso, Ecuador, and back to India, she developed an amateur interest in photography, taking pictures of potters, basket makers, weavers, and head-carriers, all the while admiring the spinal health of her subjects. She turned this interest into a professional venture, undertaking posture therapy training with Noëlle Perez, founder of the Aplomb method in Paris and one of the first Europeans to study under the Indian yoga master, B.K.S. Iyengar.[11]

That both Perez and Gokhale would ultimately find inspiration in yoga is unsurprising. After all, modern yoga—a designation that scholars use to distinguish twentieth-century yoga systems from the earlier meditative practices that were undertaken by cave-dwelling ascetic monks or fire-breathing contortionists—is unique in its focus on physical postures, or *asanas*. The evolution of yoga into a mainstream, middle-class fitness practice on a global scale has a complicated history. What is notable is that modern yoga began during the early twentieth century, partly as a backlash against British colonialism and partly as a response to the growing influence that Anglo-European physical education systems were having in Asia. As a result, certain Indian men, cast as inherently weak and feminine by colonial aggressors, created physically rigorous systems of yoga postures, blending teachings from ancient Sanskrit texts with Western notions of physical education for the purpose of developing physical strength and agility, undermining colonially bred racist depictions of India and its people.[12]

Out of all the modern forms of yoga, Iyengar developed one of the most exacting systems concerning yoga postures, emphasizing biomechanical alignment and symmetry in every pose. Largely devoid (at least at the outset) of breathing and meditative practices, Iyengar's system became one of the most culturally adaptable regimes as well. In Pune, where Iyengar developed and perfected his system of yoga, local physicians would regularly refer their patients to him, especially when these doctors felt that the limits of biomedical treatment options had been reached.[13]

When Perez opened her own studio in Paris in the 1970s, and Iyengar came to visit, he told her that her students were inherently misaligned, a comment that led Perez to study the spines of nonindustrialized people in several rural regions of Africa. There she found bodies in what she called natural "aplomb." When she returned to Paris, she created her own brand of posture-focused, yoga-inspired body work and eventually earned her doctorate in ethnophysiology from the École des Hautes Études en Sciences Sociales.[14]

Much like Lieberman, posture experts such as Gokhale and Perez claim that back pain is not so much an evolutionary flaw as a cultural one. In her published work, Gokhale contends, "I believe that the biggest risk factor for back pain, as yet unidentified and underappreciated, is posture." "Most known risk factors," she continues, "can be mitigated by good posture."[15]

Steeped in evolutionary biology and ethnophysiology, Gokhale maintains that people who sit in a slouched position for prolonged periods of time, day in and day out, develop a retroverted pelvis that puts unnatural pressure on the spinal disc that occupies the space between the fifth lumbar vertebra and the sacrum, otherwise known as the lumbosacral joint. Over time, she contends, a retroverted pelvis will squeeze the disc, leading it to bulge or herniate, creating chronic low back pain. Her exercise regime teaches students and patients to maintain an anteverted pelvis, a healthy pelvic position that is found, she contends, in "babies, indigenous peoples, and your ancestors."[16]

The simplistic evolutionary explanation of back pain—and its insistence that a "natural" posture exists in other times and places—appeals to largely white, upper- and middle-class consumers, the target demographic of today's multi-billion-dollar worldwide fitness industry. Gokhale's explanation for back pain falls easily in line with the fitness industry, an enterprise known for creating slogans such as "sitting is the new smoking," and insisting that by mimicking our evolutionary ancestors in eating (paleo dieting) and exercise (paleo fitness), mismatch diseases will be cured.[17] The emphasis on primal posture as a cure and preventive for low back pain offers the ideal mix of panic on a population level (instilling fear in billions of people who sit for hours on end, telling them they are all doomed to develop back pain), a quick and fairly straightforward fix (focused exercise on one part of the body) that is amenable to further body technology developments (ergonomic chairs, standing desks, and wearable correctors) while at the same time offering an idealized fitness utopia and hope (look at all those people living more simple, pure, and pain-free lives elsewhere).

Similar to the early years of the poor posture epidemic, the evolutionary approach to understanding human posture—and now, by extension, low back pain—has continued to inform the design and construction of everyday technologies and clothing. Whereas posture innovators of the past focused on undergarments and shoes, today wearable technologies and posture-tracking devices are considered to be the next frontier of advancement. Some of the major companies producing such devices are Marakym, BackJoy, VIBO Care, and Lumo Bodytech in the United States, along with UPRIGHT in Israel and Swedish Posture in Sweden.[18] These companies sell products that range from low-tech back supports and molded seat cushions to wearable body sensors that, when paired with your smartphone, reprimand you when you deviate from ideal spinal alignment. According to market analysts, posture correction technologies are expected to grow approximately 5.7 percent over the next five years, especially with rising demands due to the COVID-19 pandemic, with more at-home workers complaining of back pain.[19]

The field of forensics has also taken an interest in the development of wearable technologies, hoping to use posture tracking for the purpose of surveillance. Researchers in India, China, and Vietnam are at the forefront of developing software that can read and analyze individual human postures captured on security cameras. The hope, at least for those in forensic medicine and psychology, is that posture recognition technology will one day be as effective as facial recognition technology, and be able to be used alongside it. Ostensibly, this kind of posture recognition technology could be used in criminal investigations as well. Building on the work of American psychologist Paul Ekman, who has spent his career quantifying nonverbal communication of the face and body, certain intelligence agencies are busy developing an algorithm based on Ekman's six-point microexpressions of human posture.[20] To these researchers, posture construed as too straight, for example, connotes arrogance and provocation.

Although the work of constitutionalist William Sheldon has been largely denounced by psychologists and physical educators alike, the ways in which he attempted to quantify the human body for the purpose of understanding an individual's character and personality still draw interest among certain physiologists and psychologists working today. Take, for example, Amy Cuddy, social psychologist and former professor of business administration at Harvard University. Cuddy became a minor celebrity shortly after a massively popular 2012 TED Talk in which she explained to viewers that if they wished to succeed in the workplace, they needed to cultivate "power poses"—postures that were expansive rather than diminutive.[21] In her view, the difference between "high-power" and "low-power" postures could mean the difference between landing or losing a job, earning a promotion or career stagnation. In other words, if a worker—say a female employee who faces structural pay inequities—wants a raise, she needs to bolster her presence, establish a good posture, and make eye contact. Upright posture, in Cuddy's mind, is the path to power.

On the face of it, posture improvement campaigns—whether for workplace success or physical and mental health—may seem rather innocuous. What is the harm, after all, of engaging in posture exercise programs? Of buying chairs, shoes, and devices that help to encourage it? There is nothing overtly medically invasive or particularly risky about these things, right?

On an individual level, it is entirely possible that an enhanced sense of wellness can come from taking up yoga or purchasing an ergonomic chair. Of course, even posture-based exercise regimes pose some risk. Injuries from yoga, Pilates, and other fitness regimes are relatively commonplace, and are one reason why people end up needing analgesics, orthopedic consults, and physical therapy treatments.[22]

But when looking at the long history of posture improvement campaigns from a structural standpoint, it becomes evident how value-laden they are, how the presumed morality and cultural meaning of upright stance often masquerades as scientific and medical facts. In the preceding pages, I have offered a critical analysis of the posture sciences, mainly by uncovering how and why the poor posture epidemic came about, situating its rise and fall firmly in the realm of politics and culture. A similar approach needs to be taken with contemporary posture promoters who

claim that upright stance should and can be universally adopted and applied by all people, regardless of context or attention to differences in
regionality, race, ability, or gender. In recent years, for example, researchers have found that, even today, prosecutors cite poor posture as a reason
to deny African American men jury selection.[23] In his autobiography
about growing up in South Africa, comedian and political commentator
Trevor Noah succinctly addresses the extra vigilance required of Black
men and their posture in a white supremist society. "For centuries," he
writes, "colored people were told: Blacks are monkeys. Don't swing from
the trees like them. Learn to walk upright like the white man."[24]

Perhaps Cuddy would respond that Black men need to practice their
power posing in order to solve the problem. Or maybe the Liebermans
and Gokhales of the world would have these men study the habits and
postures of hunter-gatherers in sub-Saharan Africa. Neither approach,
of course, addresses the root cause of the disenfranchisement of Black
men in America's courtrooms or apartheid in South Africa. Nor would
the overly simplified solution of teaching them a Cuddy-like power
pose necessarily bring about the desired results. Most African American
men know all too well that if they adopt a powerful or aggressive posture, they are more likely to become victims of unjustified state violence. In short, both the Black men being denied jury participation and
the African men who serve as anatomical ideals for posture promoters
are disempowered, regardless of their physical form.

There are other compelling reasons to be leery of the claim that a
universal, "natural" posture exists, to be found in less developed regions
of the world, and that adopting such a posture would help slow the
rising rates of low back pain across the globe. For starters, not all paleoanthropologists working today believe that back pain is a new condition
brought about by industrialization. Some evidence suggests that degen-
erative changes can be found in the fossilized bones of prehistoric
spines, which would mean that the first upright hominids to roam the
earth likely experienced back pain, or would have been predisposed to
such a condition if they had lived long enough.[25] If this is true, the claim
that back pain is a "mismatch" disease or that there is an unadulterated
"natural" posture for industrialized people to acquire seems suspect.

Yet one does not need to rely on human fossil remains to cast doubt on the back pain as mismatch disease theory, for there are living populations of people in less developed parts of the world who offer proof to the contrary. Working at a health clinic in rural Nepal in the 1980s, medical anthropologist Robert Anderson noted that local villagers who trekked two days with heavy loads on their back to get to his facility would seek care for respiratory problems, diarrhea, and eye problems, but that "hardly anyone complain[ed] about their back." Yet when Anderson began to use a translator and asked more pointed and culturally appropriate questions, he found that "the prevalence of back pain was very high," but that the Nepali people "simply considered it a fact of life and worked through it."[26] Or consider a more recent ethnography that looks at how those living in rural Botswana convey back pain. The authors were struck by the fact that the participants in their study "rarely referred to 'low back pain' or even 'back pain.'" Instead, they used the Setswana terms *dinokeng, dinoka, noka,* or at other times *letheka*—all of which, the authors point out, literally translate into "waist pains." "And yet," the authors observed, "these waist pains were depicted by placing their hands across the low back, sometimes covering their sacroiliac joints bilaterally."[27]

Historians of medicine and anthropologists have, for some time, been aware of the contingencies of the clinical encounter, and more specifically how, under the systems of slavery and colonialism, white men of science frequently assumed that Black and other non-white peoples could not feel pain, or if they did, it was felt less acutely compared to whites.[28] This legacy still plays out in health care delivery today, with Black patients regularly denied or under-prescribed pain management therapies.[29] Knowing this, one cannot help but wonder if the same bias has informed the work of today's paleoanthropologists and ethnophysiologists, experts who observe hunter-gatherers in Africa and deem such lifestyles to be pain free.

Over the last century, health—and especially preventive health—has become increasingly commercialized: a product to be bought and sold, with the responsibility placed on individual consumers, making it a good that only those with a certain income can afford, rather than an ensured right for all. As some scholars have pointed out, wellness programs are part and parcel to neoliberal regimes, wherein ideals such as autonomy, personal growth, self-reliance, and self-regulation get mapped onto health maintenance. A complex array of economic, political, and cultural contingencies has brought this about. From the perspective of the state, preventive health and wellness programs are construed as cost-saving measures, for they place the financial burden of health maintenance onto private individuals. In turn, those who cannot participate in the market for preventive health are viewed as leading mismanaged lives, and when they sustain an injury that leads to permanent physical disability, are blamed for their condition. As long as posture surveillance is believed to assist in low back pain prevention, many posture and back health wellness programs are liable to create even greater health inequalities rather than mitigate them.[30]

A recent coauthored study published by physical therapists working in Qatar, Australia, Ireland, and the United Kingdom speaks to the urgent need of the profession to dispel the medicalized myth that poor posture leads to bad health. The authors point to the many ways in which such beliefs are not founded in scientific evidence, and are instead largely "reinforced by long-standing stereotypes" and "fear-inducing messages in the mainstream media." "People come in different shapes and sizes," they write, "with natural variation in spinal curvatures." In short, there is no single, correct posture.[31] Nor does posture correction necessarily ensure future health.

This is not news to most physical therapists who have been trained within the last several decades. But it is noteworthy that certain contemporary practitioners feel that it is necessary to combat a message that their professional forebears and fellow travelers set into motion over one hundred years ago. While the American poor posture epidemic as it was originally construed and documented is largely a historical artifact, many of the health practices, aesthetic beliefs, and assumptions associated with it have remained. The power, it turns out, does not lie so much with human posture as such, but rather with the institutions and individual actors who infuse it with such consequential meaning in the first place.

Acknowledgments

IN SOME WAYS, books are solitary endeavors, with an author toiling alone and in silence before glowing screens, amid piles of books, articles, notes, coffee mugs, and power bar wrappers. But in other respects, books are profoundly collective undertakings, pursued in tandem with colleagues and fellow scholars, foundation program officers and university administrators, librarians and archivists, friends and family members. It's deeply gratifying to have the chance to acknowledge publicly all of those who have made *Slouch* possible.

Thanks to funding agencies that continue to support the humanities, I was able to conduct deep archival research and take much-needed sabbaticals to brainstorm and write. *Slouch* was funded by the National Institutes of Health/National Library of Medicine Grant for Scholarly Works in Biomedicine and Health (1G13LM012781-01), the National Endowment for the Humanities Fellowship (FEL 257639-18), as well as the American Council of Learned Societies. These awards provided not only material assistance but also, and equally importantly, confirmed the value of my research and work.

I also received financial and intellectual support from various schools and programs at the University of Pennsylvania, including the Barbara Bates Center for the Study of the History of Nursing, the Trustees' Council of Penn Women, the Robert Wood Johnson Health and Society Scholars Program, The Penn Humanities Forum, and the School of Arts and Sciences. Thanks to a Pearce Fellowship in the History of Medicine and the support of my colleague Christopher Crenner, I was able to visit the University of Kansas Medical Center archives to access letters written by people and patients with scoliosis.

Most of the research, writing, and thinking for *Slouch* took place in the Department of the History and Sociology of Science, where I am fortunate to be a faculty member. Thanks to my brilliant colleagues (current and former)—Robert Aronowitz, David Barnes, Etienne Benson, Sebastián Gil-Riaño, B. Harun Küçük, Andi Johnson, Susan Lindee, Amy Lutz, Katherine Mason, Jessica Martucci, Ramah McKay, Jonathan Moreno, Projit Mukharji, John Tresh, Elly Truitt, Beans Velocci, and Heidi Voskuhl—for the direct and indirect ways that you have shaped my thinking about the history of science, broadly construed. I am especially grateful to Robby, Projit, Andi, Elly, and Heidi for providing shrewd feedback on earlier drafts. My historian colleagues in the School of Nursing—Cynthia Connolly, Patricia D'Antonia, and Julie Fairman—have supported this book project since its inception and played a significant role in helping me frame the project. The Penn annual faculty writing retreat, which launched just as I was embarking on this project, introduced me to an invaluable group of scholars who have become my year-round confidantes for writing support and much more. Katie Barott, Linda Chance, Julie Nelson Davis, Ayako Kano, Lisa Mitchell, Jennifer Moore, Janine Remillard, Katie Schuler, and Heather Sharkey deserve special credit—let's keep it going!

Over the years, I have benefited from an incredibly talented group of undergraduate and graduate research assistants. I hope all the enthusiasm that these young scholars showed for the book project comes through in these pages. Lea Eisenstein and Lyla Rose put innumerable hours into scouring journals, visiting archives, and securing image permissions—they know the book nearly as well as I do. Taylor Dysart, Chloe Getrajdman, Whitney Laemmli, Elaine LaFay, Mary Mitchell, Talia Moss, Sara Ray, and Sophie Qi also provided essential research help and important insights along the way.

A book of history is only as good as the archivists and librarians behind it. The Penn interlibrary loan and faculty express staff deserves special credit for managing years of my book and article requests. Nick Okrent is a dream librarian who has answered all of my queries in record time and somehow manages to consistently dig up obscure yet crucial source material. David Azzolina has been equally swift and supportive.

This book relies on a variety of archival holdings and the excellent staff who preserve the historical record and make it accessible to researchers such as myself. I'm grateful to the all those who assisted me at the Bancroft Library, University of California, Berkeley, Center for the History of Medicine at Harvard's Francis A. Countway Library of Medicine, the Hampton University Museum and Archives, the Harvard University Archives, the Hoover Institution Library and Archives, the Howard University Archives, Kansas University Medical Center Archives, the Massachusetts General Hospital Archives, the National Anthropological Archives at the Smithsonian Institute, Princeton University Archives, the Radcliff College Archives at the Harvard Radcliff Institute, Special Collections at Smith College, the University Archives at Stanford University Library, and the University Archives, University of Pennsylvania.

Several archivists deserve special mention. Margery Sly, who was head of the Smith College collections in the 1990s and was required to destroy posture photos, has been an incredible source for understanding the complexities of what was happening on the ground, as well as the daily professional challenges that most archivists face. She also put me in contact with Nanci Young, the current Smith archivist, who, along with her staff, helped me recover the school's history of posture exams and training. Their dedication came at a time when I was doubting the viability of the project, but they showed me that through creative searching and researching, not all of the records had been destroyed, as some had been preserved in less obvious places. Ben Primer had the foresight to preserve his own correspondence regarding the incineration of the Princeton posture photos and was more than happy to share these enlightening materials with me. Finally, Gina Rappaport took time at several points in this project to discuss the William Sheldon papers with me, especially regarding the possible fate of all the photos that remain in the archives but are still restricted.

Some of my best thinking happens when I am in conversation with others. Luckily, I am in a profession that fosters such dialogue through conference presentations, public lectures, and colloquiums. I presented parts of *Slouch* at the Barbara Bates Center for the History of Nursing, the Consortium for the History of Science, Technology, and Medicine, Harvard University's Department of the History of Science, Johns Hopkins University's Institute of Medicine, the Massachusetts Historical Society, Memorial University School of Medicine, Penn's Medical Sociology Workshop, the Penn Humanities Forum, Princeton University's Program in the History of Science, the Science History Institute, the University of Kansas Department of History and Philosophy of Medicine, the University of Wisconsin–Madison Department of the History of Medicine and Bioethics, and Yale University's Program in the History of Science and Medicine. A collective thank you goes out to all of my colleagues at these various forums and institutions.

I have particularly appreciated my ongoing conversations and email exchanges with Michele Eodice, Mary Fissell, Nancy Hirschmann, Susan Lederer, Elizabeth Lee, Deborah Levine, Naomi Rogers, Rosemary Stevens, Jonathan Sadowsky, Thomas Schlich, Jacob Steere-Williams, Sarah Tracy, Arleen Tuchman, Keith Wailoo, John Warner, Elizabeth Watkins, and Bess Williamson. For written responses to various chapters presented in this book, I am grateful to Jonathan Marks, Joan Jacobs Brumberg, Dominque Tobbell, Michael Rembis, Jenifer Barclay, Brigid Prial, and Carla Bittel. Mara Mills, Jaipreet Virdi, and Sarah Rose, co-editors of a forthcoming *Osiris* issue, gave me the chance to test run my thinking on epidemics and disability. Christine von Oertzen, Elaine Leong, and Carla hosted me at the Max Planck Institute for the History of Science in Berlin, providing comments on multiple rounds of an article detailing Clelia Mosher's posture measuring device.

In the project's early stages, Susan Reverby read the first four chapters of the manuscript and offered me essential feedback and words of encouragement. More recently, Wendy Kline and Jaipreet helped me bring the book across the finish line, taking the time to read the entire manuscript and assuring me that it was good to go.

I have had the good fortune to be represented by an incredible agent, Jessica Papin. Thank you for taking a chance on me. My editor, Eric Crahan, expressed a great deal of enthusiasm for the project since day one and has expertly stewarded it through peer review and production. Thanks also to Whitney Rauenhorst, Alyssa Sanford, David Heath, and to the team at Princeton University Press.

Family and friends, near and far, have fed my soul, showing me that there is more to life than "the book." My Sunday phone conversations with my mother, Rose Anne O'Donnell, keep me righted; I am blessed to call her Mom. Jess Frey, Jeanne Alvaré Goodwin, Anjali Shaw, and Audrey Yu are steadfast confidantes who have seen me through thick and thin. I would be remiss if I did not thank my physical therapists Marc McShane and Amy Lesher, who kept me not only physically intact, but also up to date on the latest theories regarding posture testing and training in the clinic.

My greatest debt goes to my own family. I faced some life-threatening health events during the course of writing this book. At one point, I wasn't even sure that I would recuperate enough to be able to research and write again. My spouse, Damon Linker, stood by my side through it all, helping to get me back on my feet, and once I was, did everything in his power to make sure that I had the space and time (and meals!) to see this project through. He is my intellectual companion, life partner, and love. The book is dedicated to my children, Mark and Katie Rose, who are my life's finest works. This project has matured alongside them as they have grown into discerning young adults whose love for culture, ideas, music, and the human psyche makes me indescribably proud. These pages are for you.

Illustration Credits

FIGURE 1.   From Dudley J. Morton, "Human Origin: Correlation of Previous Studies of
Primate Feet and Posture with Other Morphologic Evidence," *American Journal
of Physical Anthropology* 10, no. 1 (1927): 183.

FIGURE 2.   K. Frances Scott, *A College Course in Hygiene* (New York: Macmillan Co., 1939), 15.

FIGURE 3.   "Other 25—National Negro Health Week," *Afro-American* (Baltimore, MD),
April 3, 1937, 18.

FIGURE 4.   Courtesy of the R. Tait McKenzie Papers, University of Pennsylvania, Box 4,
Folder 76.

FIGURE 5.   Jessie H. Bancroft, *The Posture of School Children: With Its Home Hygiene and
New Efficiency Methods for School Training* (New York: Macmillan Co., 1913), 99.

FIGURE 6.   Jessie H. Bancroft, *The Posture of School Children: With Its Home Hygiene and
New Efficiency Methods for School Training* (New York: Macmillan Co., 1913),
162.

FIGURE 7.   "Incorrect Sitting Position for Postural Deformity and Dorsal Curvature
Cases," photograph, Library of Congress Catalogue, https://lccn.loc.gov
/2018677185.

FIGURE 8.   "Posture and Tuberculosis," 1920, National Child Welfare Association,
cooperating with National Association for the Study and Prevention of
Tuberculosis; from Library of Congress, https://www.loc.gov/item
/2014647542/.

FIGURE 9.   Courtesy of Stanford University Archives and Special Collections.

FIGURE 10.   From Lillian C. Drew, *Individual Gymnastics: A Handbook of Corrective and
Remedial Gymnastics* (Philadelphia: Lea and Febiger, 1922), 80.

FIGURE 11.   Roger I. Lee and Lloyd T. Brown, "A New Chart for the Standardization of
Body Mechanics," *Journal of Bone and Joint Surgery* 21, no. 5 (1923): 754.

FIGURE 12.   Image from John Daly McCarthy, *Health and Efficiency* (New York: Henry Holt
and Company, 1922), 53.

FIGURE 13.   Theresa Wolfson, "Seating Survey in the Garment Industry," in *Seating and Posture: An Inquiry Made by the Joint Board of Sanitary Control into the Seating Conditions in the Women's Garment Industry* (1923), 45.

FIGURE 14.   From M. J. Pullman, *Foot Hygiene and Posture: For Adults and Children* (Los Angeles: 1933), 22.

FIGURE 15.   Photograph in Herman W. Marshall, "What Do You Know about Feet?," *Boot and Shoe Recorder* (April 15, 1922): 110, courtesy of the Anglican Church of Melanesia and the British Museum.

FIGURE 16.   From M. J. Pullman, *Foot Hygiene and Posture: For Adults and Children* (Los Angeles: 1933), 28.

FIGURE 17.   Display Advertisement for B. F. Goodrich Rubberized "Posture Foundation" Sport Shoes, *Ladies' Home Journal* 54, no. 6 (June 1937): 94.

FIGURE 18.   Display Advertisement for Buster Brown Shoes, *Parents Magazine* 5, no. 10 (October 1930): 75.

FIGURE 19.   Antioch College, *The Effects of Modern Shoes upon Proper Body Mechanics, 1924–1931* (Yellow Springs, OH: Antioch College, 1931), 5.

FIGURE 20.   Antioch College, *The Effects of Modern Shoes upon Proper Body Mechanics, 1924–1931* (Yellow Springs, OH: Antioch College, 1931), 12.

FIGURE 21.   *Pittsburgh Post-Gazette*, October 11, 1935, 11.

FIGURE 22.   "Expert Adjustment with One Movement," Display Advertisement, *Corset and Underwear Review* 14, no. 5 (February 1920): 52.

FIGURE 23.   "Camco Corset," *Corset and Underwear Review* 19, no. 5 (August 1922): cover.

FIGURE 24.   From *A Manual of Camp Physiological Supports*, 6th ed. (Jackson, MI: S. H. Camp and Company, n.d.): 33.

FIGURE 25.   "Camp," Display Advertisement, *Rochester Democrat and Chronicle*, May 4, 1941, 16A.

FIGURE 26.   "Fitness Program Seen Adding Importance to Posture Week," *Women's Wear Daily* 70, no. 57 (March 22, 1945): 17.

FIGURE 27.   William Blaikie, *How to Get Strong and How to Stay So* (New York: Harper & Brothers, 1879), 225.

FIGURE 28.   Jessie H. Bancroft, *The Posture of School Children: With Its Home Hygiene and New Efficiency Methods for School Training* (New York: Macmillan Co., 1913), 235.

FIGURE 29.   Courtesy of the Massachusetts General Hospital Archives and Special Collections (subject file Orthopedic Department), Boston, Massachusetts.

FIGURE 30.   From Bess M. Mensendieck, *The Mensendieck System of Functional Exercises* (Portland, ME: Southworth-Anthoensen Press, 1937), plate X.

FIGURE 31.   L. Joseph Cahn, "Use of a Museum in Hygiene Class," *Journal of Health and Physical Education* 12, no. 1 (1941): 6.

FIGURE 32.   Courtesy of Smith College Archives, Physical Education Department Records, Dorothy Ainsworth files, unprocessed, restricted, Box 1.

FIGURE 33.   *Rochester Democrat and Chronicle*, December 20, 1927, 21, courtesy of the Rochester Museum and Science Center.

FIGURE 34.   *W.A.C. Field Manual: Physical Training* (Washington, DC: U.S. Government Printing Office, 1943), 119.

FIGURE 35.   Courtesy of Smith College Special Collections, Northampton, Massachusetts.

FIGURE 36.   "Finishing School: Wealthiest Families Send Children in Highly-Rated Palmer to Become Ladies and Gentlemen," *Ebony Magazine*, October 1947, 22.

FIGURE 37.   Courtesy of the Hennepin County Library.

FIGURE 38.   *Vassar Chronicle*, April 15, 1950, 7.

FIGURE 39.   Courtesy of Princeton University Archives, Department of Special Collections, Princeton University Library.

Notes

Introduction

1. Ron Rosenbaum, "The Great Ivy League Nude Posture Photo Scandal: How Scientists Coaxed America's Best and Brightest out of Their Clothes," *New York Times Magazine*, January 15, 1995, 26, 28–29, 30–31, 40, 46, 55–56.

2. Important exceptions include Sander L. Gilman, *Stand Up Straight!: A History of Posture* (London: Reaktion Books, 2018), as well as Patricia Vertinsky, "Physique as Destiny: William H. Sheldon, Barbara Honeyman Heath and the Struggle for Hegemony in the Science of Somatotyping," *Canadian Bulletin of Medical History* 24, no. 2 (Fall 2007): 291–316.

3. Rosenbaum, "The Great Ivy League Nude Posture Photo Scandal."

4. Patricia Marx, "Stand Up Straight!," *New Yorker*, March 29, 2021, 30. This number indicates the global market, not just the United States. See "Posture Correction Market Size, Share & Trends Analysis Report by Distribution Channel (Pharmacies & Retail Stores, E-Commerce), by Product (Sitting Support Devices, Kinesiology Tape), by End Use, by Region, and Segment Forecasts, 2022–2030," accessed April 11, 2023, https://www.grandviewresearch.com/industry-analysis/posture-correction-market-report, and "Posture Corrector Market: Global Industry Analysis and Forecast (2022–2029)," accessed April 11, 2023, https://www.maximizemarketresearch.com/market-report/posture-corrector-market/146092/.

5. Lloyd T. Brown, "The Harvard Slouch," *New York Times*, March 18, 1917, T8.

6. For more on the pre–nineteenth-century understanding of posture, especially in Europe, see Gilman, *Stand Up Straight!*

7. Tom Gundling, "Stand and Be Counted: The Neo-Darwinian Synthesis and the Ascension of Bipedalism as an Essential Hominid Synapomorphy," *History and Philosophy of the Life Sciences* 34, no. 1/2 (December 2012): 185–210. For a history of skull collecting, see Ann Fabian, *The Skull Collectors: Race, Science, and America's Unburied Dead* (Chicago: University of Chicago Press, 2010).

8. Bert Theunissen, *Eugène Dubois and the Ape-Man from Java: The History of the First Missing Link and Its Discoverer* (Dordrecht: Kluwer Academic Publishers, 1989).

9. For one example, see Arthur Keith, "Hunterian Lectures on Man's Posture: Its Evolution and Disorders (Lecture I)," *British Medical Journal* 1, no. 3246 (March 1923): 451–54.

10. John F. Kasson, *Rudeness and Civility: Manners in Nineteenth-Century Urban America* (New York: Hill and Wang, 1990).

11. Nancy Tomes, "The Making of a Germ Panic, Then and Now," *American Journal of Public Health* 90, no. 2 (February 2000): 191–98; Nancy Tomes, *The Gospel of Germs: Men, Women, and the Microbe in American Life* (Cambridge, MA: Harvard University Press, 1998).

12. Joel E. Goldthwait, "An Anatomic and Mechanistic Concept of Disease," *Boston Medical and Surgical Journal* 172, no. 24 (July 1915): 888. Views such as Goldthwait's were relatively common. For the history of holism in U.S. medicine, see Sarah W. Tracy, "George Draper and American Constitutional Medicine, 1916–1946: Reinventing the Sick Man," *Bulletin of the History of Medicine* 66, no. 1 (Spring 1992): 53–89.

13. R. Tait McKenzie, "The Regulation of Physical Instruction in Schools and Colleges from the Standpoint of Hygiene," *Science* 29, no. 743 (March 1909): 482.

14. Beth Linker, "Toward a History of Ableness," *All of Us*, June 1, 2021, http://allofusdha.org/research/toward-a-history-of-ableness/.

15. Gail Bederman, *Manliness and Civilization: A Cultural History of Gender and Race in the United States, 1880–1917* (Chicago: University of Chicago Press, 1995); Beth Linker, *War's Waste: Rehabilitation in World War I America* (Chicago: University of Chicago Press, 2011).

16. Douglas C. Baynton, "Disability and the Justification of Inequality in American History," in *The New Disability History: American Perspectives*, ed. Paul K. Longmore and Lauri Umansky (New York: New York University Press, 2001), 33–57; Douglas C. Baynton, *Defectives in the Land: Disability and Immigration in the Age of Eugenics* (Chicago: University of Chicago Press, 2016).

17. Michael Rembis, "Disability and the History of Eugenics," in *The Oxford Handbook of Disability History*, ed. Michael Rembis, Catherine Kudlick, and Kim E. Nielsen (New York: Oxford University Press, 2018), 85–103. See also David Mitchell and Sharon Snyder, "The Eugenic Atlantic: Race, Disability, and the Making of an International Eugenic Science, 1800–1945," *Disability & Society* 18, no. 7 (December 1, 2003): 843–64.

18. Jonathan Katz, *The Invention of Heterosexuality* (New York: Dutton, 1995); George Lakoff and Mark Johnson, *Metaphors We Live By* (Chicago: University of Chicago Press, 1980).

19. Nancy Leys Stepan, *"The Hour of Eugenics": Race, Gender, and Nation in Latin America* (Ithaca, NY: Cornell University Press, 1991); Alexandra Minna Stern, "From Mestizophilia to Biotypology: Racialization and Science in Mexico, 1920–1960," in *Race and Nation in Modern Latin America*, ed. Nancy P. Appelbaum, Anne S. Macpherson, and Karin Alejandra Rosemblatt (Chapel Hill: University of North Carolina Press, 2003), 187–210. For more on the history of euthenics, see Susan Currell, "Eugenic Decline and Recovery in Self-Improvement Literature of the Thirties," in *Popular Eugenics: National Efficiency and American Mass Culture in the 1930s*, ed. Susan Currell and Christina Cogdell (Athens: Ohio University Press, 2006), 44–69.

20. The history of posture improvement campaigns served much the same purpose as other hygiene efforts in the American colonial project. See, for example, Warwick Anderson, *Colonial Pathologies: American Tropical Medicine, Race, and Hygiene in the Philippines* (Durham, NC: Duke University Press, 2006).

21. This was a trend found in other nations as well. See, for example, Ruth Rogaski, *Hygienic Modernity: Meanings of Health and Disease in Treaty-Port China* (Berkeley: University of California Press, 2004).

22. Beth Linker and Emily K. Abel, "Integrating Disability, Transforming Disease History: Tuberculosis and Its Past," in *Civil Disabilities*, ed. Nancy J. Hirschmann and Beth Linker (Philadelphia: University of Pennsylvania Press, 2015), 83–102.

23. Riva Lehrer, *Golem Girl: A Memoir* (New York: One World, 2020); Sunaura Taylor, *Beasts of Burden: Animal and Disability Liberation* (New York: New Press, 2017).

24. The scholarship on the history of obesity is extensive. For some examples, see Sabrina Strings, *Fearing the Black Body: The Racial Origins of Fat Phobia* (New York: New York University Press, 2019); Peter N. Stearns, *Fat History: Bodies and Beauty in the Modern West* (New York: New York University Press, 2002); Sander L. Gilman, *Fat: A Cultural History of Obesity* (Cambridge: Polity, 2008); and Anna Mollow, "Disability Studies Gets Fat," *Hypatia* 30, no. 1 (Winter 2015): 199–216.

25. Roy Porter, "Diseases of Civilization," in *Companion Encyclopedia of the History of Medicine*, ed. W. F. Bynum and Roy Porter (London: Taylor & Francis Group, 1993), 585; Charles E. Rosenberg, "Pathologies of Progress: The Idea of Civilization as Risk," *Bulletin of the History of Medicine* 72, no. 4 (Winter 1998): 714–30.

26. Charles E. Rosenberg, *Explaining Epidemics and Other Studies in the History of Medicine* (New York: Cambridge University Press, 1992); Priscilla Wald, *Contagious: Cultures, Carriers, and the Outbreak Narrative* (Durham, NC: Duke University Press, 2008). For a critical take on the narrative logic of epidemics, see Mary E. Fissell et al., "Introduction: Reimagining Epidemics," *Bulletin of the History of Medicine* 94, no. 4 (Winter 2020): 543–61, and the articles contained in the remainder of this themed issue that problematize such a view.

27. For more on slavery and disability, see Stefanie Hunt-Kennedy, *Between Fitness and Death: Disability and Slavery in the Caribbean* (Urbana: University of Illinois Press, 2020); Jenifer L. Barclay, *The Mark of Slavery: Disability, Race, and Gender in Antebellum America* (Urbana: University of Illinois Press, 2021); Dea Boster, *African American Slavery and Disability: Bodies, Property, and Power in the Antebellum South, 1800–1860* (New York: Routledge, 2012).

28. Shaun Dreisbach, "How to Turn Off Your Bitch Switch," *Glamour*, February 2016, 96.

29. A version of Brody's article appeared in print on December 29, 2015, on page D5 of the New York edition, with the headline: "Good Posture May Better Your Position." For Black jurors, see Adam Liptak, "Exclusion of Blacks from Juries Raises Renewed Scrutiny," *New York Times*, August 16, 2015, https://www.nytimes.com/2015/08/17/us/politics/exclusion-of-blacks-from-juries-raises-renewed-scrutiny.html.

Chapter One: The Making of a Posture Science

1. Bert Theunissen, "Marie Eugène Francois Thomas Dubois," in *New Dictionary of Scientific Biography*, ed. Noretta Koertge (Detroit: Thomas Gale, 2008), 313.

2. Peter J. Bowler, *Theories of Human Evolution: A Century of Debate, 1844–1944* (Baltimore: Johns Hopkins University Press, 1986), 5.

3. Tom Gundling, "Human Origins Studies: A Historical Perspective," *Evolution: Education and Outreach* 3 (2010): 314–21; Peter J. Bowler, *Evolution: The History of an Idea* (Berkeley: University of California Press, 1984); Bowler, *Theories of Human Evolution*.

4. Misia Landau, *Narratives of Human Evolution* (New Haven, CT: Yale University Press, 1991).

5. S. V. Clevenger, "Disadvantages of the Upright Position," *American Naturalist* 18, no. 1 (January 1884): 7.

6. The use of more "primitive" peoples as model organisms is still part of today's public health practices in the West. See, for example, H. Pontzer, B. M. Wood, and D. A. Raichlen, "Hunter-Gatherers as Models in Public Health," *Obesity Reviews* 19, no. S1 (December 2018): 24–35.

7. Shari M. Huhndorf, *Going Native: Indians in the American Cultural Imagination* (Ithaca, NY: Cornell University Press, 2015), 4.

8. The ways in which indigenous peoples thought about human posture and how they responded to these claims foisted upon them requires further study. Following the research of Jenny Reardon and Kim TallBear, it seems clear that the depiction of Native Americans as possessing the ancestral backbone of the United States was a white construct used to claim ownership over Native American lands and identity. See Jenny Reardon and Kim TallBear, "'Your DNA Is Our History': Genomics, Anthropology, and the Construction of Whiteness as Property," *Current Anthropology* 53, no. S5 (April 2012): S233–45. For other accounts of how physical anthropology played a significant role in the dispossession and oppression of non-white Americans, see Lee D. Baker, *Anthropology and the Racial Politics of Culture* (Durham, NC: Duke University Press, 2010); Maile Arvin, *Possessing Polynesians: The Science of Settler Colonial Whiteness in Hawai'i and Oceania* (Durham, NC: Duke University Press, 2019); Susan Burch, *Committed: Remembering Native Kinship in and beyond Institutions* (Chapel Hill: University of North Carolina Press, 2021); Alexandra Widmer and Veronika Lipphardt, eds., *Health and Difference: Rendering Human Variation in Colonial Engagements* (New York: Berghahn Books, 2016); Jonathan Marks, *Tales of the Ex-Apes: How We Think about Human Evolution* (Berkeley: University of California Press, 2015).

9. Constance Areson Clark, *God—or Gorilla: Images of Evolution in the Jazz Age* (Baltimore: Johns Hopkins University Press, 2008).

10. Tom Gundling, "Stand and Be Counted: The Neo-Darwinian Synthesis and the Ascension of Bipedalism as an Essential Hominid Synapomorphy," *History and Philosophy of the Life Sciences* 34, no. 1/2 (December 2012): 187. See also Bowler, *Theories of Human Evolution*, 151.

11. Bert Theunissen, *Eugène Dubois and the Ape-Man from Java: The History of the First Missing Link and Its Discoverer* (Dordrecht: Kluwer Academic Publishers, 1989), 4.

12. Bowler, *Theories of Human Evolution*, 4.

13. Stephen Jay Gould, *The Mismeasure of Man* (New York: Norton, 1981). Many of Gould's historical actors believed that upright posture was only a consequence of higher brain development.

14. On this, see Stephen Jay Gould, "Posture Maketh the Man," in *Ever Since Darwin: Reflections in Natural History* (New York: Norton, 1977), 207–13. In this piece he discusses how Smith saw posture and speech as incidental manifestations of brain development. See also Bowler, *Theories of Human Evolution*, 161–73; Gundling, "Stand and Be Counted," 190.

15. Bowler, *Theories of Human Evolution*, 151.

16. Alfred Russel Wallace, *Darwinism: An Exposition of the Theory of Natural Selection, with Some of Its Applications* (New York: Humboldt, 1889), 308, quoted in Gundling, "Stand and Be Counted," 188.

17. German evolutionist Ernst Haeckel was an example of this. See Theunissen, *Eugène Dubois*.

18. Theunissen, *Eugène Dubois*, 73–75. See also Bowler, *Evolution*, 201–2.

19. For more on this, see Thomas DiPiero, "Missing Links: Whiteness and the Color of Reason in the Eighteenth Century," *Eighteenth Century* 40, no. 2 (Summer 1999): 155–74; Harriet Ritvo, "At the Edge of the Garden: Nature and Domestication in Eighteenth- and Nineteenth-Century Britain," *Huntington Library Quarterly* 55, no. 3 (Summer 1992): 363–78.

20. Theunissen, "Marie Eugène Francois Thomas Dubois."

21. Bowler, *Evolution*, 299.

22. Nancy J. Parezo and Don D. Fowler, *Anthropology Goes to the Fair: The 1904 Louisiana Purchase Exposition* (Lincoln: University of Nebraska Press, 2007), 49 (emphasis mine). See also Robert W. Rydell, *All the World's a Fair: Visions of Empire at the American International Expositions, 1876–1916* (Chicago: University of Chicago Press, 1985); Sadiah Qureshi, *Peoples on Parade: Exhibitions, Empire, and Anthropology in Nineteenth-Century Britain* (Chicago: University of Chicago Press, 2011).

23. "A Monkey Girl: The Missing Link on Exhibition in Chicago," *Chicago Daily Tribune*, December 27, 1884, 8. See also "Krao—A Missing Link," *The Continent: An Illustrated Weekly Magazine*, February 20, 1884, 5, 106. For more on the American press and the popularity of the missing link in the late nineteenth century, see David R. Angerhofer, "The American Response to *Pithecanthropus erectus*: The Missing Link and the General Reader, 1860–1920" (master's thesis, University of Maryland, 2008).

24. "A Monkey Girl."

25. "A Monkey Girl."

26. For "raw" data and a history of paleoanthropology, see Gundling, "Human Origins Studies," 314.

27. Theunissen, "Marie Eugène Francois Thomas Dubois," 313.

28. Theunissen, "Marie Eugène Francois Thomas Dubois."

29. Theunissen, *Eugène Dubois*, 9–11.

30. Theunissen, *Eugène Dubois*, 11.

31. Theunissen, *Eugène Dubois*, 33.

32. Bowler, *Theories of Human Evolution*, 5.

33. Arthur Keith, *Man: A History of the Human Body* (New York: Holt, 1912), 75. See also Arthur Keith, "Hunterian Lectures on Man's Posture: Its Evolution and Disorders (Lecture I)," *British Medical Journal* 1, no. 3246 (March 1923): 452.

34. Keith, *Man*; Frank Spencer, "Sir Arthur Keith," in *History of Physical Anthropology*, ed. Frank Spencer (New York: Garland, 1997), 560–62.

35. Arthur Keith, "*Pithecanthropus erectus*—A Brief Review of Human Fossil Remains," *Science Progress* 3, no. 17 (1895): 368.

36. Keith, "*Pithecanthropus erectus*," 369.

37. Keith would go on to privilege brain development—although never to the exclusion of posture primacy—with the so-called "Piltdown Man" discovery of 1912. The discovery of the Piltdown Man's remains in East Sussex, England, suggested that the first evolution of human

intelligence came about in Europe. To this day, much controversy surrounds the Piltdown discovery, with accusations that British paleoanthropologists planted the remains. There is extensive scholarship on the Piltdown discovery and Keith's role in it. See, for example, Jonathan Sawday, "'New Men, Strange Faces, Other Minds': Arthur Keith, Race and the Piltdown Affair (1912–53)," in *Race, Science and Medicine, 1700–1960*, ed. Waltraud Ernst and Bernard Harris (London: Routledge, 1999), 259–88.

38. O. C. Marsh, "The Ape-Man from the Tertiary of Java," *Science* 3, no. 74 (May 1896): 793.

39. Marsh, "The Ape-Man."

40. Aleš Hrdlička, "Dr. Eugene Dubois, 1858–1940," *Scientific Monthly* 52, no. 6 (1941): 578–80. For the American context, see also Gundling, "Stand and Be Counted," 196.

41. C. Loring Brace, "'Physical' Anthropology at the Turn of the Last Century," in *Histories of American Physical Anthropology in the Twentieth Century*, ed. Michael A. Little and Kenneth A. R. Kennedy (Lanham, MD: Lexington Books, 2010), 44.

42. Earnest A. Hooton, "Where Did Man Originate?," *Antiquity* 1 (January 1927): 133–50. See also Gundling, "Stand and Be Counted," 196.

43. Dudley J. Morton, "Human Origin: Correlation of Previous Studies of Primate Feet and Posture with Other Morphologic Evidence," *American Journal of Physical Anthropology* 10, no. 1 (1927): 195.

44. "Learning to Stand Up No Simple Task: Took Man Centuries, Says Yale Prof.," *Chicago Defender*, February 6, 1926, A1. Morton was quoted here by a journalist.

45. Beth Linker, "Feet for Fighting: Locating Disability and Social Medicine in First World War America," *Social History of Medicine* 20, no. 1 (April 2007): 91–109.

46. Keith, *Man*, 76.

47. F.M.R. Walshe, "The Work of Sherrington on the Physiology of Posture," *Proceedings of the Royal Society of Medicine* 17 (1924): 4–6. Sherrington's postural reflex ushered in studies of proprioception.

48. Leonard Hill, "The Influence of the Force of Gravity on the Circulation," *The Lancet* 145, no. 3728 (February 1895): 338–39. See also Keith, "Hunterian Lectures (Lecture I)," 454.

49. Dudley J. Morton, "The Relation of Evolution to Medicine," *Science* 64, no. 1660 (1926): 394–96. See also Earnest A. Hooton, "The Relation of Physical Anthropology to Medical Science," *Medical Review of Reviews* 22, no. 4 (April 1916): 260–64.

50. Critical histories of evolutionary medicine are scant. Medical historian W. F. Bynum offers the beginnings in his article, "Darwin and the Doctors: Evolution, Diathesis, and Germs in 19th-Century Britain," *Gesnerus* 40, no. 1–2 (1983): 43–53. Certain textbooks in evolutionary medicine provide historical overviews as well. See, for example, Robert Perlman, *Evolution and Medicine* (Oxford: Oxford University Press, 2013).

51. Keith, "Hunterian Lectures (Lecture I)," 452.

52. Keith, "Hunterian Lectures (Lecture I)." Many experts at the time were convinced of a link between disease and poor posture. For a couple of representative examples, see Armitage Whitman, "Postural Deformities in Children," *New York State Journal of Medicine* 24, no. 19 (October 1924): 871–74; Augustus Grote Pohlman, "Some of the Disadvantages of the Upright Position," *American Medicine* 1, no. 9 (December 1906): 541–46.

53. I am thinking here of Henry F. Osborn's theory of organic selection—i.e., that individuals could acquire new adaptive characteristics through exercise, but that these characteristics did not need to be inherited. See Bowler, *Evolution*, 262.

54. For more on Keith and his conversion, see Brace, "'Physical' Anthropology at the Turn of the Last Century," 25–53. For the racial implications of his conversion, see Amanda Rees, "Stories of Stones and Bones: Disciplinarity, Narrative and Practice in British Popular Prehistory, 1911–1935," *British Journal for the History of Science* 49, no. 3 (September 2016): 433–51; Nancy Leys Stepan, "Nature's Pruning Hook: War, Race, and Evolution, 1914–1918," in *The Political Culture of Modern Britain: Studies in Memory of Stephen Koss*, ed. J.M.W. Bean (London: Hamish Hamilton, 1987), 129–45; Marianne Sommer, "Ancient Hunters and Their Modern Representatives: William Sollas's (1849–1936) Anthropology from Disappointed Bridge to Trunkless Tree and the Instrumentalisation of Racial Conflict," *Journal of the History of Biology* 38, no. 2 (2005): 327–65; James J. Harris, "The 'Tribal Spirit' in Modern Britain: Evolution, Nationality, and Race in the Anthropology of Sir Arthur Keith," *Intellectual History Review* 30, no. 2 (2020): 273–94.

55. Bowler, *Theories of Human Evolution*, 15.

56. Clevenger, "Disadvantages of the Upright Position," 8. This was an early articulation of the burgeoning posture sciences, an article which later U.S. posture scientists of the twentieth century would reference.

57. Clevenger, "Disadvantages of the Upright Position," 7.

58. For more on the racial science of pelvimetry, see Elizabeth O'Brien, "Pelvimetry and the Persistence of Racial Science in Obstetrics," *Endeavour* 37, no. 1 (March 2013): 21–28. For examples of pelvimetry and posture, see Frank Hinchey, "Pelvic Changes of Quadrupedal Mammals on Assuming the Erect Posture," *Journal of the Missouri Medical Association* 22, no. 1 (1925): 298–303, and William Jackson Merrill, "Distortion of the Pelvis from Posture," *American Journal of Orthopedic Surgery* 16, no. 12 (December 1918): 492–94. Many scientists interested in pelvimetry distinguished four main shapes of human pelvises—oval, round, square, and cuneiform— and believed that progress along the evolutionary scale could be deduced from these shapes, with oval pelvises being characteristic of the most evolutionarily advanced Europeans and cuneiform of the "least advanced" Black races (Native Americans possessed round, and "Mongols" square).

59. Clevenger, "Disadvantages of the Upright Position," 7.

60. Arthur Keith, "Hunterian Lectures on Man's Posture: Its Evolution and Disorders (Lecture V)," *British Medical Journal* 1, no. 3250 (April 1923): 624.

61. Stepan, "Nature's Pruning Hook," 143. See also Bowler, *Evolution*, 289.

62. Arthur Keith, "Creating a New American Race," *New York Times*, June 2, 1929, 112–13, 121.

63. Arthur Keith, *Ethnos: Or the Problem of Race Considered from a New Point of View* (London: Kegan Paul, 1931): 16, as quoted in Harris, "The 'Tribal Spirit,'" 14.

64. Bowler, *Evolution*, 258. For an example of neo-Lamarckism and posture in America, see Frank Baker, "The Ascent of Man," *American Anthropologist* 3, no. 4 (October 1890): 309. Baker, a physician who founded the Anthropological Society of Washington, DC, in 1897 and edited the *American Anthropologist*, believed that certain aspects of human posture—particularly the spinal column, hands, and pelvis—were of a "recent origin" and thus "imperfectly differentiated

and liable to return to their primitive state" if not attended to. Espousing neo-Lamarckian views, which were relatively common at the time, he wrote: "the erect posture has been gradually acquired. Since gravity plays an important part in the functions of the visceral and circulatory systems, any marked change in the line of equilibrium must necessarily be accompanied by disturbances. These disturbances, to a certain extent, conflict with the acquirement of the position; as they weaken the animal. In the course of time the body may perhaps become adapted to the changed conditions, but before the perfect adaptation takes place there is a period of struggle. There is abundant evidence that such a struggle has occurred and is yet going on; the adaptation being as yet far from complete" (309).

65. For other examples, see Philip Lewin, "The Ten Commandments of Good Posture," *Hygeia* 6 (January 1928): 3–5. In his article, Lewin argues: "Man was not intended to walk upright. Many human disorders are penalties for his having assumed the upright posture" (3). This argument can be found in journalistic reports on evolution and posture well into the twentieth century. For example, see "To Breed a Brain," *Newsweek* 35, no. 19 (May 8, 1950): 54, wherein one reporter wrote: "It can be argued that man's ancestor was unwise to rise from all fours and stand erect. Physiologists are inclined to blame flat feet, varicose veins, drooping paunches, and some cardiac disorders . . . on man's upright stance, for which the human body is ill designed."

66. "Why We Can't Stand Still," *Literary Digest* 99, no. 10 (December 8, 1928): 23.

67. K. Frances Scott, *A College Course in Hygiene* (New York: Macmillan, 1939), 15.

68. For more on the history of paleoanthropology and the human–animal binary, see Murray Goulden, "Boundary-Work and the Human-Animal Binary: Piltdown Man, Science, and the Media," *Public Understanding of Science* 18, no. 3 (2009): 275–91.

69. Robin Veder, "Seeing Your Way to Health: The Visual Pedagogy of Bess Mensendieck's Physical Culture System," *International Journal of the History of Sport* 28, no. 8–9 (May 2011): 1336–52. See also Robin Veder, *The Living Line: Modern Art and the Economy of Energy* (Hanover, NH: Dartmouth College Press, 2015).

70. Marguerite Agniel, "New Ways to Correct Posture," *Parents Magazine* 4, no. 10 (October 1929): 51.

71. Frank H. Richardson, "The Runabout Child and His Problems," *Hygeia* 6 (December 1928): 691. See also Jean Grissom, "Sugar-Coated Calisthenics," *Hygeia* 5 (June 1927): 276. As she writes, "All parents want their children to have perfect posture, strong, useful feet. . . . Some mothers and fathers persist in the idea that these things all will be developed by ordinary play activities. There seems to be something wrong with this theory, in views of the many adults with wretched postures." The article features white children trotting like a horse, crawling on all fours pretending to be an elephant, and others using their feet like monkeys picking up marbles from the floor.

72. Mel Y. Chen, *Animacies: Biopolitics, Racial Mattering, and Queer Affect* (Durham, NC: Duke University Press, 2012); Jenifer L. Barclay, *The Mark of Slavery: Disability, Race, and Gender in Antebellum America* (Urbana: University of Illinois Press, 2021); Sunaura Taylor, *Beasts of Burden: Animal and Disability Liberation* (New York: New Press, 2017).

73. Aleš Hrdlička, "Quadruped Progression in the Human Child," *American Journal of Physical Anthropology* 10, no. 3 (1927): 347–54.

74. "Why Babies Move upon All Fours," *New York Times*, June 17, 1928, 120. This article was an interview with Hrdlička. See also Aleš Hrdlička, *Children Who Run on All Fours: And Other Animal-Like Behaviors of Young Children* (New York: Whittlesey House, McGraw-Hill, 1931).

75. "Why Babies Move upon All Fours."

76. For an example of this, see Baker, "The Ascent of Man." A physician-anthropologist, Baker wrote that "savages when ill-fed and living in unfavorable conditions may simulate the habits of anthropoids, and this has an effect upon their physical structure," 319.

77. Arthur Keith, "An Address on the Nature of Man's Structural Imperfections," *The Lancet* 206, no. 5334 (November 1925): 1047.

78. Keith, "An Address," 1047.

79. Arthur Keith, "National Physique and Public Health," *The Observer* (London), October 20, 1918, 3. See also Keith, "An Address," 107.

80. Keith, "An Address," 107.

81. The historical literature on empire, race, and the concept of civilization is vast. For some examples from U.S. history, see Gail Bederman, *Manliness and Civilization: A Cultural History of Gender and Race in the United States, 1880–1917* (Chicago: University of Chicago Press, 1995); Matthew Frye Jacobson, *Barbarian Virtues: The United States Encounters Foreign Peoples at Home and Abroad, 1876–1917* (New York: Hill and Wang, 2001); Rydell, *All the World's a Fair*; Walter T. K. Nugent, *Habits of Empire: A History of American Expansion* (New York: Alfred A. Knopf, 2008). For the way in which physician-anthropologists applied notions of civilization to deformity and disease, see Linker, "Feet for Fighting."

82. Keith, "An Address," 1047.

83. J. Albright Jones, "Effect of Posture Work on the Health of Children," *American Journal of Diseases of Children* 46, no. 1 (July 1933): 148–49.

84. Helen Durham and Barbara Beattie, "By Their Heads Ye Shall Know Them," *Ladies' Home Journal* 50, no. 5 (May 1933): 69–70.

85. Writing for *Collier's* magazine, Ruth Moore advised, "An easy way to test yourself for correct alignment . . . is to put a book on your head . . . and try to sit and rise, go up and down stairs, stoop to the floor and get up, all without dislodging your book" (49). See Ruth C. Moore, "Somebody Ought to Tell Them," *Collier's* 90 (September 24, 1932): 48–49. For more examples, see also Doris Lee Ashley, "Heads Up! If You Would Have a Beautiful Body," *The Delineator* 120, no. 3 (March 1932): 38; Mary Bayley Noel, "Improving Your Children's Posture," *Parents Magazine* 9, no. 6 (June 1934): 33–34, 75; Jessie H. Bancroft, *The Posture of School Children: With Its Home Hygiene and New Efficiency Methods for School Training* (New York: Macmillan, 1913).

86. These quotations come from an interview that journalist Marie Beynon Ray conducted with Joseph Pilates. See Ray, "Cutting a Fine Figure," *Collier's* 94 (August 18, 1934): 30.

87. "Zulu Girl's Noble Carriage: It Comes of Their Habit of Carrying Heavy Burdens upon Their Heads," *Chicago Defender*, July 5, 1913, 7.

88. "Photo Standalone 11—No Title," *Chicago Defender*, July 23, 1968, 14. The photo is of American beauty contestant Norma Sherer, "Miss Miami," head-carrying baskets with native women in Haiti.

89. For an excellent account of museum exhibits and other visual displays of evolution in the United States, see Clark, *God—or Gorilla*, especially chapter 7. The ape imagery became

strengthened when an increasing number of evolutionists became convinced by Raymond Dart's 1925 discovery of the first non-modern hominid from the African continent and his contention that Africa, not Asia, was the place of the beginning of the human lineage. For more, see Gundling, "Human Origins Studies," 316.

90. For more on eugenics and African Americans, see Ayah Nuriddin, "Engineering Uplift: Black Eugenics as Black Liberation," in *Nature Remade: Engineering Life, Envisioning Worlds*, ed. Luis A. Campos, Michael R. Dietrich, Tiago Saraiva, and Christian C. Young (Chicago: University of Chicago Press, 2021), 186–202; Daylanne K. English, *Unnatural Selections: Eugenics in American Modernism and the Harlem Renaissance* (Chapel Hill: University of North Carolina Press, 2004); Dorothy E. Roberts, *Killing the Black Body: Race, Reproduction, and the Meaning of Liberty* (New York: Pantheon Books, 1997).

91. Algernon B. Jackson, "Afro Health Talk: What Posture Means," *Afro-American* (Baltimore), October 30, 1937, 15. For more on the National Negro Health Movement, see Susan L. Smith, *Sick and Tired of Being Sick and Tired: Black Women's Health Activism in America, 1890–1950* (Philadelphia: University of Pennsylvania Press, 1995), 58–82. See also Paul Braff, "Saving the Race from Extinction: African Americans and National Negro Health Week," *New York Academy of Medicine*, February 27, 2018, https://nyamcenterforhistory.org/tag/african-american-history/.

92. Algernon B. Jackson, "Afro Health Talk: Posture and Purpose," *Afro-American* (Baltimore), November 2, 1940.

93. Jackson, "Afro Health Talk: Posture and Purpose."

94. Dr. A. Wilberforce Williams, the first African American health columnist who wrote for the *Chicago Defender*, was a posture crusader. His successor at the *Chicago Defender*, Henrine Ward, was a physical educator who taught at Fisk University, served as an administrator for the Chicago YMCA, and would go on to become dean of women at Bethune-Cookman College in Florida. See Williams, "Dr. A. Wilberforce Williams Talks on Preventive Measures, First Aid Remedies Hygienics and Sanitation: Posture—A Factor in Health," *Chicago Defender*, March 5, 1921, 12 and Ward, "Your Posture—Told by Miss Henrine E. Ward," *Chicago Defender*, April 27, 1935, 6.

Chapter Two: Posture Epidemic

1. Londa Schiebinger, *Nature's Body: Gender in the Making of Modern Science* (Boston: Beacon Press, 1993); Londa Schiebinger, "Skeletons in the Closet: The First Illustrations of the Female Skeleton in Eighteenth-Century Anatomy," *Representations* 14 (1986): 42–82.

2. Robert B. Osgood, *Body Mechanics: Education and Practice; Report of the Subcommittee on Orthopedics and Body Mechanics* (New York: Century, 1932), 51.

3. Nancy Tomes, *The Gospel of Germs: Men, Women, and the Microbe in American Life* (Cambridge, MA: Harvard University Press, 1998).

4. Judith Walzer Leavitt, *Typhoid Mary: Captive to the Public's Health* (Boston: Beacon Press, 1996).

5. For more on the new public health movement and how it emphasized prevention, see Dorothy Porter, ed., *The History of Public Health and the Modern State* (Amsterdam: Rodopi,

1994). See also Cynthia A. Connolly, *Saving Sickly Children: The Tuberculosis Preventorium in American Life, 1909–1970* (New Brunswick, NJ: Rutgers University Press, 2008).

6. The literature concerning these historical trends is quite expansive. For a few examples, see Kathleen Lynne Norman, "'Biological Living': The Redemption of Women and America through Healthy Living, Dress, and Eugenics" (PhD dissertation, Claremont Graduate University, 2000); Jill Fields, "'Fighting the Corsetless Evil': Shaping Corsets and Culture, 1900–1930," *Journal of Social History* 33, no. 2 (Winter 1999): 355–84; Mark Aldrich, *Safety First: Technology, Labor, and Business in the Building of American Work Safety, 1870–1939* (Baltimore: Johns Hopkins University Press, 1997); Jennifer Klein, *For All These Rights: Business, Labor, and the Shaping of America's Public-Private Welfare State* (Princeton, NJ: Princeton University Press, 2003); Alan M. Kraut, *Silent Travelers: Germs, Genes, and the "Immigrant Menace"* (New York: Basic Books, 1994); Amy L. Fairchild, *Science at the Borders: Immigrant Medical Inspection and the Shaping of the Modern Industrial Labor Force* (Baltimore: Johns Hopkins University Press, 2003); Michael B. Katz, *Reconstructing American Education* (Cambridge, MA: Harvard University Press, 1987); Beth Linker, *War's Waste: Rehabilitation in World War I America* (Chicago: University of Chicago Press, 2011); Beth Linker, "Feet for Fighting: Locating Disability and Social Medicine in First World War America," *Social History of Medicine* 20, no. 1 (April 2007): 91–109; Laila Haidarali, *Brown Beauty: Color, Sex, and Race from the Harlem Renaissance to World War II* (New York: New York University Press, 2018).

7. Many early posture studies were conducted on individuals who had little agency—public schoolchildren, immigrants, military draftees, and unskilled laborers. The few sites where subjects consented were in the life insurance industry—where individuals who wanted a policy were required to be examined—and in upper-middle-class private schools and fitness gyms. For more on the history of physical examinations, see Dan Bouk, *How Our Days Became Numbered: Risk and the Rise of the Statistical Individual* (Chicago: University of Chicago Press, 2015); Angela Nugent, "Fit for Work: The Introduction of Physical Examinations in Industry," *Bulletin of the History of Medicine* 57, no. 4 (Winter 1983): 578–95; Audrey B. Davis, "Life Insurance and the Physical Examination: A Chapter in the Rise of American Medical Technology," *Bulletin of the History of Medicine* 55, no. 3 (Fall 1981): 392–406.

8. R. W. Lovett, "The Occurrence and Prevention of Flat-Foot among City Hospital Nurses," *Medical and Surgical Report of Boston City Hospital* 7, no. 1 (1896): 193–201.

9. Notably, Swartz discusses how workers tinkered with the built environment to make standing desks. Nelle Swartz, "Industrial Posture and Seating," in *Proceedings of the National Safety Council, Tenth Annual Safety Congress* (Chicago: National Safety Conference, 1921), 845.

10. For employment discrimination based on posture deficiencies, see Beth Linker, "Spines of Steel: A Case of Surgical Enthusiasm in Cold War America," *Bulletin of the History of Medicine* 90, no. 2 (Summer 2016): 222–49. For immigration restrictions based on posture exams, see Douglas C. Baynton, *Defectives in the Land: Disability and Immigration in the Age of Eugenics* (Chicago: University of Chicago Press, 2016); Fairchild, *Science at the Borders*.

11. Jessie H. Bancroft, "Pioneering in Physical Training—An Autobiography," *Research Quarterly, American Association for Health, Physical Education and Recreation* 12, supplement 3 (1941): 666.

12. Bancroft, "Pioneering in Physical Training," 667.

13. James Allen Young, "Height, Weight, and Health: Anthropometric Study of Human Growth in Nineteenth-Century American Medicine," *Bulletin of the History of Medicine* 53, no. 2 (Summer 1979): 214–43. See also Martha H. Verbrugge, *Able-Bodied Womanhood: Personal Health and Social Change in Nineteenth-Century Boston* (New York: Oxford University Press, 1988).

14. Bancroft, "Pioneering in Physical Training," 675. Famed philosopher and educator John Dewey undertook posture training himself, and found that the techniques used "provided demonstration of the unity of the mind and body"; he encouraged the nation's schools to take up physical education that promoted a "sensory consciousness" of habit formation. John Dewey, "Introduction," in *Constructive Conscious Control of the Individual*, by F. Matthias Alexander (New York: E. P. Dutton, 1923), xxi–xxxiii.

15. Michael R. Huxley, "F. Matthias Alexander and Mabel Elsworth Todd: Proximities, Practices and the Psycho-Physical," *Journal of Dance and Somatic Practices* 3, no. 1/2 (January 2012): 25–36.

16. William James, *The Energies of Men* (New York: Moffat, Yard, 1914), 28.

17. Ruth Harris, *Guru to the World: The Life and Legacy of Vivekananda* (Cambridge, MA: Harvard University Press, 2022). James's interest in physical vigor and the strenuous life can be found in his other writings, such as "Moral Equivalent of War" and "The Experience of Activity." See Mark Dyreson, "Nature by Design: Modern American Ideas about Sport, Energy, Evolution, and Republics, 1865–1920," *Journal of Sport History* 26, no. 3 (Fall 1999): 447–69.

18. Roberta J. Park, "Setting the Scene—Bridging the Gap between Knowledge and Practice: When Americans Really Built Programmes to Foster Healthy Lifestyles, 1918–1940," *International Journal of Sports History* 25, no. 11 (September 2008): 1427–52.

19. Jessie H. Bancroft, *The Posture of School Children: With Its Home Hygiene and New Efficiency Methods for School Training* (New York: Macmillan, 1913), 4.

20. Bancroft, *The Posture of School Children*, 263.

21. Bancroft, *The Posture of School Children*, 263.

22. "Military Drills in the Schools," *School Review* 24, no. 4 (April 1916): 312–20. The superintendent of public instruction for Pennsylvania concurred: "We must conclude that in case of any malformation, local weakness or constitutional debility [military] drill tends, by its strain upon the nerves and prolonged tension on the muscle, to increase the defects rather than to relieve them." From Nathan C. Schaeffer, "Should Our Educational System include Activities Whose Special Purpose Is Preparation for War?," *School and Society* 1, no. 9 (February 1915): 290.

23. Schaeffer, "Should Our Educational System Include," 290.

24. For the history of hygiene, including evidence of posture work, in the Philippines, see Frederick O. England, *Physical Education: A Manual for Teachers* (Manila: Bureau of Printing, 1919); Warwick Anderson, *Colonial Pathologies: American Tropical Medicine, Race, and Hygiene in the Philippines* (Durham, NC: Duke University Press, 2006). For the imperialist effort to physically train Native Americans, see Jacqueline Fear-Segal and Susan D. Rose, eds., *Carlisle Indian Industrial School: Indigenous Histories, Memories, and Reclamations* (Lincoln: University of Nebraska Press, 2016).

25. Bancroft, "Pioneering in Physical Training," 674.

26. Bancroft, "Pioneering in Physical Training," 674.

27. Bancroft, "Pioneering in Physical Training," 675–76.

28. See R. Tait McKenzie, "The Proper Development of Physical Power," Lecture at YMCA Montreal, February 20, 1894, Box 6, Folder 22, R. Tait McKenzie Papers, University of Pennsylvania Archives, Philadelphia. See also R. Tait McKenzie, "The Development of Physical Efficiency among College Men," Speech at Queen's University, 1907, Box 6, Folder 43, R. Tait McKenzie Papers, University of Pennsylvania Archives, Philadelphia.

29. McKenzie, "The Development of Physical Efficiency among College Men." For quotations, see Schaeffer, "Should Our Educational System Include Activities," 289–90.

30. Emmett A. Rice, John Hutchinson, and Mabel Lee, *A Brief History of Physical Education*, 4th ed. (New York: Ronald Press, 1958), 244.

31. "How Prehistoric Woman Solved the Problem of Her Waist Line," *Current Opinion* (March 1, 1914): 201. "The strain of the erect posture, when our prehistoric ancestors abandoned life in the tree-tops, was far more serious to the female than to the male. Physiologically, the female is adapted to progression on all fours. She's too complex in structure to walk erect with the ease of the male. Woman, consequently, demands just such an aid as is afforded by the brassiere, the stays, the corset. Perhaps as time progresses she may abandon any process of reinforcement for her figure." For the history of sex segregation in late nineteenth-century physical education, see Verbrugge, *Able-Bodied Womanhood*.

32. Claire McRee, "The Debutante Slouch: Fashion and the Female Body in the United States, 1912–1925" (master's thesis, Bard Graduate Center: Decorative Arts, Design History, Material Culture, 2015).

33. J. Hamilton Moore, *The Young Gentleman and Lady's Monitor, and English Teacher's Assistant* (New York: Daniel D. Smith, 1824), 170.

34. Daniel E. Bender, *Sweated Work, Weak Bodies: Anti-Sweatshop Campaigns and Languages of Labor* (New Brunswick, NJ: Rutgers University Press, 2004). See also Albert H. Freiberg, "Some Effects of Improper Posture in Factory Labor," in *The Child Workers of the Nation: Proceedings of the Fifth Annual Conference on Child Labor* (New York: National Child Labor Committee, 1909), 104–10.

35. Matthew Frye Jacobson, *Whiteness of a Different Color: European Immigrants and the Alchemy of Race* (Cambridge, MA: Harvard University Press, 1998).

36. Bender, *Sweated Work, Weak Bodies*, 7.

37. William Skarstrom, "Gymnastic Teaching," *American Physical Education Review* 13, no. 1 (January 1913): 101; Susan Burch, *Committed: Remembering Native Kinship in and beyond Institutions* (Chapel Hill: University of North Carolina Press, 2021).

38. Dudley Allen Sargent, "The Physical Test of a Man," *American Physical Education Review* 26, no. 1 (April 1921): 188.

39. Shari M. Huhndorf, *Going Native: Indians in the American Cultural Imagination* (Ithaca, NY: Cornell University Press, 2015), 43.

40. Marguerite M. Marshall, "No Debutante Slouch for High School Girls," *Pittsburgh Press*, May 27, 1914, 18.

41. Katharine Capshaw Smith, "Childhood, the Body, and Race Performance: Early 20th-Century Etiquette Books for Black Children," *African American Review* 40, no. 4 (Winter 2006): 795–811.

42. E. Azalia Hackley, *The Colored Girl Beautiful* (Kansas City, MO: Burton, 1916), 106.

43. Hackley, *The Colored Girl Beautiful*, 75.

44. "Mother" Stoner, "Store Life: Wait upon Your Customers with a Sprightly Step," *Women's Wear* 19, no. 19 (July 23, 1919): 53.

45. Barbara A. Schreier, *Becoming American Women: Clothing and the Jewish Immigrant Experience, 1880–1920* (Chicago: Chicago Historical Society, 1994). See also McRee, "The Debutante Slouch."

46. There is much scholarship on the changing beauty standards in the U.S. during this time period. For a few examples, see Lois W. Banner, *American Beauty . . . Through Two Centuries of the American Idea, Ideal, and Image of the Beautiful Woman* (New York: Alfred A. Knopf, 1983); Kathy Peiss, *Hope in a Jar: The Making of America's Beauty Culture* (New York: Metropolitan Books, 1998); Blain Roberts, *Pageants, Parlors, and Pretty Women: Race and Beauty in the Twentieth-Century South* (Chapel Hill: University of North Carolina Press, 2014).

47. Other scientists idealized the poor white body in terms of posture and foot health as well. Linker, "Feet for Fighting."

48. Hired in 1908 by the National Child Labor Committee, which aimed to end child labor and establish free, compulsory education for all youth, Hine set out to visually document the effects that unrelenting work and poor living conditions had on children's bodies. Visiting the textile mills of Fall River, Massachusetts, known at the time as "the Spindle City," Hine captured a young female bookkeeper who posed with her back bare to the camera, and included the caption: "Postural deformity. Correct position in work most essential. . . . Picture shows extremely bad working position for this physical defect. A hunched, stooped-over position. Need for advice of examining physician." Lewis Wicks Hine, "Incorrect Sitting Position for Postural Deformity and Dorsal Curvature Cases," Photograph, January 1917, Library of Congress Catalogue, https://lccn.loc.gov/2018677185. Another image indicates sewing work: Lewis Wicks Hine, "Elizabeth Rudensky. Right Dorsal Curve," Photograph, January 1917, Library of Congress Catalogue, https://lccn.loc.gov/2018677177.

49. Sargent was known for his belief that a "straight line was a physical sign of health and longevity, of perfect structure and harmony of function." Sargent, as quoted in Verbrugge, *Able-Bodied Womanhood*, 135. Original quotation in Dudley Allen Sargent, "The Physical Proportions of the Typical Man," *Scribner's Magazine* 2 (July 1887): 16.

50. Paul Emmons, "The 'Right' Angles: Constructing Upright Posture and the Orthographic View," *Proceedings of the 87th ACSA Annual Meeting* (Fall 1999): 331–36.

51. Robin Veder, *The Living Line: Modern Art and the Economy of Energy* (Hanover, NH: Dartmouth College Press, 2015).

52. Carma R. Gorman, "Educating the Eye: Body Mechanics and Streamlining in the United States, 1925–1950," *American Quarterly* 58, no. 3 (September 2006): 839–68.

53. For more on the history of streamlining and machine–body aesthetics at the time, see Emmons, "The 'Right' Angles"; Gorman, "Educating the Eye"; Christina Cogdell, *Eugenic Design: Streamlining America in the 1930s* (Philadelphia: University of Pennsylvania Press, 2004).

54. Much like the paleoanthropologists who gave bipedalism priority over brain development, posture experts like Bancroft emphasized understanding the lines of the entire body, not just the head, as did the traditions of phrenology and physiognomy.

55. Veder, *The Living Line*, 132.

56. Veder, *The Living Line*, 132–33.

57. Anson Rabinbach, *The Human Motor: Energy, Fatigue, and the Origins of Modernity* (New York: Basic Books, 1990). Historian Richard Gillespie contends that after World War I, industrialists remained unconvinced by the physiologists' ways of measuring fatigue. As I have argued in my article, "Feet for Fighting," biomechanics and ergonomics increasingly replaced physiologists after World War I, the former being a less expensive method since it placed the work self-monitoring and self-disciplining on the individual. Richard Gillespie, "Industrial Fatigue and the Discipline of Physiology," in *Physiology in the American Context, 1850–1940*, ed. Gerald L. Geison (Bethesda, MD: American Physiological Society, 1987), 249. See also Linker, "Feet for Fighting."

58. Bancroft, *The Posture of School Children*, 108. At Bancroft's invitation, Frank and Lillian Gilbreth became active members of the APL.

59. Bancroft, "Pioneers in Physical Training," 674.

60. Bancroft, "Pioneers in Physical Training," 675.

61. Fairchild, *Science at the Borders*, 83.

62. Alexandra Minna Stern et al., "'Better Off in School': School Medical Inspection as a Public Health Strategy during the 1918–1919 Influenza Pandemic in the United States," *Public Health Reports* 125, supplement 3 (April 2010): 67.

63. Bancroft, *The Posture of School Children*, 284.

64. Bancroft, *The Posture of School Children*, 189.

65. Bancroft, *The Posture of School Children*, specifically 188–200 for a detailed description of her study and results.

66. Bancroft, *The Posture of School Children*, 5.

67. Bancroft, *The Posture of School Children*, 1.

68. Bancroft, *The Posture of School Children*, 3.

69. Joel E. Goldthwait, "An Anatomic and Mechanistic Concept of Disease," *Boston Medical and Surgical Journal* 172, no. 24 (July 1915): 888.

70. McKenzie, "The Development of Physical Efficiency among College Men." McKenzie also wrote that the students who were "recruited [to college] from the farm, the shop, the office, the factory and the night school" universally suffered from "round or crooked backs, narrow, flat chests and flabby muscles." R. Tait McKenzie, "The Regulation of Physical Instruction in Schools and Colleges from the Standpoint of Hygiene," *Science* 29, no. 743 (March 1909): 482.

71. Hackley, *The Colored Girl Beautiful*, 104. For more on the history of tuberculosis, especially as it affected African Americans, see Samuel Roberts, *Infectious Fear: Politics, Disease, and the Health Effects of Segregation* (Chapel Hill: University of North Carolina Press, 2009).

72. Irving Fisher, *How to Live: Rules for Healthful Living, Based on Modern Science* (New York: Funk & Wagnalls, 1915), 57.

73. Bancroft, *The Posture of School Children*, 185.

74. Bancroft, *The Posture of School Children*, 207–8. There are several histories about the rise of medical statistics and the establishment of numerical norms during this time period. For one of the best recent examples of such scholarship, see Bouk, *How Our Days Became Numbered*.

75. Bancroft, *The Posture of School Children*, 208, emphasis in original.

76. For a few examples, see Eliza M. Mosher, "Habits of Posture: A Cause of Deformity and Displacement of the Uterus," *New York Journal of Gynaecology and Obstetrics* 2, no. 13

(November 1893): 962–77; Eliza M. Mosher, "Faulty Habits of Posture, a Cause of Enteroptosis," *Woman's Medical Journal* 25, no. 2 (February 1915): 27–30. Pelvimetry became popular at this time, as well. For an account of this history and its role in perpetuating race science, see Elizabeth O'Brien, "Pelvimetry and the Persistence of Racial Science in Obstetrics," *Endeavour* 37, no. 1 (March 2013): 21–28.

77. For demographics of the APL, see Jessie Bancroft, New York, NY, to R. Tait McKenzie, Philadelphia PA, May 25, 1917, Box 4, Folder 76, R. Tait McKenzie Papers, University of Pennsylvania Archives, Philadelphia. Many of the physicians involved with the APL worried about the surgicalization of medical care in addition to the limits of the new laboratory sciences. In the absence of highly effective pharmaceuticals, surgery became a more popular curative modality. Abdominal surgeries surged in popularity at the time. Appendectomies and ovariectomies were also common, and so too were "pexies," otherwise known as suspensions whereby the kidneys, uterus, liver, and stomach would be lifted into their "proper" positions, locations dictated by the medical belief in organ fixity at the time. For more on the rise of abdominal surgeries, see Regina Morantz-Sanchez, *Conduct Unbecoming a Woman: Medicine on Trial in Turn-of-the-Century Brooklyn* (New York: Oxford University Press, 1999); Dale C. Smith, "Appendicitis, Appendectomy, and the Surgeon," *Bulletin of the History of Medicine* 70, no. 3 (Fall 1996): 414–41.

78. Clelia Duel Mosher, "The Schematogram: A New Method of Graphically Recording Posture and Changes in the Contours of the Body," *School and Society* 1, no. 18 (May 1915): 642–45.

79. For more about Mosher and her commitment to the posture sciences, see Beth Linker, "Tracing Paper, the Posture Sciences, and the Mapping of the Female Body," in *Working with Paper: Gendered Practices in the History of Knowledge*, ed. Carla Bittel, Elaine Leong, and Christine von Oertzen (Pittsburgh: University of Pittsburgh Press, 2019), 124–39.

80. Linker, "Tracing Paper." While scholars have addressed practices of mapping epidemic diseases, there has been very little work on how visual mapping operates in the context of non-infectious epidemics. One of the classic examples of disease mapping is John Snow's Broad Street pump. Or consider Alice Hamilton's maps of typhoid fever in Chicago in 1903. In these examples, incidence and outbreak are represented by dots on a city map. The city becomes the bounded "body" within which disease is represented; the logic motivating the creation of these maps is to track down vectors, to evaluate disease spread, and to contain, if possible, further spread of the contagion. Mosher's schematograms operate slightly differently, but with a similar end goal. With the poor posture epidemic, the pathology is picked up by seeing bad posture in other people and then taking up that faulty comportment for oneself. The schematogram thus mapped the contours and topography of the human form, correcting it on paper and in the clinic to stop the social contagion of bad posture from spreading.

81. Lloyd T. Brown, "A Combined Medical and Postural Examination of 746 Adults," *American Journal of Orthopedic Surgery* 15, no. 11 (November 1917): 775.

82. Clelia and Eliza corresponded with one another about the details of their posture work. See, in particular, Eliza Mosher, Yonkers, NY, to Clelia Mosher, Stanford, California, August 31, 1921, Box 4, vol. 4, Clelia D. Mosher Papers, Stanford University Special Collections, Stanford.

83. H. Ling Taylor (Sec. APL), New York, to Members of the American Posture League, January 10, 1916, B MS c 81.4, File "American Posture League, Report of the Executive Committee," Zabdiel Boylston Adams (1874–1940) Papers, Harvard University, Countway Library of Medicine, Boston.

84. By the 1930s, several articles and advertisements appeared in the *Journal of Health and Physical Education* that indicated widespread usage of Clelia's device. For example, one advertisement claimed that the schematograph was used at the University of Colorado, University of Oregon, Iowa State University, University of California, Barnard College, Vassar College, Springfield High School, Syracuse University, and Columbia University Teachers College. "The Mosher-Lesley Schematograph," Advertisement, *Journal of Health and Physical Education* 4, no. 3 (March 1933): 71.

85. Lloyd T. Brown, "The Harvard Slouch," *New York Times*, March 18, 1917, T8. Before this study, Brown published on the treatment of so-called floating kidneys with posture exercises. See Hugh Cabot and Lloyd T. Brown, "Treatment of Movable Kidney, with or without Infection, with Posture," *Boston Medical and Surgical Journal* 171, no. 10 (September 1914): 369–73. Brown was also part of the Posture Committee at Smith College at this time.

86. Brown, "A Combined Medical and Postural Examination of 746 Adults," 776.

87. Clarence L. Cole, E. W. Loomis, and Eugie A. Campbell, "A Report of Physical Examinations of Twenty Thousand Volunteers," *Military Surgeon* 43, no. 1 (July 1918): 45–64.

88. Cole, Loomis, and Campbell, "A Report of Physical Examinations of Twenty Thousand Volunteers." See also Linker, "Feet for Fighting."

89. James T. Rugh, "Foot Prophylaxis in the Soldier," *American Journal of Orthopedic Surgery* 16, no. 8 (August 1918): 529–37; James T. Rugh, "The Foot of the American Soldier and Its Care," *Pennsylvania Medical Journal* 22, no. 1 (January 1919): 198–205.

90. Joel E. Goldthwait, *The Division of Orthopaedic Surgery in the A.E.F.* (Norwood, MA: Plimpton Press 1941), 47.

91. It was with the Great War that flat feet became a disability that could preclude a man from service. See Linker, "Feet for Fighting."

92. For a description of the camps, see Leah C. Thomas and Joel E. Goldthwait, *Body Mechanics and Health* (Boston: Houghton Mifflin, 1922), 61.

93. "Fashionable Slouch to Go if Posture League Succeeds," *New York Times*, April 5, 1914, SM5.

94. For an excellent account of the federal attempt to improve children's health at this time, see Cynthia A. Connolly and Janet Golden, "'Save 100,000 Babies': The 1918 Children's Year and Its Legacy," *American Journal of Public Health* 108, no. 7 (July 2018): 902–7.

95. "Alert Doctor Examines 1,200 Immigrants a Day: How Dr. J. W. Schereschewsky Accomplishes a Seemingly Impossible Task," *The Sun*, December 22, 1907, 15.

96. "Alert Doctor Examines 1,200 Immigrants a Day," 15.

97. Joseph W. Schereschewsky, "Some Physical Characteristics of Male Garment-Workers of the Cloak and Suit Trades," *American Journal of Public Health* 5, no. 7 (July 1915): 602. The study was conducted on 1,000 female and 2,000 male workers, but he only tabulated the results of the male workers. See "Corsets," *Women's Wear* 11, no. 57 (September 1915): 3. See also "Labor Notes," *Women's Wear* 11, no. 93 (October 1915): 2.

98. Joseph W. Schereschewsky, "Medical Inspection of Schools," *Public Health Reports* 28, no. 35 (August 1913): 1791.

99. United States Public Health Service (USPHS), *Studies in Physical Development and Posture*, Bulletin no. 199 (Washington, DC: Government Printing Office, 1931), 2.

100. USPHS, *Studies in Physical Development and Posture* (1931), 2.

101. USPHS, *Studies in Physical Development and Posture* (1931), 3. See also USPHS, *Studies in Physical Development and Posture*, Bulletin no. 179 (Washington, DC: Government Printing Office, 1928), iv.

102. USPHS, *Studies in Physical Development and Posture* (1931), 7.

103. USPHS, *Studies in Physical Development and* Posture (1931), 10. The specific results of a sample of 215 subjects were: 6 excellent (2.79 percent), 35 good (16.27 percent), 77 fair (35.81 percent), 75 poor (34 percent), 21 very poor (9.7 percent).

104. E. Blanche Sterling, "The Posture of School Children in Relation to Nutrition, Physical Defects, School Grade, and Physical Training," *Public Health Reports (1896–1970)* 37, no. 34 (August 1922): 2043–49; E. Blanche Sterling, "Health Studies of Negro Children: II. The Physical Status of the Urban Negro Child: A Study of 5,170 Negro School Children in Atlanta, Ga.," *Public Health Reports (1896–1970)* 43, no. 42 (October 1928): 2713–74.

105. For a sample of his take on posture, see Williams, "Dr. A. Wilberforce Williams Talks on Preventive Measures, First Aid Remedies Hygienics and Sanitation: Posture—A Factor in Health," *Chicago Defender*, March 5, 1921, 12. For a brief account of Williams's work at the newspaper, see Mary Stovall, "The 'Chicago Defender' in the Progressive Era," *Illinois Historical Journal* 83, no. 3 (Autumn 1990): 159–72.

106. For a history of the Children's Bureau, see Kriste Lindenmeyer, *A Right to Childhood: The U.S. Children's Bureau and Child Welfare, 1912–46* (Urbana: University of Illinois Press, 1997).

107. Armin Klein and Leah C. Thomas, *Posture and Physical Fitness*, United States Department of Labor, Children's Bureau Publication no. 205 (Washington, DC: Government Printing Office, 1931); Armin Klein and Leah C. Thomas, *Posture Exercises: A Handbook for Schools and for Teachers of Physical Education*, United States Department of Labor, Children's Bureau Publication no. 165 (Washington, DC: Government Printing Office, 1926); Armin Klein, *Posture Clinics: Organization and Exercises*, United States Department of Labor, Children's Bureau Publication no. 164 (Washington, DC: Government Printing Office, 1926); United States Children's Bureau, "No. 6. Your Child's Posture," in *Lesson Material on Care of the Preschool Child: No. 1–9* (Washington, DC: Government Printing Office, 1928), which was succeeded by *Good Posture in the Little Child*, United States Department of Labor, Children's Bureau Publication no. 219 (Washington, DC: Government Printing Office, 1933).

108. While conducting their research in Chelsea, Klein and Thomas helped create the education film "Posture," Worchester Film Corporation, U.S. Children's Bureau, 1926.

109. Klein and Thomas, *Posture and Physical Fitness*, 41.

110. For example, see Jeanette L. Sturges, Redlands Children's Postural Clinic, Redlands, California, to Grace Abbott, Chief of U.S. Children's Bureau, January 18, 1927; Harvey L. Long, Director of Physical Education of Public Schools, Lincoln, Nebraska, to Grace Abbott, January 7, 1927, Box 272, Record Group 102, File 1925–28, 46711, National Archives and Records Administration, College Park, Maryland.

111. Beaufort S. Parsons, East Falls Church, Virginia, to Grace Abbott, Chief of U.S. Children's Bureau, April 20, 1925, Box 272, Record Group 102, File 1925–28, 46711, National Archives and Records Administration, College Park, Maryland.

112. United States Children's Bureau, "No. 6. Your Child's Posture," and *Good Posture in the Little Child*.

113. Osgood, *Body Mechanics*, 41.

114. Osgood, *Body Mechanics*, 24.

115. Herbert Hoover, "Address of President Hoover," in *White House Conference 1930: Address and Abstract of Committee Reports* (New York: Century, 1930), 10.

116. Martha H. Verbrugge, *Active Bodies: A History of Women's Physical Education in Twentieth-Century America* (New York: Oxford University Press, 2012), 3. See also Rachel Louise Moran, *Governing Bodies: American Politics and the Shaping of the Modern Physique* (Philadelphia: University of Pennsylvania Press, 2018).

117. For more on the history of university students as research subjects, see Heather Munro Prescott, *Student Bodies: The Influence of Student Health Services in American Society and Medicine* (Ann Arbor: University of Michigan Press, 2007). For the growth of university physical education department, see Verbrugge, *Active Bodies*, and Moran, *Governing Bodies*.

118. Popular magazines and journals often made reference to Metropolitan Life's posture exam. For one example, see Helen Durham and Barbara Beattie, "Are You Graceful or Awkward?," *Ladies' Home Journal* 49, no. 1 (January 1932): 24, 62. For some examples of Met Life brochures on posture, see *The Importance of Posture* (New York: Metropolitan Life Insurance Co., 1927); *Posture from the Ground Up* (New York: Metropolitan Life Insurance Co., 1939); *Standing Up to Life: Good Posture and Foot Health* (New York: Metropolitan Life Insurance Co., 1954). Met Life relied heavily on nurses to conduct many of these physical exams. See Diane Hamilton, "Cost of Caring: The Metropolitan Life Insurance Company's Visiting Nurse Service, 1909–1953," *Bulletin of the History of Medicine* 63, no. 3 (Fall 1989): 414–34; Patricia D'Antonio, *Nursing with a Message: Public Health Demonstration Projects in New York City* (New Brunswick, NJ: Rutgers University Press, 2017).

Chapter Three: Posture Commercialization

1. Charles E. Rosenberg, "What Is an Epidemic? AIDS in Historical Perspective," in *Explaining Epidemics and Other Studies in the History of Medicine* (Cambridge: Cambridge University Press, 1992), 278–92. See also Priscilla Wald, *Contagious: Cultures, Carriers, and the Outbreak Narrative* (Durham, NC: Duke University Press, 2008.

2. Rosenberg, "What Is an Epidemic?," 279.

3. See Elizabeth Toon, "Managing the Conduct of the Individual Life: Public Health Education and American Public Health, 1910 to 1940" (PhD dissertation, University of Pennsylvania, 1998); Nancy Tomes, "Merchants of Health: Medicine and Consumer Culture in the United States, 1900–1940," *Journal of American History* 88, no. 2 (September 2001): 519–47; Nancy Tomes, *Remaking the American Patient: How Madison Avenue and Modern Medicine Turned Patients into Consumers* (Chapel Hill: University of North Carolina Press, 2016).

4. For more on Foucault's theory about docile bodies—and more specifically the practice of docility-utility, see Michel Foucault, *Discipline and Punish: The Birth of the Prison,* translated by

Alan Sheridan (New York: Pantheon Books, 1977), especially 135ff. According to Foucault, body work began with in the eighteenth-century military, an en masse pursuit with military officers disciplining the soldier's body, molding a piece of formless clay into a machine with high heads and erect posture. By the nineteenth century, he argues, body work came under the purview of "the disciplines," professionals who engaged in so-called "docility-utility," coercively manipulating bodies to become more economically efficient and to promote "internal organization."

5. "Finds Chairs Add to the Ills of Man: Dr. Eliza M. Mosher Tells Posture League Few Persons Sit in Correct Positions," *New York Times*, March 14, 1915, 9.

6. Witold Rybczynski, *Now I Sit Me Down: From Klismos to Plastic Chair: A Natural History* (New York: Farrar, Straus, and Giroux, 2016), 54.

7. Marcel Mauss, "Techniques of the Body," *Economy and Society* 2, no 1 (February 1973): 77.

8. Dudley Allen Sargent, "The Physical Test of a Man," *American Physical Education Review* 26, no.1 (April 1921): 188.

9. George J. Fisher, "The American Posture League," *Journal of Health and Physical Education* 6, no. 8 (October 1935): 16–17.

10. The most detailed account of the working relationship between industry and the APL can be found in B MS c 81.2, File "American APL," Zabdiel Boylston Adams Papers, Countway Library of Medicine, Harvard University, Boston. Adams was in charge of the "shoe committee" for the APL, and his papers contain correspondence with both the League and dozens of shoe manufacturers who sought the League's endorsement.

11. E. H. Bradford and J. S. Stone, "The Seating of School Children," *Transactions of the American Orthopedic Association: Thirteenth Session*, vol. 12 (Philadelphia: American Orthopedic Association, 1899), 170–83. The authors provide a history of physician-based research into school seating dating back to the early nineteenth century.

12. Robin Veder, *The Living Line: Modern Art and the Economy of Energy* (Hanover, NH: Dartmouth College Press, 2015), 190–93. See also M. Uribe y Troncoso, "Influence of Different Kinds of Handwriting on the Hygienic Posture and Deformities of School Children," *Public Health Papers and Reports* 31, pt. 1 (1905): 182–86. For more on the history of handwriting, see Tamara Plakins Thornton, *Handwriting in America: A Cultural History* (New Haven, CT: Yale University Press, 1996).

13. Bradford and Stone, "The Seating of School Children," 175.

14. R. Tunstall Taylor, "Lateral Curvature of the Spine," *American Physical Education Review* 9, no. 3 (September 1904): 192.

15. Rodris Roth, "Nineteenth-Century American Patent Furniture," in *Innovative Furniture in America from 1800 to the Present*, ed. David A. Hanks (New York: Horizon Press, 1981), 23–46.

16. Sigfried Giedion, *Mechanization Takes Command: A Contribution to Anonymous History* (New York: Oxford University Press, 1948), 393.

17. And yet it is important to unmask such histories since, as disability and design historian Bess Williamson demonstrates, accessible design is inherently inclusive. Adaptive features often enhance user experience, allowing for individual variations, assisting disabled and nondisabled bodies alike. Bess Williamson, *Accessible America: A History of Disability and Design* (New York:

New York University Press, 2019). See also Christina Cogdell, *Eugenic Design: Streamlining America in the 1930s* (Philadelphia: University of Pennsylvania Press, 2004).

18. Frank Bunker Gilbreth and Lillian Moller Gilbreth, *Fatigue Study, The Elimination of Humanity's Greatest Unnecessary Waste* (New York: Sturgis and Walton, 1916), 100.

19. Gilbreth and Gilbreth, *Fatigue Study*, 99.

20. Daniel E. Bender, *Sweated Work, Weak Bodies: Anti-Sweatshop Campaigns and Languages of Labor* (New Brunswick, NJ: Rutgers University Press, 2004).

21. Henry Moskowitz, "The Joint Board of Sanitary Control in the Cloak, Suit and Skirt Industry of New York City," *Annals of the American Academy of Political and Social Science* 44 (November 1912): 39–58.

22. Nancy Tomes, *The Gospel of Germs: Men, Women, and the Microbe in American Life* (Cambridge, MA: Harvard University Press, 1998), 212–20. See also Bender, *Sweated Work, Weak Bodies*.

23. George Price, "Joint Board of Sanitary Control: Ten Years of Progress," in *Ten Years of Industrial Sanitary Self Control: Tenth Annual Report of the Joint Board of Sanitary Control in the Cloak, Suit, and Skirt and Dress and Waist Industries* (New York: Joint Board of Sanitary Control, 1921), 13.

24. Theresa Wolfson, "Health Education," in *Eleventh Annual Report of the Joint Board of Sanitary Control in the Cloak, Suit, and Skirt and Dress and Waist Industries* (New York: Joint Board of Sanitary Control, 1922), 19–21.

25. Theresa Wolfson, "Health Education," 21.

26. "Posture at Work: Influence of Occupation and Height of Furniture upon the Chest and Spine," *Iron Age* 99, no. 15 (April 1917): 930.

27. Harold Mestre, "Seating of Women and Minors in the Fruit and Vegetable Canning Industry of California," *Industrial Welfare Commission*, State of California, Bulletin no. 2a (Sacramento: California Printing Office, 1919), 141.

28. Hilles, Edith, and Wilhelmina Conger. "Attempts to Standardize Seating in Industry." In *Industrial Posture and Seating*, prepared by the Bureau of Women in Industry, State of New York Department of Labor, Special Bulletin 104 (April 1921): 40.

29. Hilles and Conger, "Attempts to Standardize Seating in Industry," 42.

30. Minnesota, Ohio, Kansas, and eventually New York were among the first to require seat backs. See Hilles and Conger, "Attempts to Standardize Seating in Industry," 36.

31. Frances Perkins, foreword to *Industrial Posture and Seating*, prepared by the Bureau of Women in Industry, State of New York Department of Labor, Special Bulletin 104 (April 1921): 1.

32. Hilles and Conger, "Attempts to Standardize Seating in Industry."

33. Only three states at the time—Minnesota, Ohio, and New York—had laws that required work chairs to have back rests. See Nelle Swartz, "Industrial Posture and Seating," in *Proceedings of the National Safety Council, Tenth Annual Safety Congress* (Chicago: National Safety Council, 1921), 845–51.

34. George Price, "Defective Seating, Faulty Posture Health," in *Twelfth Annual Report of the Joint Board of Sanitary Control in the Cloak, Suit, and Skirt and Dress and Waist Industries* (New York: Joint Board of Sanitary Control, 1923), 35–36.

35. For more on the American Windsor chair, see Rybczynski, *Now I Sit Me Down*, 87–100.

36. Joel E. Goldthwait, "The Importance of Correct Furniture to Assist in the Best Body Function, as Recognized by the Massachusetts Institute of Technology and Smith College," *Journal of Bone and Joint Surgery* 5, no. 2 (April 1923): 179–84. See also "Corrective Posture Chair, Plimpton Scofield Co.," *Official Gazette of the United States Patent Office* 295 (February 1922): 457.

37. Goldthwait, "The Importance of Correct Furniture."

38. Arthur B. Emmons and Joel E. Goldthwait, "A Work Chair," *Journal of Industrial Hygiene* 3, no. 5 (September 1921): 154.

39. Theresa Wolfson, "Seating Survey in the Garment Industry," *Nation's Health* 5, no. 3 (March 1923): 165–68.

40. The German Bureau of Economy and Efficiency came to similar conclusions, designing work chairs similar to that designed by Goldthwait and his colleagues. For more, see Jennifer Karns Alexander, "Efficiency and Pathology: Mechanical Discipline and Efficient Worker Seating in Germany, 1929–1932," *Technology and Culture* 47, no. 2 (April 2006): 286–310.

41. John Daly McCarthy, *Health and Efficiency* (New York: Henry Holt, 1922), 50.

42. "Gives Benches for Parks: APL Presents Seats Built upon Hygienic Lines," *New York Times*, March 11, 1917, 14.

43. Henry Eastman Bennett, an efficiency engineer who began his career at the College of William and Mary, was hired by the American Seating Company in the early 1920s. Bennett offers a description of his and the company's work in Henry Eastman Bennett, "Some Requirements of Good School Seating," *Elementary School Journal* 23, no. 3 (November 1922): 203–14.

44. Arthur B. Emmons, "Organized Preventive Medicine Is Nowhere More Effectual than as Applied to Industrial Groups," *Nation's Health* 6, no. 1 (January 1924): 8–9, 74.

45. Emmons and Goldthwait, "A Work Chair."

46. Alexander, "Efficiency and Pathology," 309.

47. Edward Tenner, *Our Own Devices: How Technology Remakes Humanity* (New York: Alfred A. Knopf, 2003), especially chapter 5.

48. For more on the history of flat feet, see Beth Linker, "Feet for Fighting: Locating Disability and Social Medicine in First World War America," *Social History of Medicine* 20, no. 1 (April 2007): 91–109.

49. Henry Ling Taylor, American APL, New York, to Zabdiel Boylston Adams, Boston, November 26, 1913 in Box "American APL," File "1913–1914," Zabdiel Boylston Adams Papers, Countway Library of Medicine, Harvard University, Boston.

50. "Inventor's Profile: Charles F. Brannock," Smithsonian Institution, accessed August 13, 2020, https://invention.si.edu/sites/default/files/Kid-friendly-Inventor-Profile-Charles-Brannock.pdf.

51. The superintendent of Boston City Hospital had begun to keep records on the number of nursing workdays lost due to foot complaints. In 1892, an aggregate of forty-two days was lost. In 1893, 125 days, and so on.

52. Robert W. Lovett, "The Occurrence and Prevention of Flat-Foot among City Hospital Nurses," *Medical and Surgical Reports of the Boston City Hospital* 7, no. 1 (1896): 94.

53. Despite all his research, Lovett could not pin down what caused foot disability. He concluded that "the only reliable information obtained in these cases was given by the imprints seen through glass. A foot with a well distributed pressure area seemed rather less likely to give trouble than one resting on two islands. . . . A flat foot may be perfectly serviceable, as may also a severely pronated

one, while an apparently well-balanced foot may become painful." Robert W. Lovett, "Occurrence of Painful Affections of the Feet among Trained Nurses," *Journal of Bone and Joint Surgery* s2-1, no. 1 (August 1903): 60. For a history of how flat feet became a disability, see Linker, "Feet for Fighting."

54. Lovett, "Occurrence of Painful Affections of the Feet," 54.

55. William H. Mulligan Jr., "Mechanization and Work in the American Shoe Industry: Lynn, Massachusetts, 1852–1883," *Journal of Economic History* 41, no. 1 (March 1981): 59–63.

56. Frank P. Aborn, Aborn and Co., Lynn, MA, to Zabdiel Boylston Adams, Massachusetts General Hospital, Boston, March 18, 1914, File "American APL," Zabdiel Boylston Adams Papers, Countway Library of Medicine, Harvard University, Boston.

57. Aborn to Boylston Adams, March 18, 1914.

58. For an excellent discussion of the dialectic between decadence and primitivism, see Elisa F. Glick, "Harlem's Queer Dandy: African-American Modernism and the Artifice of Blackness," *Modern Fiction Studies* 49, no. 3 (Fall 2003): 414–42.

59. Joe Marr, "Growing Interest in Foot Remedies," *Boot and Shoe Recorder* (January 22, 1916): 64.

60. "The Orthopedic Department," *Boot and Shoe Recorder* (January 22, 1916): 75–76.

61. See, for example, a series of articles authored by orthopedic surgeon, Dr. Herman W. Marshall, the first of which was "What Do You Know about Feet?," *Boot and Shoe Recorder* (April 15, 1922): 108–10.

62. Nancy Rexford, *Women's Shoes in America, 1795–1930* (Kent, OH: Kent State University Press, 2000).

63. "Something New and Vital in Shoes," Display Advertisement, *New York Times*, March 30, 1919, 41.

64. "At Last—a 'Human' Shoe!," Display Advertisement, *American Physical Education Review* 25, no. 1 (January 1920): 127.

65. Glick, "Harlem's Queer Dandy."

66. Henry B. Scates, Boston, to Zabdiel Boylston Adams, Boston, March 16, 1914, Box "American APL," File "1913 and 1914," Zabdiel Boylston Adams Papers, Harvard University, Countway Library of Medicine, Boston.

67. For an example of Buster Brown and its promise of posture exams in the African American press, see "Buster Brown," Display Advertisement 10, *New York Amsterdam News*, September 16, 1950, 3. For an account of labor conditions at the Brown Shoe Company, see "Brown Shoe Company, Inc. History," *International Directory of Company Histories*, vol. 68 (London: St. James Press, 2005), http://www.fundinguniverse.com/company-histories/brown-shoe-company-inc-history /. For the Brown Shoe Company's employment and exploitation of little people, see Jerry Maren, who played one of the "munchkins" in *The Wizard of Oz* film and worked for the Brown Shoe Company. See Stephen Cox, *The Munchkins of Oz*, 3rd ed. (Nashville: Cumberland House, 2002), 107. For more on how shoe stores at this time appealed to science and children, especially through the fluoroscope (X-ray technology applied in shoe fitting), see Jacalyn Duffin and Charles R. R. Hayter, "Baring the Sole: The Rise and Fall of the Shoe-Fitting Fluoroscope," *Isis* 91, no. 2 (June 2000): 260–82.

68. Edith Abbott, "Women in Industry: The Manufacture of Boots and Shoes," *American Journal of Sociology* 15, no. 3 (November 1909): 335–60.

69. "Retail Shoe Women's Symposium," *Boot and Shoe Recorder* (May 13, 1922): 78–80, quote on 79. For percent increase in women workers, see "Retail Shoe Women's Symposium," 80.

70. Antioch College, *The Effects of Modern Shoes upon Proper Body Mechanics, 1924–1931* (Yellow Springs, OH: Antioch College, 1931). According to this report, Antioch Shoes became a mainstay in several hospitals where superintendents of nursing reported improved efficiency and less foot trouble among student nurses. The cost of the shoes ranged from $7.25 to $10.50. Antioch was concerned with making an "economical" shoe, but not a "cheap" one that would break down and need frequent replacement. The study estimated that student nurses needed to replace their Antioch Shoes approximately twice per year.

71. See "Forget Your Feet Because Antioch Didn't," Display Advertisement, *Saturday Review,* October 24, 1942, 47. See also Antioch College's pamphlet "Walk in Beauty" (Yellow Springs, OH: Antioch College, 1941).

72. L. E. La Fetra, "The Relation of Clothing to Posture," *Proceedings of the Ninth Congress of the American School Hygiene Association*, vol. 6 (New York: American School Hygiene Association, 1917), 117–21.

73. Michael Hau, *The Cult of Hygiene and Beauty in Germany: A Social History, 1890–1930* (Chicago: University of Chicago Press, 2003).

74. "Boys' Right Posture Suits Reduced," Display Advertisement, *Chicago Daily Tribune,* May 29, 1920, 28.

75. Julia Grant, *The Boy Problem: Educating Boys in Urban America, 1870–1970* (Baltimore: Johns Hopkins University Press, 2014).

76. Ingrid Loschek, "Twentieth Century Fashion," in *Encyclopedia of Clothing and Fashion*, ed. Valerie Steele, vol. 3 (Detroit: Charles Scribner's Sons, 2005), 348.

77. JoAnne Olian, ed., *Children's Fashions, 1900–1950s, as Pictured in Sears Catalogs* (Mineola, NY: Dover Publications, 2003), especially the introduction.

78. For but one example of this, see Dr. Luther Emmett Holt, who was the author of one of the best-selling parent advice books at the time. Holt, *Food, Health, and Growth: A Discussion of the Nutrition of Children* (New York: MacMillan, 1922), 226.

79. See Jane Farrell-Beck and Colleen Gau, *Uplift: The Bra in America* (Philadelphia: University of Pennsylvania Press, 2002).

80. Farrell-Beck and Gau, *Uplift.*

81. Alice S. Cutler, "True Principles of Scientific Corset Fitting," *Corset and Underwear Review* 10, no. 5 (February 1918): 55.

82. Emphasis in original. Albert M. Judd, "Body Posture and Poise—Its Effect on the General Health," *Long Island Medical Journal* 18, no. 2 (February 1924): 41–47, quote on 47.

83. Regina Morantz-Sanchez, *Conduct Unbecoming a Woman: Medicine on Trial in Turn-of-the-Century Brooklyn* (New York: Oxford University Press, 1999).

84. Dale C. Smith, "Appendicitis, Appendectomy, and the Surgeon," *Bulletin of the History of Medicine* 70, no. 3 (Fall 1996): 414–41.

85. Eliza M. Mosher, "Faulty Habits of Posture, a Cause of Enteroptosis," *Woman's Medical Journal* 25, no. 2 (February 1915): 27–30.

86. "Dr. Joel Goldthwait Speaks on Dress: Physical Director of Smith College Urges Health through Proper Clothing," *Wellesley College News* 29, no. 24 (April 20, 1921): 1, 4.

87. Valerie Steele, *The Corset: A Cultural History* (New Haven, CT: Yale University Press, 2001), especially chapter 6.

88. "College Girl Corsets," Display Advertisement, *Corset and Underwear Review* 17, no. 1 (April 1921): 39.

89. "College Girl Corsets," Display Advertisement, *Corset and Underwear Review* 17, no. 2 (May 1921): 17. By 1924, according to historian Jill Fields, "the elastic step-in girdles were sold in corset departments nationwide." Jill Fields, *An Intimate Affair: Women, Lingerie, and Sexuality* (Berkeley: University of California Press, 2007), 74.

90. "Women Who Are Inventors," *New York Times*, October 19, 1913, X11.

91. "Women Who Are Inventors."

92. Jill Fields, "'Fighting the Corsetless Evil': Shaping Corsets and Culture, 1900–1930," *Journal of Social History* 33, no. 2 (Winter 1999): 355–84, 369. See also Steele, *The Corset*, 153.

93. "'Stand Up Straight' Theme of Booklet Issued by H. W. Gossard—Hints to Growing Girls," *Women's Wear*, January 13, 1923, 6. For more on the history of menstruation and the commercialization of hygiene products, see Joan Jacobs Brumberg, *The Body Project: An Intimate History of American Girls* (New York: Random House, 1997); Lara Freidenfelds, *The Modern Period: Menstruation in Twentieth-Century America* (Baltimore: Johns Hopkins University Press, 2009).

94. "What Paris Really Thinks about Corsets," *Corset and Underwear Review* 18, no. 1 (October 1921): 77–78.

95. Fields, "'Fighting the Corsetless Evil,'" 370.

96. Alphonsus P. Haire, "Medical Corsetry and Scientific Fitting," *Corset and Underwear Review* 18, no. 2 (November 1921): 71–72.

97. Fields, "'Fighting the Corsetless Evil,'" 372.

98. For public trust, see Paul Starr, *The Social Transformation of American Medicine* (New York: Basic Books, 1982). For a brief history of chiropractic and the AMA, see Steve Agocs, "Chiropractic's Fight for Survival," *AMA Journal of Ethics* 13, no. 6 (June 2011): 384–88.

99. For the National Association of Negro Tailors, see "Fashion Show at Dunbar," *Philadelphia Tribune*, April 16, 1921, 2.

100. Amelia Wilson, "Presto! Bad Posture Gone," *Philadelphia Tribune*, March 14, 1940, 9.

101. "City Shopper: You Can Improve Your Posture!," *New York Amsterdam News*, November 6, 1943, 11B.

102. Jeffrey T. Schnapp, "Crystalline Bodies: Fragments of a Cultural History of Glass," *West 86th: A Journal of Decorative Arts, Design History, and Material Culture* 20, no. 2 (Fall–Winter 2013): 173–94. For a contemporary perspective on the statue's arrival to North America, see John Lardner, "You Can See through Her," *Los Angeles Times*, August 29, 1936, 12.

103. "Fitness Program Seen Adding Importance to Posture Week," *Women's Wear Daily* 70, no. 57 (March 1945): 17. Other industrialists attributed Samuel Camp's success to his insistence on an "education before sales" approach to marketing and selling. See Baert D. Brand, "Firm Marks 50th Year Fighting Appearance Defects," *Women's Wear Daily* 97, no. 17 (July 1958): 25.

104. "Samuel Higby Camp Institute for Better Posture Formed," *Women's Wear Daily* 63, no. 76 (October 1941): 16.

105. "Television Show Launches National Posture Week," *Women's Wear Daily* 70, no. 92 (May 1945): 44.

106. Matthew H. Hersch, "High Fashion: The Women's Undergarment Industry and the Foundations of American Spaceflight," *Fashion Theory* 13, no. 3 (2009): 345–70.

107. "Four Million Men Wearing Girdles," *New Journal and Guide*, January 20, 1951, B1.

Chapter Four: Posture Queens and Fitness Regimes

1. For more on Victorian- and Progressive-era fitness, see Harvey Green, *Fit for America: Health, Fitness, Sport, and American Society* (New York: Pantheon Books, 1986); John F. Kasson, *Houdini, Tarzan, and the Perfect Man: The White Male Body and the Challenge of Modernity in America* (New York: Hill and Wang, 2001); Clifford Putney, *Muscular Christianity: Manhood and Sports in Protestant America, 1880–1920* (Cambridge, MA: Harvard University Press, 2001); Martha H. Verbrugge, *Able-Bodied Womanhood: Personal Health and Social Change in Nineteenth-Century Boston* (New York: Oxford University Press, 1988).

2. Nikolas Rose, "The Neurochemical Self and Its Anomalies," in *Risk and Morality*, ed. Richard V. Ericson and Aaron Doyle (Toronto: University of Toronto Press, 2003), 407–37.

3. Carolyn Thomas de la Peña, *The Body Electric: How Strange Machines Built the Modern American* (New York: New York University Press, 2003), 55.

4. Thomas de la Peña, *The Body Electric*, 56.

5. Dudley Allen Sargent, "Is the Teaching of Physical Training a Trade or a Profession?," *Proceedings of the American Association for the Advancement of Physical Education*, Sixth Annual Meeting (Ithaca, NY: Andrus & Church, 1891), 6–24; pages 19–24 include commentary from Sargent's colleagues.

6. Both Hartwell and Wood come from *Proceedings of the American Association for the Advancement of Physical Education*, 1891, 22–24, as quoted in Roberta J. Park, "Science, Service, and the Professionalization of Physical Education, 1885–1905," *International Journal of the History of Sport* 24, no. 12 (December 2007): 1677.

7. Park, "Professionalization of Physical Education," 1679.

8. See Kasson, *Houdini, Tarzan, and the Perfect Man*; Elizabeth Toon and Janet Golden, "Rethinking Charles Atlas," *Rethinking History: Journal of Theory and Practice* 4, no. 1 (2000): 80–84.

9. Jessie H. Bancroft, *School Gymnastics, Free Hand: A System of Physical Exercises for Schools* (New York: E. L. Kellogg, 1896), 72.

10. For the toe-in and -out exercise, see Jessie H. Bancroft, *School Gymnastics with Light Apparatus* (Boston: D. C. Heath, 1900), 7.

11. Lillian Curtis Drew, *Individualized Gymnastics: A Handbook of Corrective and Remedial Gymnastics* (Philadelphia: Lea and Febiger, 1922), 22.

12. Drew, *Individualized Gymnastics*, 27.

13. William James, *The Energies of Men* (New York: Moffat, Yard, 1914). See Mark Dyreson, "Nature by Design: Modern American Ideas about Sport, Energy, Evolution, and Republics, 1865–1920," *Journal of Sport History* 26, no. 3 (Fall 1999): 447–69.

14. Judith P. Swazey, *Reflexes and Motor Integration: Sherrington's Concept of Integrative Action* (Cambridge, MA: Harvard University Press, 1969). See also Roger Smith, "Representations of Mind: C. S. Sherrington and Scientific Opinion, c. 1930–1950," *Science in Context* 14, no. 4 (December 2001): 511–39.

15. J. Wayne Lazar, "Problems of Consciousness in Nineteenth Century British and American Neurology," in *Brain, Mind, and Consciousness in the History of Neuroscience*, ed. C.U.M. Smith and Harry Whitaker (New York: Springer, 2014), 158.

16. Lazar, "Problems of Consciousness," 150.

17. W.E.B. Du Bois, "Strivings of the Negro People," *Atlantic Monthly* 80 (August 1897): 194.

18. Mabel Elsworth Todd, "The Balancing of Forces in the Human Being: Its Application to Postural Patterns," in *Early Writings, 1920–1934* (New York: Dance Horizons, 1977), 5. For more on Todd, James, Dewey, and Alexander, see Matt Zepelin, "From Esotericism to Somatics: A History of Mind-Body Theory and Practice Across the Divide of Modernism, 1820s to 1950s" (PhD dissertation, University of Colorado at Boulder, 2018), especially chapter 6.

19. Mabel Elsworth Todd, "Principles of Posture, with Special Reference to the Mechanics of the Hip-Joint," *Boston Medical and Surgical Journal* 184, no. 25 (June 1921): 667–73. See Robin Veder, "The Expressive Efficiencies of American Delsarte and Mensendieck Body Culture," *Modernism/Modernity* 17, no. 4 (November 2010): 819–38.

20. This recollection is not directly attributed. It may come from one of Todd's well-known students, Barbara Clark. See John Rolland, "Mabel Todd: An Introduction to Her Work," *Contact Quarterly* (Fall 1979): 6–7.

21. John Dewey, "Introduction," in *Constructive Conscious Control of the Individual*, by F. Matthias Alexander (New York: E. P. Dutton, 1923), xxi–xxxiii.

22. Michael Huxley, "F. Matthias Alexander and Mabel Elsworth Todd: Proximities, Practices and the Psycho-Physical," *Journal of Dance and Somatic Practices* 3, no. 1/2 (January 2012): 25–36.

23. Zepelin, "From Esotericism to Somatics," 300.

24. F. Matthias Alexander, *Man's Supreme Inheritance: Conscious Guidance and Control in Relation to Human Evolution in Civilization* (New York: E. P. Dutton, 1918), 22.

25. John Dewey, "Reply to a Reviewer," *New Republic* 15 (1918): 55.

26. Joseph Hubertus Pilates, *Your Health: A Corrective System of Exercising that Revolutionizes the Entire Field of Physical Education* (New York: Joseph H. Pilates, 1934), 49.

27. Pilates, *Your Health*, 2.

28. Robin Veder, "Seeing Your Way to Health: The Visual Pedagogy of Bess Mensendieck's Physical Culture System," *International Journal of the History of Sport* 28, no. 8–9 (May 2011): 1336–52.

29. Joel E. Goldthwait, Boston, to William Allan Neilson, Northampton, Box 11, Folder "Goldthwait, Joel, 1918–1935," February 25, 1924, Office of the President, William Allan Neilson Files, Smith College Archives, Smith College, Northampton, Massachusetts.

30. Jean S. McGill, *The Joy of Effort: A Biography of R. Tait McKenzie* (Bewdley, ON: Clay, 1980), 57. After graduating from medical school in 1892, McKenzie continued to study anatomy and perfect his practice in anthropometry by working first as a medical examiner for the Montreal branch of the New York Metropolitan Insurance Company, then as a lecturer in anatomy at McGill, and finally as the first director of physical education at the University of Pennsylvania—a post which he held from 1904 to 1929.

31. See R. Tait McKenzie, "The Proper Development of Physical Power," Lecture at YMCA Montreal, February 1894, Box 6, Folder 22, R. Tait McKenzie Papers, University of Pennsylvania

Archives, Philadelphia. See also R. Tait McKenzie, "The Development of Physical Efficiency among College Men," Speech at Queen's University 1907, Box 6, Folder 43, R. Tait McKenzie Papers, University of Pennsylvania Archives, Philadelphia.

32. "Posture" (Cambridge: Massachusetts Institute of Technology, 1928; London: Wellcome Trust, 2008), video, 13 minutes, https://archive.org/details/Posture-wellcome. For another black-and-white film from the time, see "Educated Feet for Correct Posture," produced by Marcia Midleton and Edward J. Hummel (Beverly Hills, CA: Beverly Hills Unified Schools, n.d.), video, 16 minutes, https://archive.org/details/0973_Educated_Feet_for_Correct_Posture_01_26_19_19.

33. Marie C. Harrington, dir., "Posture and Personality" (Hardcastle Film Associates, 1949), video, 8 minutes, https://archive.org/details/2244_Posture_and_Personality; "Health: Your Posture" (Centro Corporation, 1953), video, 10 minutes, https://archive.org/details/HealthY01953.

34. Much of the information about poster weeks can be found in Drew, *Individualized Gymnastics*. I have also learned a great deal from the Smith Archives. See, for example, "Student Activities Release on Jan 16, 1933," Box 52, File "Hygiene—Posture," Smith College Archives, Smith College, Northampton, Massachusetts.

35. Drew, *Individual Gymnastics*, especially chapter 7.

36. Jessie H. Bancroft, *The Posture of School Children: With Its Home Hygiene and New Efficiency Methods for School Training* (New York: Macmillan, 1913), 208.

37. Mabel Lee, "Views of Parents on the Physical Education Program for Their Daughters," *Journal of Health and Physical Education* (hereafter *JHPE*) 4, no. 4 (1933): 13.

38. Lee, "Views of Parents." By the mid-1930s certain physical educators argued that the stigmatization did grave psychological harm to those children who could not attain A-grade posture. See, for example, E. F. Voltmer, "A Plea for the Less Gifted Physically," *JHPE* 6, no. 5 (1935): 28–29, 61.

39. For firsthand accounts of the stigma that children with curved spines and other visible physical defects felt, see Beth Linker and Emily Abel, "Integrating Disability, Transforming Disease History: Tuberculosis and Its Past," in *Civil Disabilities: Citizenship, Membership, Belonging*, ed. Nancy Hirschmann and Beth Linker (Philadelphia: University of Pennsylvania Press, 2015), 83–102, and Beth Linker, "Spines of Steel: A Case of Surgical Enthusiasm in America," *Bulletin of the History of Medicine* 90, no. 2 (Summer 2016): 222–49. The first half of the twentieth century was a time when cities across the nation adopted municipal laws prohibiting "unsightly" persons from the streets (otherwise known as "ugly laws"). Susan M. Schweik, *The Ugly Laws: Disability in Public* (New York: New York University Press, 2009).

40. A. R. Shands Jr. et al., "End-Result Study of the Treatment of Idiopathic Scoliosis: Report of the Research Committee of the American Orthopaedic Association," *Journal of Bone and Joint Surgery* 23, no. 4 (October 1941): 963–77.

41. D. K (Pennsylvania) to P. R. Harrington (Houston, Texas), December 3, 1960, in "Time Article, 1960, Patient Inquiries," in the Paul R. Harrington Archives (HA), University of Kansas Medical Center Archives, Kansas City, Kansas [hereafter TPI/HA]. For privacy and protection, I have de-identified letter writers. I also removed the letter writers' cities of origin but retained states of origin.

42. J. K. (Oregon) to P. R. Harrington (Houston, Texas), July 28, 1965, in TPI/HA.

43. E. S. (Pennsylvania) to Paul R. Harrington (Houston, Texas), October 23, 1960, in TPI/HA.

44. All quotes come from Faye F. Nixon, "Pushing Posture," *Hygeia* 13 (May 1935): 463–64.

45. Dorothye E. Brock, "Some Practical Ideas about Posture Training," *American Physical Education Review* 28, no. 8 (1923): 366–74, student responses on 371–72.

46. Brock, "Some Practical Ideas," 372.

47. For the ending of posture contests at Barnard, see "Perfect Posture Goes with Brains at Girls' College," *Philadelphia Inquirer*, February 20, 1962, 13.

48. "Unrouged Girl Is Barnard's Best," *New York Times*, December 10, 1925, 5.

49. Lois W. Banner, *American Beauty: A Social History . . . Through Two Centuries of the American Idea, Ideal, and Image of the Beautiful Woman* (New York: Alfred A. Knopf, 1983). See also Elwood Watson and Darcy Martin, eds., *"There She Is, Miss America": The Politics of Sex, Beauty, and Race in America's Most Famous Pageant* (New York: Palgrave Macmillan, 2004).

50. "YWCA Opens War on Beauty Contests," *New York Times*, April 18, 1924, 21.

51. "Unrouged Girl Is Barnard's Best."

52. "Health Is a Game for Modern Girls," *New York Times*, February 26, 1928, 136.

53. Joel E. Goldthwait (Boston) to William Allan Neilson (Northampton), December 6, 1926, Box 11, Folder "Goldthwait, Joel, 1918–1935," Office of the President, William Allan Neilson Files, Smith College Archives, Smith College, Northampton, Massachusetts.

54. Goldthwait to Neilson, February 25, 1924.

55. "Japanese Girl Winner in College Contest," *Chicago Defender*, March 20, 1926, A1.

56. "Photo Standalone 2" in "Indian Maid Wins Contest," *China Press*, September 17, 1930, A2.

57. Maxine Leeds Craig, *Ain't I a Beauty Queen? Black Women, Beauty, and the Politics of Race* (New York: Oxford University Press, 2002), 53.

58. Martha H. Verbrugge, *Active Bodies: A History of Women's Physical Education in Twentieth-Century America* (New York: Oxford University Press, 2012), 87. See chapter 4 for more on Allen.

59. Maryrose Reeves Allen to Mr. Goodman, February 5, 1957, 5, Box 160-7, Folder 13, Allen Papers, Howard University, Washington, DC, as quoted in Verbrugge, *Active Bodies*, 87.

60. Craig, *Ain't I a Beauty Queen*, 46.

61. Ava Purkiss, *Fit Citizens: A History of Black Women's Exercise from Post-Reconstruction to Postwar America* (Chapel Hill: University of North Carolina Press, 2023). See also Ava Purkiss, "'Beauty Secrets: Fight Fat': Black Women's Aesthetics, Exercise, and Fat Stigma, 1900–1930s," *Journal of Women's History* 29, no. 2 (Summer 2017): 14–37.

62. Purkiss, "'Beauty Secrets: Fight Fat,'" 21.

63. Verbrugge, *Active Bodies*, 88.

64. "A Study of the Health Problems of the Teacher and the Child," 23–24, Health Workshop, Hampton Institute Summer School, 1944, Box 160-9, Folder 13, Allen Papers, as cited in Verbrugge, *Active Bodies*, 88.

65. A. Wilberforce Williams, "Dr. A. Wilberforce Williams Talks on Preventive Measures, First Aid Remedies, Hygienics, and Sanitation: Posture—A Factor in Health," *Chicago Defender*, March 5, 1921, 12.

66. Algernon Jackson, "Afro Health Talk: What Posture Means," *Afro-American* (Baltimore), October 30, 1937, 15.

67. Algernon Jackson, "Afro Health Talk: Posture and Purpose," *Afro-American* (Baltimore, November 2, 1940, 7.

68. "19 Girls Win in Walk Right Contest," *Afro-American* (Baltimore), December 24, 1927, 18.

69. "Bennett College Gives Correct Posture Contest," *Chicago Defender*, April 20, 1940, 12.

70. "Get High Rating in Posture Contest," *Chicago Defender*, January 25, 1936, 21.

71. "Norman Graduation," *Afro-American* (Baltimore), July 11, 1931, 5.

72. James P. Comer, *Maggie's American Dream: The Life and Times of a Black Family* (New York: New American Library, 1988), 69–70.

73. Posture Committee of the American Physical Education Association, "Symposium of Preventative and Corrective Physical Education," *JHPE* 2, no. 4 (1931): 50.

74. Helen Durham and Barbara Beattie, "Hiding Your Hips," *Ladies' Home Journal* 48, no. 11 (November 1931): 24, 136.

75. Durham and Beattie, "Hiding Your Hips," 136.

76. Helen Durham, "Changing Chins," *Collier's*, March 7, 1931, 29, 56. Durham interviewed Beattie for this article.

77. Ruth F. Wadsworth, *Charm by Choice* (New York: Woman's Press, 1930), 63. See also Ruth F. Wadsworth, "Sitting Pretty," *Collier's*, March 30, 1929, 22, 48.

78. There is much scholarship on the history of obesity. For some of the more recent examples, see Sabrina Strings, *Fearing the Black Body: The Racial Origins of Fat Phobia* (New York: New York University Press, 2019); Deborah I. Levine, "Corpulence and Correspondence: President William H. Taft and the Medical Management of Obesity," *Annals of Internal Medicine* 160, no. 8 (October 2013): 580–81. For obesity through the lens of disability studies, see Sander L. Gilman, "Defining Disability: The Case of Obesity," *PMLA* 120, no. 2 (March 2005): 514–17. For a history of the medicalization of aging, see Cara Kiernan Fallon, "Forever Young: The Social Transformation of Aging in America Since 1900" (PhD dissertation, Harvard University, 2018).

79. "Good Posture Is a Real Health Asset," *Hygeia* 14 (January 1936): 92. These same admonitions appeared in other nonmedical magazines. For example, writing for the *American Magazine* in 1918, George McClellan, then sixty-one years old and married to a woman five years his junior, bragged that because of his regular exercise and attention to good posture, people often mistook him for being his wife's son. Good posture, he insisted, could "reduce a distended abdomen" as well as "the superfluous flesh on the face and cut down a double chin." George H. McClellan, "Mistaken for My Wife's Son," *American Magazine* 85 (February 1918): 104–5.

80. "Making Posture Exercises Attractive," *Hygeia* 9 (February 1931): 172.

81. "Chins Go Up and Heads High as Six Win in Posture Contest of Madison Junior," *Rochester Democrat and Chronicle*, December 20, 1927, 21.

82. "Lynchburg School Has Posture Contest," *Afro-American*, February 3, 1934, 12.

83. Posture Committee of the American Physical Education Association, "Symposium of Preventative and Corrective Physical Education."

84. Roger J. R. Hynes, "The Most Beautiful Spines in America: The History of the Posture Queens," *Chiropractic History* 22, no. 2 (2002): 65–71, 68.

85. P. Reginald Hug details the history of the Alabama State Chiropractic Association's posture queen contest. A practitioner himself, Hug presents this history without acknowledging the entrenched racism and sexism inherent to the contest, especially as it was practiced in Alabama. P. Reginald Hug, "Posture Queen Contests in Alabama," *Journal of Chiropractic Humanities* 15 (2008): 67–80.

86. Louise Goodwin, "How I Won the National Most Perfect Back Contest," *Chiropractic Journal* 1, no. 9 (September 1933): 5. See also Hynes, "The Most Beautiful Spines in America."

87. Douglas C. Baynton, "Disability and the Justification of Inequality in American History," in *The New Disability History: American Perspectives*, ed. Paul Longmore and Lauri Umansky (New York: New York University Press, 2001), 33–57.

Chapter Five: The Geopolitics of Posture

1. George Weisz, *Chronic Disease in the Twentieth Century: A History* (Baltimore: Johns Hopkins University Press, 2014).

2. Beth Linker, "A Dangerous Curve: The Role of History in America's Scoliosis Screening Programs," *American Journal of Public Health* 102, no. 4 (April 2012): 606–16; Davis Yosifon and Peter N. Stearns, "The Rise and Fall of American Posture," *American Historical Review* 103, no. 4 (October 1998): 1057–95.

3. Heather Munro Prescott, *Student Bodies: The Influence of Student Health Services in American Society and Medicine* (Ann Arbor: University of Michigan Press, 2007). When she surveyed the field in 1942, physiatrist Frances Hellebrandt wrote that the number of scientific publications on the subject of posture could fill a "very large book." F. A. Hellebrandt and Elizabeth Brogdon Franseen, "Physiological Study of the Vertical Stance of Man," *Physiological Reviews* 23, no. 3 (July 1943): 220–55.

4. Rachel Louise Moran, *Governing Bodies: American Politics and the Shaping of the Modern Physique* (Philadelphia: University of Pennsylvania Press, 2018).

5. The scholarship on the history of the Cold War in America is extensive. For several excellent examples of histories that touch on culture, bodies, race, and the early Cold War years, see Elaine Tyler May, *Homeward Bound: American Families in the Cold War Era* (New York: Basic Books, 1988); Mary L. Dudziak, *Cold War Civil Rights: Race and the Image of American Democracy* (Princeton, NJ: Princeton University Press, 2000); Jeffrey Montez de Oca, "The 'Muscle Gap': Physical Education and US Fears of a Depleted Masculinity, 1954–1963," in *East Plays West: Sport and the Cold War*, ed. Stephen Wagg and David Andrews (London: Routledge, 2006), 137–62; Toby C. Rider and Kevin B. Witherspoon, "Introduction," in *Defending the American Way of Life: Sport, Culture, and the Cold War*, ed. Toby C. Rider and Kevin B. Witherspoon (Fayetteville: University of Arkansas Press, 2018), 3–10; David Serlin, *Replaceable You: Engineering the Body in Postwar America* (Chicago: University of Chicago Press, 2004).

6. The news coverage on Kennedy's long history of back pain may have been comparatively sparse, but it was nevertheless extant. For the most comprehensive piece of reporting, including the president's Ottawa incident, see Frank E. Carey, "Kennedy's Back Injury Dates from 1942 Football Mishap," *Tampa Tribune*, June 25, 1961, 14-A.

7. A copy of Kraus's handwritten exam notes can be found in the Hans Kraus Personal Papers. See "1961: 4 July–13 December," Medical File: K, 1961–1993, HKPP-001-001, John F. Kennedy Presidential Library and Museum, Boston, Massachusetts. For more on Kraus's work with the president, see Janet G. Travell, interview by Theodore C. Sorensen, January 20, 1966, page 18, John F. Kennedy Library Oral History Program.

8. Adolf Lorenz, *My Life and Work: The Search for a Missing Glove* (New York: C. Scribner's Sons, 1936). For more on the Viennese orthopedic tradition, see Thomas Schlich, "The Perfect Machine: Lorenz Böhler's Rationalized Fracture Treatment in World War I," *Isis* 100, no. 4 (December 2009): 758–91.

9. Susan E. B. Schwartz, *JFK's Secret Doctor* (New York: Skyhorse, 2012), 28. The author interviewed Kraus and had access to the family papers, but the quotation is unattributed.

10. Bruce Grynbaum, "Obituaries: Hans Kraus, MD," *JAMA* 276, no. 3 (July 1996): 259.

11. Robert Flack (grandson of Sonya Escherich-Eisenmenger-Weber) in discussion with the author, July 13, 2021. The life and work of Sonya Escherich-Eisenmenger-Weber has received little attention from scholars. Much of the biographical information presented here comes from my correspondence with her grandson, Robert Flack, who has created a document of the family genealogy. Robert Flack, *A Brief History of the Family of Sonja Escherich-Eisenmenger-Weber* (Tryon, NC: 1990), https://flackology.com/genealogy/histories/Brief%20history%20of%20 Sonya%20Weber.pdf.

12. "Thirty-Ninth Annual Report of the Babies Hospital of the City of New York," December 1927 (page 43), Augustus C. Long Health Sciences Library Archives and Special Collections, Columbia University Irving Medical Center, New York, New York.

13. For more on gender and the history of physical therapy, see Beth Linker, "Strength and Science: Gender, Physiotherapy, and Medicine in the United States, 1918–35," *Journal of Women's History* 17, no. 3 (Fall 2005): 106–32.

14. To chart the growth and popularity of Weber's clinic, I referenced annual reports of the Babies' Hospital and Columbia University's College of Physicians and Surgeons Vanderbilt Clinic. For the 1930s, see "Forty-Second Annual Report of the Babies Hospital," December 1930: 69, 71. For the popularity of the Posture Clinic at Vanderbilt, see, for example, "Seventy-Fourth Annual Report of the Presbyterian Hospital in the City of New York," December 1942: 83–84, Augustus C. Long Health Sciences Library Archives and Special Collections, Columbia University Irving Medical Center, New York, New York, https://www.library -archives.cumc.columbia.edu/digital-collections-columbia-university-medical-center -affiliated-hospitals.

15. Patricia Spain Ward, *Simon Baruch: Rebel in the Ranks of Medicine, 1840–1921* (Tuscaloosa: University of Alabama Press, 1994).

16. For a historic overview of physiatry and allied health professions, see Glenn Gritzer and Arnold Arluke, *The Making of Rehabilitation: A Political Economy of Medical Specialization, 1890–1980* (Berkeley: University of California Press, 1989), 99.

17. Frank H. Krusen, "The Future of Physical Medicine," *JAMA* 125, no. 16 (August 1944): 1093–97.

18. Hans Kraus and S. Eisenmenger-Weber, "Evaluation of Posture Based on Structural and Functional Measurements," *Physical Therapy* 25, no. 6 (November 1945): 267–71; Hans Kraus

and S. Eisenmenger-Weber, "Fundamental Considerations of Posture Exercises: Guided by Qualitative and Quantitative Measurements and Tests," *Physical Therapy* 27, no. 6 (November 1947): 361–68; Hans Kraus and S. Eisenmenger-Weber, "Quantitative Tabulation of Posture Evaluation: Based on Structural and Functional Measurements," *Physical Therapy* 26, no. 5 (September 1946): 235–42.

19. John Steinbeck, *Vanderbilt Clinic* (New York: Presbyterian Hospital, 1947). The famed novelist, whose family received care at the Vanderbilt Clinic, wrote promotional copy for this pamphlet, which was intended to attract donations.

20. Ernest Havemann, "Mystery and Misery of the Backache," *Life*, June 11, 1956, 151–52, 155–56, 158, 160, 163, 165–66. For a more detailed account of the increased clinical attention to pain and its management at this time, see Keith Wailoo, *Pain: A Political History* (Baltimore: Johns Hopkins University Press, 2014).

21. Kelly R. Haisley et al., "Major Barbara Stimson: A Historical Perspective on the American Board of Surgery through the Accomplishments of the First Woman to Achieve Board Certification," *Annals of Surgery* 267, no. 6 (June 2018): 1000–1006.

22. Barbara B. Stimson, "The Low Back Problem," *Psychosomatic Medicine* 9, no. 3 (May 1947): 210–12.

23. Several years later, she raised the number to 50 percent. See Barbara B. Stimson, "Backache," *American Journal of Nursing* 51, no. 11 (November 1951): 672–74.

24. Stimson, "The Low Back Problem."

25. Kraus and Eisenmenger-Weber, "Evaluation of Posture." Hans Kraus, "Diagnosis and Treatment of Low Back Pain," *General Practitioner* 5, no. 4 (April 1952): 55–60.

26. Hans Kraus and Ruth P. Hirschland, "Muscular Fitness and Orthopedic Disability," *New York State Journal of Medicine* 54, no. 2 (January 1954): 212–15.

27. Robert H. Boyle, "The Report that Shocked the President," *Sports Illustrated*, August 15, 1955, 31–32, 72.

28. To complete the battery of posture tests on children in Europe and eventually Guatemala, Kraus hired two assistants, Ruth Hirschland Schlesinger and her sister-in-law Bonnie Prudden Hirschland. Kraus came to know the Hirschland family through his expat networks, especially those revolving around skiing and mountaineering in the Adirondack Mountains. Similar to Kraus, the Hirschlands fled Germany in 1933 when Ruth's father, Kurt M. Hirschland, the well-known senior partner of the Jewish banking firm Simon Hirschland Bank, was forcibly removed by the Nazis, who "Aryanized" the company. Ruth and her husband Rudolf died together by committing suicide in 1996. See Wolfgang Saxon, "Rudolf Schlesinger, 87, Expert on the World's Legal Systems," *New York Times*, November 22, 1996, D19. See also Schwartz, *JFK's Secret Doctor*, and Gretchen Rous Besser, "Hans Kraus: The Skiers' Doctor," *Skiing Heritage* 21, no. 4 (December 2009): 28–31.

29. Moran, *Governing Bodies.*

30. Moran, *Governing Bodies*, 93.

31. John F. Kennedy, "The Soft American," *Sports Illustrated*, December 26, 1960, 15–22.

32. Kennedy, "The Soft American."

33. "Let's Get the Kinks out of the Kids!," Display Advertisement 46, *New York Times*, February 5, 1962, 12; "Could This Be Our Deadliest Disease?," Display Advertisement 17, *Washington*

Post, June 24, 1963, A5; "Astronaut Glenn Discusses a Down to Earth Problem," Display Advertisement 11, *New York Times*, April 29, 1963, A4; "The Silent Epidemic," Display Advertisement 45, *Washington Post*, February 19, 1962, A10; Hans Kraus, *Hypokinetic Disease: Diseases Produced by Lack of Exercise* (Springfield, IL: Thomas, 1961).

34. Weber retired to North Carolina and agreed to only the occasional media interview. See Jack Havlicek, "Physical Fitness—Or Lack of It, Perturbs Doctors," *Asheville Citizen* (Asheville, NC), November 13, 1964, 19; Jean Perry, "How Two Experts Shape Up on Health Care," *Daily News* (Jacksonville, NC), July 5, 1979, 12. Kraus, by contrast, authored self-help books such as *The Causes, Prevention, and Treatment of Sports Injuries* (New York: Playboy Press, 1981) and *Backache, Stress, and Tension* (New York: Simon and Schuster, 1965).

35. Kenneth P. O'Donnell and David F. Powers, *"Johnny, We Hardly Knew Ye": Memories of John Fitzgerald Kennedy* (Boston: Little, Brown, 1972), 79, 96, as quoted in T. Glenn Pait and Justin T. Dowdy, "John F. Kennedy's Back: Chronic Pain, Failed Surgeries, and the Story of Its Effects on His Life and Death," *Journal of Neurosurgery: Spine* 27, no. 3 (September 2017): 247–55.

36. Pait and Dowdy, "John F. Kennedy's Back." See also James N. Giglio, "Growing Up Kennedy: The Role of Medical Ailments in the Life of JFK, 1920–1957," *Journal of Family History* 31, no. 4 (October 2006): 358–85.

37. "The Presidency: Back on the Course," *Time*, July 26, 1963, 17.

38. Pait and Dowdy, "John F. Kennedy's Back."

39. Judith A. Bellafaire, *The Women's Army Corps: A Commemoration of World War II Service* (Washington, DC: U.S. Army Center of Military History, 1993), https://history.army.mil/brochures/wac/wac.htm.

40. Leisa D. Meyer, *Creating GI Jane: Sexuality and Power in the Women's Army Corps during World War II* (New York: Columbia University Press, 1996).

41. Leisa D. Meyer, "Creating GI Jane: The Regulation of Sexuality and Sexual Behavior in the Women's Army Corps during World War II," *Feminist Studies* 18, no. 3 (Autumn 1992): 585.

42. Meyer, "Creating GI Jane."

43. Bellafaire, *The Women's Army Corps*.

44. *W.A.C. Field Manual: Physical Training* (Washington, DC: U.S. Government Printing Office, 1943), iv.

45. *W.A.C. Field Manual*, 8, emphasis in original.

46. Rationing of raw materials also affected the supply of girdles and foundation wear during World War II.

47. *W.A.C. Field Manual*, 10.

48. Dorothy S. Ainsworth, "Our Contribution to Morale in Times of War and Peace," *Journal of Health and Physical Education* 14, no. 2 (1943): 68.

49. Martha H. Verbrugge, *Active Bodies: A History of Women's Physical Education in Twentieth-Century America* (New York: Oxford University Press, 2012).

50. "Fundamental Movement" lecture notes and outline, December 2, 1946, Department of Exercise and Sport Studies records, Box "Departments: Physical Education," File "Courses, Reports," Smith College Archives, CA-MS-01043, Smith College Special Collections, Northampton, Massachusetts.

51. Syllabus for "Technique of Physical Examination," a course offered in 1946, Department of Exercise and Sport Studies records, Box "Departments: Physical Education," File "Courses, Reports," Smith College Archives, CA-MS-01043, Smith College Special Collections, Northampton, Massachusetts.

52. Patricia Vertinsky and Bieke Gils, "'Physical Education's First Lady of the World': Dorothy Sears Ainsworth and the International Association of Physical Education and Sport for Women and Girls," *International Journal of the History of Sport* 33, no. 13 (2016): 1500–1516.

53. Dorothy S. Ainsworth, "Report for the Dept. of Physical Education," November 1, 1934, Department of Exercise and Sport Studies records, Box "Departments: Physical Education," File "Courses and Reports, 1930s," Smith College Archives, CA-MS-01043, Smith College Special Collections, Northampton, Massachusetts.

54. Hans Bonde, "The Iconic Symbolism of Niels Bukh: Aryan Body Culture, Danish Gymnastics and Nordic Tradition," *International Journal of the History of Sport* 16, no. 4 (1999): 104–18. Bukh traveled to the United States in 1923 and 1926, and in the 1930s he traveled to South America and South Africa.

55. Niels Bukh, *Fundamental Gymnastics: The Basis of Rational Physical Development* (New York: E. P. Dutton, 1928), 4.

56. Bonde, "The Iconic Symbolism of Niels Bukh." For an example of an interwar American adaptation of Bukh's program, see Dorothy Sumption, *Fundamental Danish Gymnastics for Women* (New York: A. S. Barnes, 1929), 1–9.

57. Hazel C. Peterson, a student of Ainsworth's, completed a biography of her mentor for a doctoral dissertation. See Hazel C. Peterson, "Dorothy S. Ainsworth: Her Life, Professional Career and Contributions to Physical Education" (PhD dissertation, Ohio State University, 1968).

58. Robert L. Griswold, "'Russian Blonde in Space': Soviet Women in the American Imagination, 1950–1965," *Journal of Social History* 45, no. 4 (Summer 2012): 881.

59. Dorothy S. Ainsworth, "Contribution of Physical Education to the Social Service Agency," *Journal of the American Association for Health, Physical Education, and Recreation* 21, no. 6 (1950): 325, 367–68.

60. Vertinsky and Gils, "'Physical Education's First Lady of the World.'"

61. Jonathan Katz, *The Invention of Heterosexuality* (New York: Dutton, 1995), 96.

62. Anna G. Creadick, *Perfectly Average: The Pursuit of Normality in Postwar America* (Amherst: University of Massachusetts Press, 2010), 74.

63. This was in contrast to the zoot suit. See Kathy Peiss, *Zoot Suit: The Enigmatic Career of an Extreme Style* (Philadelphia: University of Pennsylvania Press, 2011).

64. Jennifer Terry, *An American Obsession: Science, Medicine, and Homosexuality in Modern Society* (Chicago: University of Chicago Press, 1999); Carolyn Herbst Lewis, *Prescription for Heterosexuality: Sexual Citizenship in the Cold War Era* (Chapel Hill: University of North Carolina Press, 2010); Margot Canaday, *The Straight State: Sexuality and Citizenship in Twentieth-Century America* (Princeton, NJ: Princeton University Press, 2009).

65. For more on Sheldon, see Patricia Vertinsky, "Physique as Destiny: William H. Sheldon, Barbara Honeyman Heath and the Struggle for Hegemony in the Science of Somatotyping," *Canadian Bulletin of Medical History* 24, no. 2 (Fall 2007): 291–316; Sarah W. Tracy, "An Evolving Science of Man: The Transformation and Demise of American Constitutional Medicine,

1920–1950," in *Greater than the Parts: Holism in Biomedicine, 1920–1950*, ed. Christopher Lawrence and George Weisz (New York: Oxford University Press, 1998), 161–88. For primary sources, see William H. Sheldon, "The New York Study of Physical Constitution and Psychotic Pattern," in *Sensation and Measurement*, ed. Howard R. Moskowitz, Bertram Scharf, and Joseph C. Stevens (New York: Springer, 1974), 147–55; William H. Sheldon, *Atlas of Men: A Guide for Somatotyping the Adult Male at All Ages* (New York: Harper, 1954); J. E. Lindsay Carter and Barbara Honeyman Heath, *Somatotyping: Development and Applications* (Cambridge: Cambridge University Press, 1990).

66. Quotations from Liam O'Connor, "How Your Shape Shapes Your Life," *Popular Science* 160, no. 5 (May 1952): 116–19, 228, 230, 232.

67. Vertinsky, "Physique as Destiny."

68. Vertinsky, "Physique as Destiny," and Tracy, "An Evolving Science of Man." Much of this same information, including the number of psychiatric patients that he photographed, can be found in Sheldon's Personal Papers, held at the National Anthropological Archives, Smithsonian Institution, Suitland, Maryland. See, for example, Box 5, file "New York Psychiatric," and Box 14, file Oregon State Prison.

69. Sheldon Papers, Box 10, file "Atlas of Women Inquiries," and Box 30, file "College Series Statistics," National Anthropological Archives, Smithsonian Institution, Suitland, Maryland.

70. Sheldon, *Atlas of Men*.

71. St. Clair Drake, "Reflections on Anthropology and the Black Experience," *Anthropology and Education Quarterly* 9, no. 2 (Summer 1978): 85–109.

72. Katrina Hazzard-Donald, "Dance," in *Encyclopedia of African American History, 1896 to the Present: From the Age of Segregation to the Twenty-First Century*, ed. Paul Finkelman (New York: Oxford University Press, 2009).

73. David K. Wiggins, "Charles Holston Williams: Hamptonian Loyalist and Champion of Racial Uplift through Physical Education, Dance, Recreation, and Sport," *International Journal of the History of Sport* 36, no. 17–18 (2019): 1531–51. Wiggins points out that it was also important to signal masculinity among the Black male dancers, making sure that they were not overly effete or homosexual. Wiggins, "Charles Holston Williams," 1540.

74. "Hale America Program to Keep Nation from Going Soft: Three-Fold Purpose," *Atlanta Daily World*, March 30, 1942, 5.

75. Lucius Jones, "Sports Slants," *Atlanta Daily World*, April 22, 1942, 5.

76. "Physical Education and Athletic Association Bulletin Published for the Class of 1954 in the Fall of 1958," Department of Exercise and Sport Studies records, Box "Departments: Physical Education," file "Publications, 1920–1993," Smith College Archives, CA-MS-01043, Smith College Special Collections, Northampton, Massachusetts.

77. Margaret Perkins, Smith College Press Board, to Helen Stafford, Society Editor, *Hartford Courant*, Hartford, CT, March 12, 1940, Department of Exercise and Sport Studies records, Box "Departments: Physical Education," File "Body Mechanics," Smith College Archives, CA-MS-01043, Smith College Special Collections, Northampton, Massachusetts.

78. Kathy Peiss, *Hope in a Jar: The Making of America's Beauty Culture* (New York: Metropolitan Books, 1998), 239.

79. Peiss, *Hope in a Jar*, 245.

80. Joseph Hansen, Evelyn Reed, and Mary-Alice Waters, *Cosmetics, Fashions, and the Exploitation of Women* (New York: Pathfinder Press, 1986): 39, as quoted in Peiss, *Hope in a Jar*, 256.

81. Florence C. Whipple, "An Experience in Health Motivation for College Girls," *Journal of Health and Physical Education* 18, no. 9 (November 1947): 639–88.

82. For quote, see J. Robertson, "Are You Norma, Typical Woman? Search to Reward Ohio Winners," *Plain Dealer* (Cleveland), September 9, 1945, 1, as quoted in Creadick, *Perfectly Average*, 29.

83. "Books: The Glad Hatter, Lilly Daché's Glamour Book," *Time*, March 26, 1956, 112.

84. Paisley Harris, "Gatekeeping and Remaking: The Politics of Respectability in African American Women's History and Black Feminism," *Journal of Women's History* 15, no. 1 (Spring 2003): 212–20.

85. Evelyn Brooks Higginbotham, *Righteous Discontent: The Women's Movement in the Black Baptist Church, 1880–1920* (Cambridge, MA: Harvard University Press, 1993), 187.

86. For the purpose and methodology of her dissertation research, see "An Outline: 'The Development of Beauty in College Women through Health and Physical Education,' 1938," Maryrose Reeves Allen Papers, Series B "Education," Box 160-4, Folder 4, Moorland-Spingarn Research Center, Howard University Archives, Washington, DC. For model club and concerns about superficial beauty, see Reeves Allen Papers, Series C, Box 160-7, Folder 4.

87. This quote comes from the 1939 draft of her dissertation. "An Outline," Reeves Allen Papers.

88. See Reeves Allen papers, Series C "Teaching Materials," Boxes 160-7 through 160-10.

89. "Foundations of Physical Education" handwritten lecture notes, Reeves Allen Papers, Series C "Teaching Materials," Box 160-4, Folder 4.

90. For more on Allen, see Verbrugge, *Active Bodies*; Ava Purkiss, "'Beauty Secrets: Fight Fat': Black Women's Aesthetics, Exercise, and Fat Stigma, 1900–1930s," *Journal of Women's History* 29, no. 2 (Summer 2017): 14–37; Kimberly D. Brown, "The Battle for Black Beauty: Howard University's Grooming Program for Women and African-American Activism in Redefining Aesthetic Ideology through Pageants Since 1925" (PhD dissertation, Howard University, 2013).

91. Reeves Allen Papers, Series C "Teaching Materials," Boxes 160-7 through 160-10. The most commercially visible relationship between Prudden and Kraus can be found in their trade publications, such as Bonnie Prudden, Hans Kraus, and Marjorie Morris, *Is Your Child Really Fit?* (New York: Harper, 1956).

92. "Body Aesthetics" course lecture notes, Reeves Allen Papers, Series C "Teaching Materials," Box 160-4, Folder 5.

93. Reeves Allen Papers, Series B "Education," Box 160-4, Folder 4.

94. Newspaper clipping, Reeves Allen Papers, Series C "Teaching Materials," Box 160-4, Folder 4.

95. Newspaper clipping, Reeves Allen Papers.

96. DeVore's charm school, according to historian Malia McAndrew, targeted "ordinary black women," and within a decade of opening had trained over 19,000 clients. Malia McAndrew, "Selling Black Beauty: African American Modeling Agencies and Charm Schools in Postwar America," *OAH Magazine of History* 24, no. 1 (January 2010): 32.

97. McAndrew, "Selling Black Beauty," 30. While most charm school advertisements in the Black press depict Black women as experts, I found one example of a "Prison Charm School" at Westfield State Farm in Bedford Hills, NY, with an image of Bonnie Prudden leading a group of Black women prisoners in posture and fitness exercises. See "Prison Charm School," *Ebony Magazine*, January 1960, 75–78.

98. Katharine Capshaw Smith, "Childhood, the Body, and Race Performance: Early 20th Century Etiquette Books for Black Children," *African American Review* 40, no. 4 (Winter 2006): 796.

99. Natalie Scurlock, "Newest Fall Fashions Demand Good Posture," *Afro-American* (Baltimore), August 2, 1947, M-4. For other examples of newspaper articles that promote the importance of posture to charm, see Lou Swarz, "Charm," *Philadelphia Tribune*, March 22, 1952, 5; "Buffer College Group Sponsors 'Charm' School," *Chicago Defender*, April 19, 1947, 17.

100. "Kissing Co-Ed Sent Home; Students Revolt," *Pittsburgh Courier*, February 9, 1929, 1.

101. Susan K. Cahn, "If We Got That Freedom: 'Integration' and the Sexual Politics of Southern College Women, 1940–1960," in *Connexions: Histories of Race and Sex in North America*, ed. Jennifer Brier, Jim Downs, and Jennifer L. Morgan (Champaign: University of Illinois Press, 2016): 186.

Chapter Six: The Perils of Posture Perfection

1. "Posture Photos Return to Files," *Cornell Daily Sun* (Ithaca, NY), September 19, 1950, 1.

2. Harlee Fried, "A Side View of Posture Pictures: It Gave Women a Sense of Pride," *Miscellany News* (Poughkeepsie, NY), November 11, 1977, 6.

3. Donald L. Robinson, "Posture Program Not to Be Abolished, According to Kiphuth," *Yale Daily News* (New Haven, CT), February 10, 1956, 1.

4. Robinson, "Posture Program Not to Be Abolished."

5. Robinson, "Posture Program Not to Be Abolished."

6. Ellen Kelly, "Taking Posture Pictures," *Journal of Health and Physical Education* 17, no. 8 (1946): 464–505.

7. Marcella F. Hance, interview by Betsy McKlveen, May 6, 1988, Pembroke Center Oral History Project, Brown University, Providence, Rhode Island.

8. *North v. Board of Trustees of the University*, 137 Ill. 296 (1891) at 306, quoted in P. Lee, "The Curious Life of In Loco Parentis at American Universities," *Higher Education in Review* 8 (2011): 69.

9. James D. Anderson, *Education of Blacks in the South* (Chapel Hill: University of North Carolina Press, 1988), 54.

10. Heather Munro Prescott, *Student Bodies: The Influence of Student Health Services in American Society and Medicine* (Ann Arbor: University of Michigan Press, 2007).

11. For examples of this, see Martha H. Verbrugge, *Able-Bodied Womanhood: Personal Health and Social Change in Nineteenth-Century Boston* (New York: Oxford University Press, 1988); Martha H. Verbrugge, *Active Bodies: A History of Women's Physical Education in Twentieth-Century America* (New York: Oxford University Press, 2012). See also Margaret A.

Lowe, *Looking Good: College Women and Body Image, 1875–1930* (Baltimore: Johns Hopkins University Press, 2003).

12. Pauline Ames Plimpton, "Letters and Opinions," *Smith Alumnae Quarterly* (Spring 1985): 2.

13. Mary Evans Boname, "Letters and Opinions," *Smith Alumnae Quarterly* (Spring 1985): 2–3.

14. Sylvia Plath, *Letters Home: Correspondence 1950–1963* (New York: Harper & Row, 1975), 48, quoted in Jonathan P. Lewis, "Posture Pictures and Other Tortures: The Battle to Control Esther Greenwood's Body," *Quarterly Horse* 1, no. 1 (November 2016), https://www.quarterlyhorse.org/fall16/lewis.

15. Sylvia Plath, *The Bell Jar* (New York: Bantam Winstone, 1981), 56.

16. Beth Linker, "Tracing Paper, the Posture Sciences, and the Mapping of the Female Body," in *Working with Paper*, ed. Carla Bittel, Elaine Leong, and Christine von Oertzen (Pittsburgh: University of Pittsburgh Press, 2019), 124.

17. Samuel Warren and Louis Brandeis, "The Right to Privacy," *Harvard Law Review* 4, no. 5 (December 1890): 193–220.

18. Sarah E. Igo, *The Known Citizen: A History of Privacy in America* (Cambridge, MA: Harvard University Press, 2018), 37.

19. Heidi Katherine Knoblauch, "Patients' Posture: Medical Photography, Collecting, and Privacy, 1862–1962" (PhD dissertation, Yale University, 2015), 155.

20. Knoblauch, "Patients' Posture."

21. Knoblauch, "Patients' Posture."

22. "X-Ray Examinations! Urged," *Washington Post*, October 5, 1937, 25.

23. "Photos of Nude Co-Eds Stir Up Detroit Furor," *Chicago Daily Tribune*, October 4, 1937, 1.

24. Sarah Lincoln Doyle, "Head, Shoulders, Knees, and Toes: The Origins of Posture Photographs at Wellesley College, 1875–1942" (master's thesis, Wellesley College, 1996).

25. Charlotte G. MacEwan, Elizabeth Powell, and Eugene C. Howe, "An Objective Method of Grading Posture: Its Development, Routine Procedure and Applications," *Physical Therapy* 15, no. 5 (September 1935): 169.

26. Kenneth D. Miller, "A Physical Educator Looks at Posture," *Journal of School Health* 21, no. 3 (March 1951): 89–94.

27. Louis B. Laplace and Jesse T. Nicholson, "Physiologic Effects of the Correction of Faulty Posture," *JAMA* 107, no. 13 (1936): 1009–12.

28. James Frederick Rogers, "Constitutional and Acquired Posture," *Journal of School Health* 3, no. 1 (January 1933): 8.

29. James Frederick Rogers, "The Long and Short of the Carriage Business," *Journal of Health and Physical Education* 3, no. 10 (December 1932): 11–59.

30. Dr. J. Philip Keeve would offer the final blow to school-based posture exams (with the exception of scoliosis exams) and training. Keeve, "'Fitness,' 'Posture' and Other Selected School Health Myths," *Journal of School Health* 37, no. 1 (January 1967): 8–15.

31. F. A. Hellebrandt and Elizabeth Brogdon Franseen, "Physiological Study of the Vertical Stance of Man," *Physiological Reviews* 23, no. 3 (July 1943): 220–55.

32. D. W. Schulenberg, "Letter to the Editor," *New York Times Magazine*, February 5, 1995, 14.

33. See "Letters to the Editor" in response to Rosenbaum's article. "Letters to the Editor," *New York Times Magazine*, February 2, 1995, 14.

34. A colleague shared this correspondence with me from his own family personal papers. The names have been de-identified. The letter was sent from Sharpe, Northampton, Massachusetts, January 19, 1953, to her fiancé in Washington, DC.

35. "Posture Ordeal 'Grim,' Says '52," *Hartford Courant*, October 7, 1948, Academic Departments records, Smith College Archives, CA-MS-01010, Smith College Special Collections, Northampton, Massachusetts.

36. Theresa Elizabeth Gagnon, interview by Karen Lamoree, November 29, 1988, Pembroke Center Oral History Project, Brown University; Ann Martha Chmielewski, interview by Wendy Korwin, May 25, 2013, Pembroke Center Oral History Project, Brown University; Elissa L. Beron, interview by Miriam Dale Pichey, May 18, 1992, Pembroke Center Oral History Project, Brown University.

37. Ulle Viiroja in 50th Reunion, Pembroke College class of 1966, interview by Whitney Pape, May 28, 2016, Pembroke Center Oral History Project, Brown University.

38. Gloria E. Del Papa, interview by Carol Fenimore, December 2, 1985, Pembroke Center Oral History Project, Brown University.

39. Bailey describes the setting of the exam: "There was a bowling alley in the basement and I remember going down to the bowling alley and basically stripped. I mean, we had undies on, but you had to stand against the wall and take pictures from a distance. I don't remember seeing the pictures, but they were, you know, I guess you know the whole story. They were trying to make sure your posture was perfect." See Black Alumnae at their 25th reunion, class of 1968, interview by Lydia L. English, Joyce P. Foster, Damali Patterson, and Karen Wyche, May 28, 1993, Pembroke Center Oral History Project, Brown University. See also Black Alumnae at their 50th Reunion, class of 1968, interview by Mary Murphy, May 25, 2018, Pembroke Center Oral History Project, Brown University.

40. Joanne Meyerowitz, "Women, Cheesecake, and Borderline Material: Responses to Girlie Pictures in the Mid-Twentieth-Century U.S.," *Journal of Women's History* 8, no. 3 (Fall 1996): 9–35.

41. Meyerowitz, "Women, Cheesecake, and Borderline Material."

42. Meyerowitz, "Women, Cheesecake, and Borderline Material."

43. Theodore Peterson, *Magazines in the Twentieth Century* (Urbana: University of Illinois Press, 1956), 338–40.

44. "X-Ray Examinations! Urged," 25.

45. Elissa L. Beron, interview by Miriam Dale Pichey, emphasis in original.

46. For more on the relationship between both compulsory able-bodiedness and heterosexuality, see Robert McRuer, *Crip Theory: Cultural Signs of Queerness and Disability* (New York: New York University Press, 2006). According to Martha Verbrugge, the presence and struggles of lesbians remained physical education's "open secret—the issue was everywhere and nowhere." Verbrugge, *Active Bodies*, 197–99.

47. "Posture Photos Return to Files."

48. Holly V. Scott, *Younger than That Now: The Politics of Age in the 1960s* (Amherst: University of Massachusetts Press, 2016).

49. Beth Bailey, *Sex in the Heartland* (Cambridge, MA: Harvard University Press, 1999), 46.

50. Bailey, *Sex in the Heartland*.

51. Glenn C. Altschuler and Isaac Kramnick, *Cornell: A History, 1940–2015* (Ithaca, NY: Cornell University Press, 2014).

52. Larry D. Coleman, "From Revolution to Revelation: The Howard Experience 1969–1976," *New Directions* 3, no. 4 (October 1976): 4.

53. Elissa L. Beron, interview by Miriam Dale Pichey.

54. Marjorie Alice Jones, interview by Julia Hyun, April 16, 1988, Pembroke Center Oral History Project, Brown University.

55. Wilton S. Sogg, "Letter to the Editor," *New York Times Magazine*, February 12, 1995, 12. At one particular Yale fraternity formal dinner party, women from one of the Seven Sisters schools were brought into a dining room, "where they discovered the basis on which they had been invited: around the table, as place cards, were their posture photos." See Steven D. Price, "Letter to the Editor," *New York Times Magazine*, February 5, 1995, 14.

56. Erika Lorraine Milam, "Making Males Aggressive and Females Coy: Gender across the Animal-Human Boundary," *Signs* 37, no. 4 (Summer 2012): 936.

57. "G.W. Taboos Nudity Except Behind Screens, So Coeds Pose Only for Silhouette Photos," *Washington Post*, October 5, 1937, 1.

58. Fradd also assured the reporter that he had never "heard of any complaints whatsoever" outside of "this Wayne Affair." These quotes come from a Springfield, Massachusetts, newspaper clipping. See Tom Stevenson, "Vivid Outlines Show Defects in Posture," *The Republican* (Springfield, MA), n.d., n.p. Clipping found in Physical Education Department Scrapbook, 1935–1940, Box 1252, File "Hygiene and Bacteriology," File "Hygiene—Posture," Academic Departments records, Smith College Archives, CA-MS-01010, Smith College Special Collections, Northampton, Massachusetts.

59. Bailey, *Sex in the Heartland*.

60. Bailey, *Sex in the Heartland*.

61. For example, in Alfred Kinsey's 1952 *Sexual Behavior in the Human Female* (Philadelphia: Saunders, 1953), he found that only 44 percent of the women in his sample said that they "restricted their pre-marital coitus" because of fear of pregnancy, whereas 80 percent cited "moral reasons." Kinsey, quoted in Bailey, *Sex in the Heartland*, 3. "Fifty percent of the male respondents in Kinsey's study of male sexual behavior wanted to marry a virgin. A young man who went to high school in the early 1960s recalled, 'You slept with one kind of woman, and dated the other kind, and the women you slept with, you didn't have much respect for, generally.'" See Bailey, *Sex in the Heartland*, 3.

62. Dolores Flamiano, "The (Nearly) Naked Truth: Gender, Race, and Nudity in Life, 1937," *Journalism History* 28, no. 3 (Fall 2002): 121–36. See also Meyerowitz, "Women, Cheesecake, and Borderline Material," 4; and Beth Bailey, *From Front Porch to Back Seat: Courtship in Twentieth-Century America* (Baltimore: Johns Hopkins University Press, 1988).

63. For examples of the importance of posture exams in other contexts, see William James, "A Study of the Expression of Bodily Posture," *Journal of General Psychology* 7, no. 2

(January 1932): 405; J. Ramsay Hunt, "Psychic Representations of Movement and Posture: Their Relation to Symptomatology," *Archives of Neurology and Psychiatry* 14, no. 1 (July 1925): 7–19; John MacKay, "The Epileptic Constitution: A Study of Seventy Cases of Epilepsy at Whittingham Mental Hospital, Preston, Lancs" (PhD dissertation, University of Glasgow, 1936).

64. George Thomas Stafford, *Preventive and Corrective Physical Education* (New York: A. S. Barnes, 1950).

65. Beth Linker, "A Dangerous Curve: The Role of History in America's Scoliosis Screening Programs," *American Journal of Public Health* 102, no. 4 (April 2012): 606–16.

66. Beth Linker, "Spines of Steel: A Case of Surgical Enthusiasm in Cold War America," *Bulletin of the History of Medicine* 90, no. 2 (Summer 2016): 222–49.

67. Naomi Rogers, *The Polio Wars: Sister Elizabeth Kenny and the Golden Age of American Medicine* (New York: Oxford University Press, 2014); Daniel J. Wilson, *Polio: The Epidemic and Its Survivors* (Chicago: University of Chicago Press, 2005).

68. L. B., California, to P. R. Harrington, Houston, Texas, November 4, 1960, in the Paul R. Harrington Papers, University of Kansas Medical Center Archives, KUMC-MSS-11, Kansas City, Kansas. The correspondence is contained in a bound volume titled "Time Article, 1960, Patient Inquiries." For privacy and protection, I have de-identified the letter writer but retained the state of origin. See Paul R. Harrington Papers, University of Kansas Medical Center Archives, KUMC-MSS-11, University of Kansas, Kansas City, Kansas.

69. Audra Jennings, "Mildred Scott: A Pennsylvania Woman at the Heart of the Early Disability Rights Movement," *Pennsylvania Legacies* 17, no. 2 (Fall 2017): 20–25.

70. Stafford, *Preventive and Corrective Physical Education*, 17.

71. Stafford, *Preventive and Corrective Physical Education*, 24.

72. Linker, "A Dangerous Curve."

73. See, for example, the account of "Miles" in Iris Halberstam-Mickel, *Stand Up Straight!: Personal Recollections about Scoliosis by the People Who Live with It* (Dubuque, IA: Kendall Hunt, 1989), 38. For other scoliosis memoirs, see Rosalie Griesse, *The Crooked Shall Be Made Straight* (Atlanta: John Knox Press, 1979); Louise F. Sohrabi, *The Story of a Woman's Fight against Scoliosis* (Alameda, CA: Rima Press, 1983). Because scoliosis was often a result of polio, one can find more accounts in Julie Silver and Daniel Wilson, *Polio Voices: An Oral History from the American Polio Epidemics and Worldwide Eradication Efforts* (Westport, CT: Praeger, 2007).

74. See "Eli" and "Jerry" in Halberstam-Mickel, *Stand Up Straight!*, 46 and 84, respectively.

75. Halberstam-Mickel, *Stand Up Straight!*

76. "Martha" in Halberstam-Mickel, *Stand Up Straight!*, 43.

77. "Sharon" in Halberstam-Mickel, *Stand Up Straight!*, 59. For an excellent analysis of staring from a critical disability studies standpoint, see Rosemarie Garland-Thomson, *Staring: How We Look* (New York: Oxford University Press, 2009).

78. "Frank" in Halberstam-Mickel, *Stand Up Straight!*, 44.

79. Stafford, *Preventive and Corrective Physical Education*, 10.

80. "Marlene" in Halberstam-Mickel, *Stand Up Straight!*, 35.

81. Private correspondence with SMR, in author's possession. See also Halberstam-Mickel, *Stand Up Straight!*

82. Walter P. Blount and John H. Moe, *The Milwaukee Brace* (Baltimore: Williams and Wilkins, 1973), x; Frank C. Wickers, Wilton H. Bunch, and Paul M. Barnett, "Psychological Factors in Failure to Wear the Milwaukee Brace for Treatment of Idiopathic Scoliosis," *Clinical Orthopaedics and Related Research* 126 (July–August 1977): 62–66.

83. Fred Pelka, *What We Have Done: An Oral History of the Disability Rights Movement* (Amherst: University of Massachusetts Press, 2012), 113–16.

84. Pelka, *What We Have Done*, 128.

85. Judith Heumann, *Being Heumann: An Unrepentant Memoir of a Disability Rights Activist* (Boston: Beacon Press, 2020).

86. For the intersection of students' rights and patients' rights, see Prescott, *Student Bodies*, especially chapter 7. For the rise of informed consent in clinical practice, see David J. Rothman, *Strangers at the Bedside: A History of How Law and Bioethics Transformed Medical Decision Making* (New York: Basic Books, 1991).

87. Jane Yolen, "Posture Picture on the Wall, Who's the Straightest of Us All?," *Smith Alumnae Quarterly* (Fall 1984): 19–21.

88. Andrew Letendre, "The Naked Truth about Yale's Posture Program," *Yale Alumni Magazine*, June 22, 2015.

89. Fried, "A Side View of Posture Pictures."

90. Robin Morgan, "No More Miss America! Ten Points of Protest," in *Sisterhood Is Powerful: An Anthology of Writings from the Women's Liberation Movement*, ed. Robin Morgan (New York: Random House, 1970), 521–24.

91. Catherine Prendergast and Nancy Abelmann, "Alma Mater: College, Kinship, and the Pursuit of Diversity," *Social Text* 24, no. 1 (86) (Spring 2006): 40.

92. Craig Robert Forrest, "A History of *in Loco Parentis* in American Higher Education" (PhD dissertation, University of Missouri–Columbia, 2020), 154. Cornell's athletic director, Robert Kane, remembered for his postwar glory years of building a winning football team, retired in 1976 beleaguered by student radicals in the university senate who opposed the two-year mandatory physical education requirement and the large athletics budget. Kane left the program feeling that neither the students nor the administration supported physical education and athletics. See Altschuler and Kramnick, *Cornell: A History*, 152.

93. Fried, "A Side View of Posture Pictures," 6.

94. Helen Gurley Brown, "Having It All," *Miami News*, November 17, 1982, 20.

95. Jennifer Smith Maguire, "Body Lessons: Fitness Publishing and the Cultural Production of the Fitness Consumer," *International Review for the Sociology of Sport* 37, no. 3–4 (April 2002): 449–64.

96. Fried, "A Side View of Posture Pictures," 6.

97. According to historian Elizabeth Haiken, plastic surgeons noticed an uptick in male patients beginning in the late 1960s. Facelifts appealed to businessmen who were "either starting a new job or afraid of losing [their] current job to a younger man." See Elizabeth Haiken, "Virtual Virility, or, Does Medicine Make the Man?," *Men and Masculinities* 2, no. 4 (April 2000): 393.

98. Alma Pryor, "Alma Pryor on Being Beautiful," *Chicago Defender*, September 6, 1975, 11.

99. "Using Your Posture for a Healthier Body," *Philadelphia Tribune*, August 21, 1990, 4B.

100. Jane E. Brody, "Personal Health: Breaking the Bad Habit of Poor Posture," *New York Times*, April 21, 1993, C12.

Chapter Seven: The Posture Photo Scandal

1. Ben M. Primer, "Report on Destruction of Princeton University Posture Photos" (n.d.), Nude Posture Photographs, Box 181, Folder 2, AC109, Department of Special Collections, Princeton University Library, Princeton, New Jersey.

2. Ron Rosenbaum, "The Great Ivy League Nude Posture Photo Scandal: How Scientists Coaxed America's Best and Brightest out of Their Clothes," *New York Times*, January 15, 1995, 26, 28–29, 30–31, 40, 46, 55–56.

3. For more on Sheldon, see Patricia Vertinsky, "Embodying Normalcy: Anthropometry and the Long Arm of William H. Sheldon's Somatotyping Project," *Journal of Sport History* 29, no. 1 (Spring 2002): 95–133; Patricia Vertinsky, "Physique as Destiny: William H. Sheldon, Barbara Honeyman Heath and the Struggle for Hegemony in the Science of Somatotyping," *Canadian Bulletin of Medical History* 24, no. 2 (Fall 2007): 291–316; Sarah W. Tracy, "Sheldon, William Herbert (1898–1977), Psychologist and Physician," in *American National Biography*, https://doi .org/10.1093/anb/9780198606697.article.1400964.

4. Rosenbaum, "The Great Ivy League Nude Posture Photo Scandal," 28.

5. Rekha Murthy, "Posturing for an Empty Scandal? Nude Photos Taken at Ivy League and Seven Sister Schools Are Merely Pictures of the Past for Most," *Brown Daily Herald* (Providence, RI), vol. 130, no. 20, February 23, 1995, 1, 9.

6. Rosenbaum, "The Great Ivy League Nude Posture Photo Scandal," 26, 28.

7. Mary Laprade to Jill Ker Conway, December 13, 1984, Office of the President, Jill Ker Conway Files, File Series IV, Administrative Issues, Box 5, Folder "Posture Pictures 1979–1984," Smith College Special Collections, Northampton, Massachusetts. Hereafter referred to as JKC files.

8. Gretchen Dieck to Jill K. Conway, July 25, 1979, JKC files.

9. The initial impetus for university-based IRBs came about with the establishment of the National Institutes of Health, but Tuskegee and its aftermath created even more strictures and regulations so that research subjects would be protected. For the early years of IRBs, see Laura Stark, *Behind Closed Doors: IRBs and the Making of Ethical Research* (Chicago: University of Chicago Press, 2011). For Tuskegee, see Susan Reverby, *Examining Tuskegee: The Infamous Syphilis Study and Its Legacy* (Chapel Hill: University of North Carolina Press, 2009).

10. Gerry to Jill Ker Conway, July 27, 1979, JKC files. Conway's request was written on this memo.

11. Gretchen Dieck to Jill Ker Conway, December 4, 1979, JKC files. Enclosed was Dieck's entire research proposal to be reviewed by Smith's Human Subjects Research Committee.

12. Gretchen S. Dieck, "Posture Pictures, Permission, and Privacy Protection," *IRB: Ethics & Human Research* 3, no. 10 (December 1981): 6–7.

13. Dieck, "Posture Pictures, Permission, and Privacy Protection," 6–7.

14. Gretchen S. Dieck, "The Relationship between Postural Asymmetry and Back and Neck Pain in Smith College Freshmen: A Historical Cohort Study (Massachusetts)" (PhD dissertation, Yale University, 1982).

15. Nancy Lange and Trudy Stella, Alumnae Association of Smith, to Jill Ker Conway, January 13, 1980, JKC files.

16. Dieck, "Posture Pictures, Permission, and Privacy Protection."

17. Emphasis mine. "Smith College Class of 1959 Newsletter," March 25, 1981, Box 2215, Folder 5, Class Letters, 1957–1982, Smith College Archives, CA-MS-01024, Smith College Special Collections, Northampton, Massachusetts.

18. "Smith College Class of 1959 Newsletter," March 25, 1981.

19. Marcella Frances Fagan Hance, interview by Betsy McKlveen, May 6, 1988, interview part I, transcript and recording, Pembroke Center Oral History Project, John Hay Library, Brown University, Providence, Rhode Island.

20. "Smith College Class of 1959 Newsletter," March 25, 1981, 2.

21. Dieck, "Posture Pictures, Permission, and Privacy Protection."

22. Katherine O'Bryan Canaday to Class of 1957, February 28, 1981, JKC files.

23. Dieck, "The Relationship between Postural Asymmetry and Back and Neck Pain in Smith College Freshmen."

24. Dieck, "The Relationship between Postural Asymmetry and Back and Neck Pain in Smith College Freshmen."

25. Gretchen S. Dieck et al., "An Epidemiologic Study of the Relationship between Postural Asymmetry in the Teen Years and Subsequent Back and Neck Pain," Spine 10, no. 10 (December 1985): 872–77.

26. Beth Linker, "Spines of Steel: A Case of Surgical Enthusiasm in Cold War America," Bulletin of the History of Medicine 90, no. 2 (Summer 2016): 222–49.

27. Beth Linker, "A Dangerous Curve: The Role of History in America's Scoliosis Screening Programs," American Journal of Public Health 102, no. 4 (April 2012): 606–16.

28. Amy Poehler, Yes, Please (New York: Dey Street Books, 2014), 5.

29. See, for example, S. L. Weinstein and I. V. Ponseti, "Curve Progression in Idiopathic Scoliosis," Journal of Bone and Joint Surgery (American Volume) 65, no. 4 (April 1983): 447–55. Linker, "A Dangerous Curve."

30. Jane A. Mott to Gretchen Dieck, December 4, 1981, JKC files.

31. O. Donald Chrisman to Mary Laprade, April 4, 1983, JKC files.

32. Emphasis added. Judith Marksbury to Jill Ker Conway, October 21, 1983, JKC files.

33. Jill Ker Conway to Mary Laprade, October 31, 1983, JKC files.

34. Mary Laprade to Helen, Jill Ker Conway's secretary, December 13, 1984, JKC files.

35. Dick Cavett, "Women Alone Can't Make a Just Society; 'Graduation from Hell,'" New York Times, June 12, 1992, A24.

36. Naomi Wolf, The Beauty Myth: How Images of Beauty Are Used against Women (New York: William Morrow, 1991).

37. Carolyn Bronstein, Battling Pornography: The American Feminist Anti-Pornography Movement, 1976–1986 (New York: Cambridge University Press, 2011), 323.

38. Claire Bond Potter, "Taking Back Times Square: Feminist Repertoires and the Transformation of Urban Space in Late Second Wave Feminism," Radical History Review 113 (Spring 2012): 67–80.

39. Bronstein, Battling Pornography.

40. Wolf, *The Beauty Myth*, 280.

41. C. Davis Fogg, "The Posture Pictures," September 25, 2015, https://www.fictionontheweb .co.uk/2015/09/the-posture-pictures-by-c-davis-fogg.html. According to Kris Belden-Adams, Fogg published an autobiographical account very similar to this fictionalized one on a Yale alumni blog for the class of 1959, which has since been taken down. See Kris Belden-Adams, "'We Did What We Were Told': The 'Compulsory Visibility' and De-Empowerment of US College Women in Nude 'Posture Pictures,' 1880–1940." *Miranda: Revue Pluridisciplinaire du Monde Anglophone* 25 (2022), https://doi.org/10.4000/miranda.44430.

42. Wolf, *The Beauty Myth*, 16.

43. Stephen J. Gould, "Curveball," *New Yorker*, November 28, 1994, 139. See also Joseph D. McInerney, "Why Biological Literacy Matters: A Review of Commentaries Related to *The Bell Curve: Intelligence and Class Structure in American Life*," *Quarterly Review of Biology* 71, no. 1 (March 1996): 81–96.

44. Dorothy Nelkin, "The Science Wars: Responses to a Marriage Failed," *Social Text* 46/47 (Spring–Summer 1996): 93–100.

45. Nelkin, "The Science Wars."

46. McInerney, "Why Biological Literacy Matters."

47. Rosenbaum, "The Great Ivy League Nude Posture Photo Scandal," 56.

48. Rosenbaum, "The Great Ivy League Nude Posture Photo Scandal," 28.

49. Vertinsky, "Embodying Normalcy"; Tracy, "Sheldon, William Herbert (1898–1977)."

50. In the uproar at Wayne University during the 1930s, when students protested the posture photography practice, Dr. Irvin Sander insisted that the exam was not compulsory. In other words, the student had the autonomy and agency to refuse to stand for the photos. It is doubtful, however, that this option was ever made clear to students. Instead, most students report that they were told that the exam was a requirement for both matriculation and graduation. See "Photos of Nude Co-Eds Stir Up Detroit Furor," *Chicago Daily Tribune*, October 4, 1937, 1.

51. "Somatotype Announcement," February 1950, "Somatotyping" file, Box 56, Dorothy Sears Ainsworth Papers, Smith College Archives, CA-MS-00300, Smith College Special Collections, Northampton, Massachusetts.

52. Rosenbaum, "The Great Ivy League Nude Posture Photo Scandal," 30.

53. Margery N. Sly, Director of Special Collections, Temple University, in conversation with the author, July 20, 2022.

54. Margery N. Sly, in conversation with the author.

55. Margery N. Sly, in conversation with the author.

56. Margery N. Sly, "Women's Health and the Pix Controversy: Two Decades Later." Paper delivered first at the Mid-Atlantic Regional Archives Conference, May 2, 2009, then again at the Society of American Archivists, August 2014. Sly kindly shared a copy of her conference talk with me.

57. Ben M. Primer, email to Jacquelyn Savani and Justin Harmon, January 17, 1995, Nude Posture Photographs, Box 181, Folder 2, AC109, Department of Special Collections, Princeton University Library, Princeton, New Jersey.

58. Robert Gavin, "Some Can Recall Stripping, but Not a Camera," *Syracuse Herald-Journal* (Syracuse, NY), January 27, 1995, A6.

59. Phil McDade, "UW's Good-Posture Secret," *Wisconsin State Journal* (Madison, WI), January 27, 1995, 1A. For an account of Cook's leadership in the Society for American Archivists, see Courtney Bailey, "The Blessings of Providence on an Association of Archivists," September 27, 2015, accessed July 25, 2022, https://cbaileymsls.wordpress.com/tag/j-frank-cook/.

60. About Kent State University, reporter Fran Henry writes, "Unlike Ivy Leaguers.., the KSU coeds posed in leotards. Also, the KSU students always knew the whereabouts of their photos: in their own personal class workbooks." Fran Henry, "KSU Posture Photos Used in Classwork," *Plain Dealer* (Cleveland), January 23, 1995, 3E. A similar practice of having students pose in leotards was adopted at the University of Iowa and also in the latter years of Wellesley's posture program. Sarah Lincoln Doyle, "Head, Shoulders, Knees, and Toes: The Origins of Posture Photographs at Wellesley College, 1875–1942" (master's thesis, Wellesley College, 1996).

61. "Nude Photos of Yale Graduates Are Shredded," *New York Times*, January 29, 1995, 16.

62. Ortner as quoted in "Nude Photos of Yale Graduates Are Shredded," 16.

63. Tamara L. Bray and Thomas W. Killion, eds., *Reckoning with the Dead: The Larsen Bay Repatriation and the Smithsonian Institution* (Washington, DC: Smithsonian Institution Press, 1994).

64. Ann M. Palkovich, "Anthropology and the Sheldon Nude Posture Photographs," paper delivered at the Society of American Archivists, August 2014. Palkovich kindly shared a copy of her conference talk with me.

65. Sly, "Women's Health and the Pix Controversy."

66. Margery N. Sly, in conversation with the author, July 20, 2022. She also characterized those early days after the Rosenbaum article as her "struggling to explain to people who aren't historians or archivists why there might be value in keeping them when . . . you know you're trying to speak from a point of logic, they are speaking emotionally. And yeah, that didn't go very far."

67. This quote comes from "No Nudes Good News for Ivy League Alum: Museum Blocks Access to 'Posture' Photos," *Times* (Trenton, NJ), January 21, 1995, contained in Nude Posture Photographs, Box 181, Folder 2, AC109, Department of Special Collections, Princeton University Library, Princeton, New Jersey.

68. Ben M. Primer, "Those Posture Photos," *Princeton Alumni Weekly*, June 7, 1995, 4.

69. Primer, "Those Posture Photos," 4.

70. Sly, director of special collections, Temple University, in conversation with the author, July 20, 2022.

71. The inventory list of the Sheldon papers is publicly accessible. National Anthropological Archives, Guide to the William H. Sheldon Papers, Circa 1920–Circa 1980 (Washington, DC: Smithsonian Institution), https://sova.si.edu/record/NAA.1987-39?s=20&n=10&t=C&q=Stevens%2C+Martin&i=27.

72. Molly Rogers, *Delia's Tears: Race, Science, and Photography in Nineteenth-Century America* (New Haven, CT: Yale University Press, 2010). See also "About the Research Project," Peabody Museum of Archaeology and Ethnology, accessed August 3, 2022, https://peabody.harvard.edu/ZD-about-research-project. The descendants of the Taylors have recently claimed ownership of these photographs, and events are still unfolding as of the writing of this book. See Keith

Greenwood, "Who Owns a Photograph of People Once Considered 'Property'?," *Washington Post*, November 10, 2021, https://www.washingtonpost.com/opinions/2021/11/10/harvard-photos-renty-delia-taylor-zealy-daguerreotypes/.

73. One indisputable concern with the preservation and free access of these materials is that they could be used to advance spurious claims based on scientific ableism and racism. There are many examples of this from the past and that are still happening today. One salient example comes from the 1990s and early 2000s, when the Havasupai tribe of New Mexico willingly gave blood samples and other vital data to researchers who wished to research the genetic basis of diabetes. The tribe later learned that the researchers used sample hand- and fingerprints to assert that the Havasupai people had abnormally high rates of schizophrenia and depression. See Cathy Gere, "Evolutionary Genetics and the Politics of the Human Archive," in *Science in the Archives: Pasts, Presents, Futures,* ed. Lorraine Daston (Chicago: University of Chicago Press, 2017): 203–22.

Epilogue: iPosture

1. Sheng Chen et al., "Global, Regional and National Burden of Low Back Pain 1990–2019: A Systematic Analysis of the Global Burden of Disease Study 2019," *Journal of Orthopaedic Translation* 32 (September 2021): 49–58.

2. Joseph L. Dieleman et al., "US Health Care Spending by Payer and Health Condition, 1996–2016," *JAMA* 323, no. 9 (March 2020): 863–84.

3. Robert A. Aronowitz, *Risky Medicine: Our Quest to Cure Fear and Uncertainty* (Chicago: University of Chicago Press, 2015).

4. Daniel E. Lieberman, *Exercised: Why Something We Never Evolved to Do Is Healthy and Rewarding* (New York: Pantheon Books, 2020), 299.

5. Lieberman, *Exercised*, 299.

6. Daniel E. Lieberman, *The Story of the Human Body: Evolution, Health, and Disease* (New York: Vintage Books, 2013), 159, emphasis in original.

7. Lieberman, *The Story of the Human Body*, 339.

8. Lieberman, *Exercised*, especially chapter 13.

9. Amy Schoenfeld, "The Posture Guru of Silicon Valley," *New York Times*, May 12, 2013, BU3.

10. Esther Gokhale, *8 Steps to a Pain-Free Back: Natural Posture Solutions for Pain in the Back, Neck, Shoulder, Hip, Knee, and Foot* (Stanford, CA: Pendo Press, 2008), xvii.

11. For more on Perez, see Dana Davis, "Noëlle Perez-Christiaens—A Posture Pioneer," accessed July 30, 2022, https://sonomabodybalance.com/2019/11/noelle-perez-christiaens-a-posture-pioneer/.

12. There is extensive scholarship on the history of yoga and its modern iteration. An exemplary study is Mark Singleton, *Yoga Body: The Origins of Modern Posture Practice* (New York: Oxford University Press, 2010). For a helpful survey of other recent books on the history of yoga, especially in the United States, see Sarah Schrank, "American Yoga: The Shaping of Modern Body Culture in the United States," *American Studies* 53, no. 1 (2014): 169–81.

13. Suzanne Newcombe, "The Institutionalization of the Yoga Tradition: 'Gurus' B.K.S. Iyengar and Yogini Sunita in Britain," in *Gurus of Modern Yoga*, ed. Mark Singleton and Ellen Goldberg (New York: Oxford University Press, 2013), 147–68; Frederick M. Smith and Joan

White, "Becoming an Icon: B.K.S. Iyengar as a Yoga Teacher and a Yoga Guru," in *Gurus of Modern Yoga*, ed. Mark Singleton and Ellen Goldberg (New York: Oxford University Press, 2013), 122–40; Andrea Jain, *Selling Yoga: From Counterculture to Pop Culture* (New York: Oxford University Press, 2014); Singleton, *Yoga Body*.

14. Davis, "Noëlle Perez-Christiaens."

15. See Gokhale, *8 Steps to a Pain-Free Back*, 10; Schoenfeld, "The Posture Guru."

16. Gokhale, *8 Steps to a Pain-Free Back*, 20.

17. For an excellent critique of this trend, see Gavin Weedon and Paige Marie Patchin, "The Paleolithic Imagination: Nature, Science, and Race in Anthropocene Fitness Cultures," *Environment and Planning E: Nature and Space* 5, no. 2 (June 2022): 719–39.

18. "Posture Corrector Market—Growth, Trends, COVID-19 Impact, and Forecasts (2022–2027)," accessed July 31, 2022, https://www.mordorintelligence.com/industry-reports /posture-corrector-market.

19. "Posture Corrector Market." For a critical analysis of the Lumo Lift, see Brad Millington, "'Quantify the Invisible': Notes toward a Future of Posture," *Critical Public Health* 26, no. 4 (2016): 405–17.

20. Haopeng Wang and Xianying Feng, "Automatic Recognition of the Micro-Expressions of Human Torso Posture in the Environment of Interrogation," in *Proceedings of the 2nd International Conference on Mechatronics Engineering and Information Technology (ICMEIT 2017)* (Dalian, China: Atlantis Press, 2017), https://doi.org/10.2991/icmeit-17.2017.4; M. L. Gavrilova et al., "Emerging Trends in Security System Design Using the Concept of Social Behavioural Biometrics," in *Information Fusion for Cyber-Security Analytics*, ed. Izzat M Alsmadi, George Karabatis, and Ahmed Aleroud, vol. 691 (Cham, Switzerland: Springer, 2017), 229–51.

21. Amy Cuddy, "Your iPhone Is Ruining Your Posture—and Your Mood," *New York Times*, December 12, 2015, http://www.nytimes.com/2015/12/13/opinion/sunday/your-iphone-is -ruining-your-posture-and-your-mood.html; Amy Cuddy, *Presence: Bringing Your Boldest Self to Your Biggest Challenges* (New York: Back Bay Books, 2018); Susan Dominus, "When the Revolution Came for Amy Cuddy," *New York Times Magazine*, October 22, 2017, 29. In recent years, Cuddy's work has come under serious scrutiny. See, for example, Maquita Peters, "'Power Poses' Co-Author: 'I Do Not Believe the Effects Are Real,'" *NPR*, October 1, 2016, https://www .npr.org/2016/10/01/496093672/power-poses-co-author-i-do-not-believe-the-effects-are-real.

22. For a few examples, see Young Mo Yang et al., "Spontaneous Diaphragmatic Rupture Complicated with Perforation of the Stomach during Pilates," *American Journal of Emergency Medicine* 28, no. 2 (February 2010): 259.e1–259.e3; Kelly Russell et al., "Epidemiology of Yoga-Related Injuries in Canada from 1991 to 2010: A Case Series Study," *International Journal of Injury Control and Safety Promotion* 23, no. 3 (2016): 284–90.

23. Adam Liptak, "Exclusion of Blacks from Juries Raises Renewed Scrutiny," *New York Times*, August 16, 2015, https://www.nytimes.com/2015/08/17/us/politics/exclusion-of-blacks -from-juries-raises-renewed-scrutiny.html. See also Herald Price Fahringer, "'Mirror, Mirror on the Wall . . .': Body Language, Intuition, and the Art of Jury Selection," *American Journal of Trial Advocacy* 17, no. 1 (Summer 1993): 197–204.

24. Trevor Noah, *Born a Crime: Stories from a South African Childhood* (New York: Spiegel & Gau, 2016), 120.

25. Robert Anderson, "Human Evolution, Low Back Pain, and Dual-Level Control," in *Evolutionary Medicine*, ed. Wenda R. Trevathan, E. O. Smith, and James J. McKenna (New York: Oxford University Press, 1999), 333–49.

26. Anderson was interviewed by journalist Benedict Carey. See Carey, "Back Pain Backlash," *Los Angeles Times*, July 24, 2000, D7A.

27. Maria Hondras et al., "*Botlhoko, Botlhoko!* How People Talk about Their Musculoskeletal Complaints in Rural Botswana: A Focused Ethnography," *Global Health Action* 8, no. 1 (December 2015), https://doi.org/10.3402/gha.v8.29010.

28. The literature on this is extensive. For a few noteworthy examples, see Deirdre Cooper Owens, *Medical Bondage: Race, Gender, and the Origins of American Gynecology* (Athens: University of Georgia Press, 2017); Keith Wailoo, *Drawing Blood: Technology and Disease Identity in Twentieth-Century America* (Baltimore: Johns Hopkins University Press, 1999); Keith Wailoo, *Pain: A Political History* (Baltimore: Johns Hopkins University Press, 2014); Martin S. Pernick, "The Calculus of Suffering in Nineteenth-Century Surgery," *Hastings Center Report* 13, no. 2 (April 1983): 26–36; Julie Livingston, *Improvising Medicine: An African Oncology Ward in an Emerging Cancer Epidemic* (Durham, NC: Duke University Press, 2012).

29. Deirdre Cooper Owens, "Listening to Black Women Saves Black Lives," *The Lancet* 397, no. 10276 (February 2021): 788–89. See also Samuel L. Swift et al., "Racial Discrimination in Medical Care Settings and Opioid Pain Reliever Misuse in a U.S. Cohort: 1992 to 2015," *PLoS ONE* 14, no. 12 (December 2019): e0226490.

30. Clare Patton, Marisa McVey, and Ciara Hackett, "Enough of the 'Snake Oil': Applying a Business and Human Rights Lens to the Sexual and Reproductive Wellness Industry," *Business and Human Rights Journal* 7, no. 1 (February 2022): 12–28.

31. Diane Slater et al., "'Sit Up Straight': Time to Re-Evaluate," *Journal of Orthopaedic & Sports Physical Therapy* 49, no. 8 (August 2019): 562–64.

Bibliography

Selected Archival Sources

Brown University, Providence, RI
    Pembroke Center Oral History Project
Columbia University Irving Medical Center, Augustus C. Long Health Sciences Library
    Archives and Special Collections, New York, NY
        Annual Reports
Hampton University Archives, Hampton, VA
    Physical Education Department Records
Harvard University, Francis A. Countway Library of Medicine, Boston, MA
    Lloyd T. Brown Diaries
    Zabdiel Boylston Adams Papers
Hennepin County Library Digital Collections, Minneapolis, MN
    Minneapolis Newspaper Photograph Collection
Howard University Archives, Moorland-Spingarn Research Center, Washington, DC
    Annual Reports
    Campbell Carrington Johnson Papers
    College Health Review
    Course Catalogs
    Departmental Files
    Maryrose Reeves Allen Papers
John F. Kennedy Presidential Library and Museum, Boston, MA
    Hans Kraus Personal Papers
    Oral History Program
    Paul R. Harrington Papers
Massachusetts General Hospital, Archives and Special Collections, Boston, MA
    Orthopedic Department Records
National Anthropological Archives, Smithsonian Institution, Suitland, MD
    William Herbert Sheldon Papers

National Archives and Records Administration, College Park, MD
>Children's Bureau Records

Princeton University Library, Department of Special Collections, Princeton, NJ

Smith College, Special Collections, Northampton, MA
>Academic Departments Records
>
>Alumnae Oral History Project Collection
>
>Clark Science Center Office Records
>
>Classes of 1951–1960 Records
>
>Department of Exercise and Sport Studies Records
>
>Dorothy Sears Ainsworth Papers
>
>Office of College Relations Records
>
>Office of the President Jill Ker Conway Files
>
>Smith College Archivist Reference Files
>
>William Allan Neilson Files

Stanford University Library, Department of Special Collections and University Archives, Stanford, CA
>Clelia D. Mosher Papers

University of Kansas, Medical Center Archives, Kansas City, KS
>Paul R. Harrington Papers

University of Pennsylvania, Philadelphia, PA
>R. Tait McKenzie Papers

Dissertations

Angerhofer, David R. "The American Response to *Pithecanthropus erectus*: The Missing Link and the General Reader, 1860–1920." Master's thesis, University of Maryland, 2008.

Brown, Kimberly D. "The Battle for Black Beauty: Howard University's Grooming Program for Women and African-American Activism in Redefining Aesthetic Ideology through Pageants Since 1925." PhD dissertation, Howard University, 2013.

Dieck, Gretchen S. "The Relationship between Postural Asymmetry and Back and Neck Pain in Smith College Freshmen: A Historical Cohort Study (Massachusetts)." PhD dissertation, Yale University, 1982.

Doyle, Sarah Lincoln. "Head, Shoulders, Knees, and Toes: The Origins of Posture Photographs at Wellesley College, 1875–1942." Master's thesis, Wellesley College, 1996.

Fallon, Cara Kiernan. "Forever Young: The Social Transformation of Aging in America Since 1900." PhD dissertation, Harvard University, 2018.

Forrest, Craig Robert. "A History of *in Loco Parentis* in American Higher Education." PhD dissertation, University of Missouri–Columbia, 2020.

Knoblauch, Heidi Katherine. "Patients' Posture: Medical Photography, Collecting, and Privacy, 1862–1962." PhD dissertation, Yale University, 2015.

MacKay, John. "The Epileptic Constitution: A Study of Seventy Cases of Epilepsy at Whittingham Mental Hospital, Preston, Lancs." PhD dissertation, University of Glasgow, 1936.

McRee, Claire. "The Debutante Slouch: Fashion and the Female Body in the United States, 1912–1925." Master's thesis, Bard Graduate Center: Decorative Arts, Design History, Material Culture, 2015.

Norman, Kathleen Lynne. "'Biological Living': The Redemption of Women and America through Healthy Living, Dress, and Eugenics." PhD dissertation, Claremont Graduate University, 2000.

Peterson, Hazel. "Dorothy S. Ainsworth: Her Life, Professional Career and Contributions to Physical Education." PhD dissertation, Ohio State University, 1968.

Toon, Elizabeth. "Managing the Conduct of the Individual Life: Public Health Education and American Public Health, 1910 to 1940." PhD dissertation, University of Pennsylvania, 1998.

Zepelin, Matt. "From Esotericism to Somatics: A History of Mind-Body Theory and Practice Across the Divide of Modernism, 1820s to 1950s." PhD dissertation, University of Colorado at Boulder, 2018.

Periodicals

Afro-American

American Magazine

Asheville Citizen

Atlanta Daily World

Atlantic Monthly

Boot and Shoe Recorder

Brown Daily Herald

Chicago Daily Tribune

Chicago Defender

China Press

Collier's

The Continent: An Illustrated Weekly Magazine

Cornell Daily Sun

Corset and Underwear Review

Current Opinion

Daily News

The Delineator

Ebony Magazine

Glamour

Ladies' Home Journal

Life Magazine

Literary Digest

Los Angeles Sentinel

Los Angeles Times

Miami News

Miscellany News

New Journal and Guide
New Republic
Newsweek
New York Amsterdam News
New Yorker
New York Times
New York Times Magazine
The Observer
Parents Magazine
Philadelphia Inquirer
Philadelphia Tribune
Pittsburgh Courier
Pittsburgh Post-Gazette
Pittsburgh Press
Plain Dealer
Popular Science
Princeton Alumni Weekly
Rochester Democrat and Chronicle
Saturday Evening Post
Saturday Review
Smith Alumnae Quarterly
Sports Illustrated
Syracuse Herald-Journal
The Sun
Tampa Tribune
Time
Vassar Chronicle
Washington Post
Wellesley College News
Wisconsin State Journal
Women's Wear Daily
Yale Alumni Magazine
Yale Daily News

Primary Sources

Abbott, Edith. "Women in Industry: The Manufacture of Boots and Shoes." *American Journal of Sociology* 15, no. 3 (November 1909): 335–60.

"About the Research Project." Peabody Museum of Archaeology and Ethnology. Accessed August 3, 2022. https://peabody.harvard.edu/ZD-about-research-project.

Agniel, Marguerite. "New Ways to Correct Posture." *Parents Magazine* 4, no. 10 (October 1929): 20–21, 50–52.

Ainsworth, Dorothy S. "Contribution of Physical Education to the Social Service Agency." *Journal of the American Association for Health, Physical Education, and Recreation* 21, no. 6 (1950): 325–68.

Ainsworth, Dorothy S. "The Function and Purpose of the Congress." Report of the International Congress on the Physical Education of Girls and Women (1950): 24–28.

Ainsworth, Dorothy S. "Our Contribution to Morale in Times of War and Peace." *Journal of Health and Physical Education* 14, no. 2 (1943): 67–69, 116–17.

"Alert Doctor Examines 1,200 Immigrants a Day: How Dr. J. W. Schereschewsky Accomplishes a Seemingly Impossible Task." *The Sun*, December 22, 1907.

Alexander, F. Matthias. *Man's Supreme Inheritance: Conscious Guidance and Control in Relation to Human Evolution in Civilization.* New York: E. P. Dutton, 1918.

Anderson, James D. *Education of Blacks in the South.* Chapel Hill: University of North Carolina Press, 1988.

Anderson, Robert. "Human Evolution, Low Back Pain, and Dual-Level Control." In *Evolutionary Medicine,* edited by Wenda R. Trevathan, E. O. Smith, and James J. McKenna, 333–49. New York: Oxford University Press, 1999.

Antioch College. *The Effects of Modern Shoes upon Proper Body Mechanics, 1924–1931.* Yellow Springs, OH: Antioch College, 1931.

Antioch College. "Walk in Beauty." Pamphlet. Yellow Springs, OH: Antioch College, 1941.

Ashley, Doris Lee. "Heads Up! If You Would Have a Beautiful Body." *The Delineator* 120, no. 3 (March 1932): 38.

"Astronaut Glenn Discusses a Down to Earth Problem." Display Advertisement 11. *New York Times,* April 29, 1963.

"At Last—a 'Human' Shoe!" Display Advertisement. *American Physical Education Review* 25, no. 1 (January 1920): 127.

"B. F. Goodrich Rubberized 'Posture Foundation' Sport Shoes." Display Advertisement. *Ladies' Home Journal* 54, no. 6 (June 1937): 94.

Baker, Frank. "The Ascent of Man." *American Anthropologist* 3, no. 4 (October 1890): 297–320.

Bancroft, Jessie H. "Pioneering in Physical Training—An Autobiography." *Research Quarterly. American Association for Health, Physical Education and Recreation* 12, supplement 3 (1941): 666–78.

Bancroft, Jessie H. *The Posture of School Children: With Its Home Hygiene and New Efficiency Methods for School Training.* New York: Macmillan, 1913.

Bancroft, Jessie H. *School Gymnastics, Free Hand: A System of Physical Exercises for Schools.* New York: E. L. Kellogg, 1896.

Bancroft, Jessie H. *School Gymnastics with Light Apparatus.* Boston: D. C. Heath, 1900.

Bennett, Henry Eastman. "Some Requirements of Good School Seating." *Elementary School Journal* 23, no. 3 (November 1922): 203–14.

"Bennett College Gives Correct Posture Contest." *Chicago Defender,* April 20, 1940.

Besser, Gretchen Rous. "Hans Kraus: The Skiers' Doctor." *Skiing Heritage* 21, no. 4 (December 2009): 28–31.

Bethea, Dennis. "What Causes a Backache." *Afro-American* (Baltimore), October 16, 1948.

Blaikie, William. *How to Get Strong and How to Stay So*. New York: Harper & Brothers, 1879.

Blount, Walter P., and John H. Moe. *The Milwaukee Brace*. Baltimore: Williams and Wilkins, 1973.

Boname, Mary Evans. "Letters and Opinions." *Smith Alumnae Quarterly* (Spring 1985): 2–3.

"Books: The Glad Hatter, Lilly Daché's Glamour Book." *Time*, March 26, 1956.

Boyle, Robert H. "The Report that Shocked the President." *Sports Illustrated*, August 15, 1955.

"Boys' Right Posture Suits Reduced." Display Advertisement. *Chicago Daily Tribune*, May 29, 1920.

Bradford, E. H., and J. S. Stone. "The Seating of School Children." *Transactions of the American Orthopedic Association: Thirteenth Session*, vol. 12, 170–83. Philadelphia: American Orthopedic Association, 1899.

Brand, Baert D. "Firm Marks 50th Year Fighting Appearance Defects." *Women's Wear Daily* 97, no. 17 (July 1958): 25.

Brock, Dorothye E. "Some Practical Ideas about Posture Training." *American Physical Education Review* 28, no. 8 (1923): 330–34, 367–74.

Brody, Jane E. "Good Posture May Better Your Position." *New York Times*, December 29, 2015.

Brody, Jane E. "Personal Health: Breaking the Bad Habit of Poor Posture." *New York Times*, April 21, 1993.

Brown, Lloyd T. "A Combined Medical and Postural Examination of 746 Adults." *American Journal of Orthopedic Surgery* 15, no. 11 (November 1917): 774–87.

Brown, Lloyd T. "The Harvard Slouch." *New York Times*, March 18, 1917.

"Buffer College Group Sponsors 'Charm' School." *Chicago Defender*, April 19, 1947.

Bukh, Niels. *Fundamental Gymnastics: The Basis of Rational Physical Development*. New York: E. P. Dutton, 1928.

"Buster Brown." Display Advertisement 10. *New York Amsterdam News*, September 16, 1950.

"Buster Brown Shoes." Display Advertisement. *Parents Magazine* 5, no. 10 (October 1930).

"But Arnold" Cartoon, *Vassar Chronicle* (Poughkeepsie, NY), April 15, 1950.

Cabot, Hugh, and Lloyd T. Brown. "Treatment of Movable Kidney, with or without Infection, with Posture." *Boston Medical and Surgical Journal* 171, no. 10 (September 1914): 369–73.

Cahn, L. Joseph. "Use of a Museum in Hygiene Class." *Journal of Health and Physical Education* 12, no. 1 (1941): 6.

"Camco Corset." *Corset and Underwear Review* 19, no. 5 (August 1922): cover.

"Camp." Display Advertisement. *Minnesota Medicine* 16 (1933): iii.

"Camp." Display Advertisement. *New York State Journal of Medicine* 40, no. 16 (August 15, 1940): inside cover.

"Camp." Display Advertisement. *Rochester Democrat and Chronicle*, May 4, 1941, 16A.

Camp, S. H. Corset. U.S. Patent 1,438,942, filed April 11, 1919, and issued December 19, 1922.

Camp, S. H. Corset and Abdominal Support. U.S. Patent 1,463,252, filed April 22, 1918, and issued July 31, 1923.

"Camp's Atlas on Anatomical Studies." Pamphlet, 6th ed., 1935.

Carey, Benedict. "Back Pain Backlash." *Los Angeles Times*, July 24, 2000.

Carey, Frank E. "Kennedy's Back Injury Dates from 1942 Football Mishap." *Tampa Tribune*, June 25, 1961.

Carter, J. E. Lindsay, and Barbara Honeyman Heath. *Somatotyping: Development and Applications*. Cambridge: Cambridge University Press, 1990.

Cavett, Dick. "Women Alone Can't Make a Just Society: 'Graduation from Hell.'" *New York Times*, June 12, 1992.

Chen, Sheng, Mingjue Chen, Xiaohao Wu, Sixiong Lin, Chu Tao, Huiling Cao, Zengwu Shao, and Guozhi Xiao. "Global, Regional and National Burden of Low Back Pain 1990–2019: A Systematic Analysis of the Global Burden of Disease Study 2019." *Journal of Orthopaedic Translation* 32 (September 2021): 49–58.

"Chins Go Up and Heads High as Six Win in Posture Contest of Madison Junior." *Rochester Democrat and Chronicle*, December 20, 1927.

"City Shopper: You Can Improve Your Posture!" *New York Amsterdam News*, November 6, 1943.

Clevenger, S. V. "Disadvantages of the Upright Position." *American Naturalist* 18, no. 1 (January 1884): 1–9.

Cole, Clarence L., E. W. Loomis, and Eugie A. Campbell. "A Report of Physical Examinations of Twenty Thousand Volunteers." *Military Surgeon* 43, no. 1 (July 1918): 45–64.

Coleman, Larry D. "From Revolution to Revelation: The Howard Experience 1969–1976." *New Directions* 3, no. 4 (October 1976): 4.

"College Girl Corsets." Display Advertisement. *Corset and Underwear Review* 17, no. 1 (April 1921): 39.

"College Girl Corsets." Display Advertisement. *Corset and Underwear Review* 17, no. 2 (May 1921): 17.

"Contends Negro Already Has Second Front Opened." *Atlanta Daily World*, August 14, 1942.

"Corrective Posture Chair, Plimpton Scofield Co." *Official Gazette of the United States Patent Office* 295 (February 1922): 457.

"Corsets." *Women's Wear* 11, no. 57 (September 1915): 3.

"Could This Be Our Deadliest Disease?" Display Advertisement 17. *Washington Post*, June 24, 1963.

Cuddy, Amy. *Presence: Bringing Your Boldest Self to Your Biggest Challenges*. New York: Back Bay Books, 2018.

Cuddy, Amy. "Your iPhone Is Ruining Your Posture—and Your Mood." *New York Times*, December 12, 2015. http://www.nytimes.com/2015/12/13/opinion/sunday/your-iphone-is-ruining-your-posture-and-your-mood.html.

Cutler, Alice S. "True Principles of Scientific Corset Fitting." *Corset and Underwear Review* 10, no. 5 (February 1918): 55.

Dewey, John. "Introduction." In *Constructive Conscious Control of the Individual*, by F. Matthias Alexander, xxi–xxxiii. New York: E. P. Dutton, 1923.

Dewey, John. "Reply to a Reviewer." *New Republic* 15 (1918): 55.

Dieck, Gretchen S. "Posture Pictures, Permission, and Privacy Protection." *IRB: Ethics & Human Research* 3, no. 10 (December 1981): 6–7.

Dieck, Gretchen S., Jennifer L. Kelsey, Vijay K. Goel, Manohar M. Panjabi, Stephen D. Walter, and Mary H. Laprade. "An Epidemiologic Study of the Relationship between Postural Asymmetry in the Teen Years and Subsequent Back and Neck Pain." *Spine* 10, no. 10 (December 1985): 872–77.

Dieleman, Joseph L., Jackie Cao, Abby Chapin, Carina Chen, Zhiyin Li, Angela Liu, Cody Horst, et al. "US Health Care Spending by Payer and Health Condition, 1996–2016." *JAMA* 323, no. 9 (March 2020): 863–84.

Dominus, Susan. "When the Revolution Came for Amy Cuddy." *New York Times Magazine*, October 22, 2017.

Drake, St. Clair. "Reflections on Anthropology and the Black Experience." *Anthropology and Education Quarterly* 9, no. 2 (1978): 85–109.

Drew, Lillian Curtis. *Individualized Gymnastics: A Handbook of Corrective and Remedial Gymnastics*. Philadelphia: Lea and Febiger, 1922.

Dreisbach, Shaun. "How to Turn Off Your Bitch Switch." *Glamour*, February 2016.

"Dr. Joel Goldthwait Speaks on Dress: Physical Director of Smith College Urges Health through Proper Clothing." *Wellesley College News* 29, no. 24 (April 20, 1921).

Du Bois, W.E.B. "Strivings of the Negro People." *Atlantic Monthly* 80 (August 1897): 194–98.

Durham, Helen. "Changing Chins." *Collier's*, March 7, 1931, 29, 56.

Durham, Helen, and Barbara Beattie. "Are You Graceful or Awkward?" *Ladies' Home Journal* 49, no. 1 (January 1932): 24, 62.

Durham, Helen, and Barbara Beattie. "By Their Heads Ye Shall Know Them." *Ladies' Home Journal* 50, no. 5 (May 1933): 69–70.

Durham, Helen, and Barbara Beattie. "Hiding Your Hips." *Ladies' Home Journal* 48, no. 11 (November 1931): 24, 136–37.

"Educated Feet for Correct Posture." Produced by Marcia Midleton and Edward J. Hummel. Beverly Hills, CA: Beverly Hills Unified Schools, n.d. Video, 16 minutes. https://archive.org /details/0973_Educated_Feet_for_Correct_Posture_01_26_19_19.

Ellis, Havelock. "How Prehistoric Woman Solved the Problem of Her Waist Line." *Current Opinion*, March 1, 1914, 201.

Emmons, Arthur B. "Organized Preventive Medicine Is Nowhere More Effectual than as Applied to Industrial Groups." *Nation's Health* 6, no. 1 (January 1924): 8–9, 74.

Emmons, Arthur B., and Joel Goldthwait. "A Work Chair." *Journal of Industrial Hygiene* 3, no. 5 (September 1921): 154–58.

England, Frederick O. *Physical Education: A Manual for Teachers*. Manila: Bureau of Printing, 1919.

"Expert Adjustment with One Movement." Display Advertisement. *Corset and Underwear Review* 14, no. 5 (February 1920): 52.

Fahringer, Herald Price. "'Mirror, Mirror on the Wall . . .': Body Language, Intuition, and the Art of Jury Selection." *American Journal of Trial Advocacy* 17, no. 1 (Summer 1993): 197–204.

"Fashion Show at Dunbar." *Philadelphia Tribune*, April 16, 1921.

"Fashionable Slouch to Go if Posture League Succeeds." *New York Times*, April 5, 1914.

"Finds Chairs Add to the Ills of Man: Dr. Eliza M. Mosher Tells Posture League Few Persons Sit in Correct Positions." *New York Times*, March 14, 1915.

"Finishing School: Wealthiest Families Send Children in Highly-Rated Palmer to Become Ladies and Gentlemen." *Ebony Magazine*, October 1947.

Fisher, George J. "The American Posture League." *Journal of Health and Physical Education* 6, no. 8 (October 1935): 16–17.

Fisher, Irving. *How to Live: Rules for Healthful Living, Based on Modern Science*. New York: Funk & Wagnalls, 1915.

"Fitness Program Seen Adding Importance to Posture Week." *Women's Wear Daily* 70, no. 57 (March 1945): 17.

Fogg, C. Davis. "The Posture Pictures." September 25, 2015. https://www.fictionontheweb.co.uk /2015/09/the-posture-pictures-by-c-davis-fogg.html.

"Forget Your Feet Because Antioch Didn't." Display Advertisement. *Saturday Review*, October 24, 1942.

"Four Million Men Wearing Girdles." *New Journal and Guide*, January 20, 1951.

Freiberg, Albert H. "Some Effects of Improper Posture in Factory Labor." In *The Child Workers of the Nation: Proceedings of the Fifth Annual Conference on Child Labor*, 104–10. New York: National Child Labor Committee, 1909. Pamphlet no. 102, paper presented at the Fifth Annual Conference on Child Labor in Chicago, January 21–23, 1909.

Fried, Harlee. "A Side View of Posture Pictures: It Gave Women a Sense of Pride." *Miscellany News* (Poughkeepsie, NY), November 11, 1977.

"G. W. Taboos Nudity Except behind Screens, So Coeds Pose Only for Silhouette Photos." *Washington Post*, October 5, 1937.

Gavin, Robert. "Some Can Recall Stripping, but Not a Camera." *Syracuse Herald-Journal* (Syracuse, NY), January 27, 1995.

Gavrilova, M. L., F. Ahmed, S. Azam, P. P. Paul, W. Rahman, M. Sultana, and F. T. Zohra. "Emerging Trends in Security System Design Using the Concept of Social Behavioural Biometrics." In *Information Fusion for Cyber-Security Analytics*, edited by Izzat M. Alsmadi, George Karabatis, and Ahmed Aleroud, 229–51. Cham, Switzerland: Springer, 2017.

"Get High Rating in Posture Contest." *Chicago Defender*, January 25, 1936.

Gilbreth, Frank Bunker, and Lillian Moller Gilbreth. *Fatigue Study: The Elimination of Humanity's Greatest Unnecessary Waste*. New York: Sturgis and Walton, 1916.

"Gives Benches for Parks: Posture League Presents Seats Built upon Hygienic Lines." *New York Times*, March 11, 1917.

Gokhale, Esther. *8 Steps to a Pain-Free Back: Natural Posture Solutions for Pain in the Back, Neck, Shoulder, Hip, Knee, and Foot*. Stanford, CA: Pendo Press, 2008.

Goldthwait, Joel E. "An Anatomic and Mechanistic Concept of Disease." *Boston Medical and Surgical Journal* 172, no. 24 (July 1915): 881–98.

Goldthwait, Joel E. *The Division of Orthopaedic Surgery in the A.E.F.* Norwood, MA: Plimpton Press, 1941.

Goldthwait, Joel E. "The Importance of Correct Furniture to Assist in the Best Body Function, as Recognized by the Massachusetts Institute of Technology and Smith College." *Journal of Bone and Joint Surgery* 5, no. 2 (April 1923): 179–84.

Good Posture in the Little Child. United States Department of Labor, Children's Bureau Publication no. 129. Washington, DC: Government Printing Office, 1933.

"Good Posture Is a Real Health Asset." *Hygeia* 14 (January 1936): 92.

Goodwin, Louise. "How I Won the National Most Perfect Back Contest." *Chiropractic Journal* 1, no. 9 (September 1933): 5.

Gould, Stephen Jay. "Curveball." *New Yorker*, November 28, 1994.

Greenwood, Keith. "Who Owns a Photograph of People Once Considered 'Property'?" *Washington Post*, November 10, 2021, https://www.washingtonpost.com/opinions/2021/11/10/harvard-photos-renty-delia-taylor-zealy-daguerreotypes/.

Griesse, Rosalie. *The Crooked Shall Be Made Straight*. Atlanta: John Knox Press, 1979.

Grissom, Jean. "Sugar-Coated Calisthenics." *Hygeia* 5 (June 1927): 276–77.

Grynbaum, Bruce. "Obituaries: Hans Kraus, MD." *JAMA* 276, no. 3 (July 1996): 259.

Gurley Brown, Helen. "Having It All." *Miami News*, November 17, 1982.

Hackley, E. Azalia. *The Colored Girl Beautiful*. Kansas City, MO: Burton, 1916.

Haire, Alphonsus P. "Medical Corsetry and Scientific Fitting." *Corset and Underwear Review* 18, no. 2 (November 1921): 71–72.

Haisley, Kelly R., Sabrina E. Drexel, Jennifer M. Watters, John G. Hunter, and Richard J. Mullins. "Major Barbara Stimson: A Historical Perspective on the American Board of Surgery through the Accomplishments of the First Woman to Achieve Board Certification." *Annals of Surgery* 267, no. 6 (June 2018): 1000–1006.

Halberstam-Mickel, Iris. *Stand Up Straight! Personal Recollections about Scoliosis by the People Who Live with It*. Dubuque, IA: Kendall Hunt, 1989.

"Hale America Program to Keep Nation from Going Soft: Three-Fold Purpose." *Atlanta Daily World*, March 30, 1942.

Harrington, Marie C., dir. "Posture and Personality." Hardcastle Film Associates, 1949. Video, 8 minutes. https://archive.org/details/2244_Posture_and_Personality.

Havemann, Ernest. "Mystery and Misery of the Backache." *Life*, June 11, 1956.

Havlicek, Jack. "Physical Fitness—Or Lack of It, Perturbs Doctors." *Asheville Citizen* (Asheville, NC), November 13, 1964.

"Health Is a Game for Modern Girls." *New York Times*, February 26, 1928.

"Health: Your Posture." Centro Corporation, 1953. Video, 10 minutes. https://archive.org/details/HealthYo1953.

Hellebrandt, F. A., and Elizabeth Brogdon Franseen. "Physiological Study of the Vertical Stance of Man." *Physiological Reviews* 23, no. 3 (July 1943): 220–55.

Henry, Fran. "KSU Posture Photos Used in Classwork." *Plain Dealer* (Cleveland), January 23, 1995.

Hill, Leonard. "The Influence of the Force of Gravity on the Circulation." *The Lancet* 145, no. 3728 (February 1895): 338–39.

Hilles, Edith, and Wilhelmina Conger. "Attempts to Standardize Seating in Industry." In *Industrial Posture and Seating*, prepared by the Bureau of Women in Industry, State of New York Department of Labor. Special Bulletin 104 (April 1921): 35–51.

Hinchey, Frank. "Pelvic Changes of Quadrupedal Mammals on Assuming the Erect Posture." *Journal of the Missouri Medical Association* 22, no. 1 (1925): 298–303.

Holt, L. Emmett. *Food, Health, and Growth: A Discussion of the Nutrition of Children*. New York: Macmillan, 1922.

Hondras, Maria, Corrie Myburgh, Jan Hartvigsen, and Helle Johannessen. "*Botlhoko, Botlhoko!* How People Talk about Their Musculoskeletal Complaints in Rural Botswana: A Focused Ethnography." *Global Health Action* 8, no. 1 (December 2015). https://doi.org/10.3402/gha.v8.29010.

Hooton, Earnest A. "The Relation of Physical Anthropology to Medical Science." *Medical Review of Reviews* 22, no. 4 (April 1916): 260–64.

Hooton, Earnest A. "Where Did Man Originate?" *Antiquity* 1 (January 1927): 133–50.

Hoover, Herbert. "Address of President Hoover." In *White House Conference 1930: Address and Abstract of Committee Reports*, 5–13. New York: Century, 1930.

Hrdlička, Aleš. *Children Who Run on All Fours: And Other Animal-Like Behaviors of Young Children.* New York: Whittlesey House, McGraw-Hill, 1931.

Hrdlička, Aleš. "Dr. Eugene Dubois, 1858–1940." *Scientific Monthly* 52, no. 6 (1941): 578–80.

Hrdlička, Aleš. "Quadruped Progression in the Human Child." *American Journal of Physical Anthropology* 10, no. 3 (1927): 347–54.

Hunt, J. Ramsay. "Psychic Representations of Movement and Posture: Their Relation to Symptomatology." *Archives of Neurology and Psychiatry* 14, no. 1 (July 1925): 7–19.

The Importance of Posture. New York: Metropolitan Life Insurance, 1927.

"Indian Maid Wins Contest." *China Press*, September 17, 1930.

Industrial Posture and Seating. Prepared by the Bureau of Women in Industry, State of New York Department of Labor. Special Bulletin 104 (April 1921): 36

"Inventor's Profile: Charles F. Brannock." Smithsonian Institution. Accessed August 13, 2020. https://invention.si.edu/sites/default/files/Kid-friendly-Inventor-Profile-Charles-Brannock .pdf.

Jackson, Algernon B. "Afro Health Talk: Posture and Purpose." *Afro-American* (Baltimore), November 2, 1940.

Jackson, Algernon B. "Afro Health Talk: What Posture Means." *Afro-American* (Baltimore), October 30, 1937.

James, William. *The Energies of Men.* New York: Moffat, Yard, 1914.

James, William. "A Study of the Expression of Bodily Posture." *Journal of General Psychology* 7, no. 2 (January 1932): 405–37.

"Japanese Girl Winner in College Contest." *Chicago Defender*, March 20, 1926.

Jones, J. Albright. "Effect of Posture Work on the Health of Children." *American Journal of Diseases of Children* 46, no. 1 (July 1933): 148–54.

Jones, Lucius. "Sports Slants." *Atlanta Daily World*, April 22, 1942.

Judd, Albert M. "Body Posture and Poise—Its Effect on the General Health." *Long Island Medical Journal* 18, no. 2 (February 1924): 41–47.

Keeve, J. Phillip. "'Fitness,' 'Posture' and Other Selected School Health Myths." *Journal of School Health* 37, no. 1 (January 1967): 8–15.

Keith, Arthur. "An Address on the Nature of Man's Structural Imperfections." *The Lancet* 206, no. 5334 (November 1925): 1047–51.

Keith, Arthur. "Creating a New American Race." *New York Times*, June 2, 1929.

Keith, Arthur. *Ethnos: Or the Problem of Race Considered from a New Point of View.* London: Kegan Paul, 1931.

Keith, Arthur. "Hunterian Lectures on Man's Posture: Its Evolution and Disorders. Lecture I." *British Medical Journal* 1, no. 3246 (March 1923): 451–54.

Keith, Arthur. "Hunterian Lectures on Man's Posture: Its Evolution and Disorders. Lecture V." *British Medical Journal* 1, no. 3250 (April 1923): 624.

Keith, Arthur. *Man: A History of the Human Body*. New York: Holt, 1912.

Keith, Arthur. "National Physique and Public Health." *The Observer* (London), October 20, 1918.

Keith, Arthur. "*Pithecanthropus erectus*—A Brief Review of Human Fossil Remains." *Science Progress* 3, no. 17 (1895): 348–69.

Kelly, Ellen. "Taking Posture Pictures." *Journal of Health and Physical Education* 17, no. 8 (1946): 464–505.

Kennedy, John F. "The Soft American." *Sports Illustrated*, December 26, 1960.

Kinsey, Alfred. *Sexual Behavior in the Human Female*. Philadelphia: Saunders, 1953.

"Kissing Co-Ed Sent Home; Students Revolt." *Pittsburgh Courier*, February 9, 1929.

Klein, Armin. *Posture Clinics: Organization and Exercises*. United States Department of Labor, Children's Bureau Publication no. 164. Washington, DC: Government Printing Office, 1926.

Klein, Armin, and Leah C. Thomas. *Posture and Physical Fitness*. United States Department of Labor, Children's Bureau Publication no. 205. Washington, DC: Government Printing Office, 1931.

Klein, Armin, and Leah C. Thomas. *Posture Exercises: A Handbook for Schools and for Teachers of Physical Education*. United States Department of Labor, Children's Bureau Publication no. 165. Washington, DC: Government Printing Office, 1926.

"Krao—A Missing Link." *The Continent: An Illustrated Weekly Magazine*, February 20, 1884.

Kraus, Hans. *Backache, Stress, and Tension*. New York: Simon and Schuster, 1965.

Kraus, Hans. *The Causes, Prevention, and Treatment of Sports Injuries*. New York: Playboy Press, 1981.

Kraus, Hans. "Diagnosis and Treatment of Low Back Pain." *General Practitioner* 5, no. 4 (April 1952): 55–60.

Kraus, Hans. *Hypokinetic Disease: Diseases Produced by Lack of Exercise*. Springfield, IL: Thomas, 1961.

Kraus, Hans, and S. Eisenmenger-Weber. "Evaluation of Posture Based on Structural and Functional Measurements." *Physical Therapy* 25, no. 6 (November 1945): 267–71.

Kraus, Hans, and S. Eisenmenger-Weber. "Fundamental Considerations of Posture Exercises: Guided by Qualitative and Quantitative Measurements and Tests." *Physical Therapy* 27, no. 6 (November 1947): 362–68.

Kraus, Hans, and S. Eisenmenger-Weber. "Quantitative Tabulation of Posture Evaluation: Based on Structural and Functional Measurements." *Physical Therapy* 26, no. 5 (September 1946): 235–42.

Kraus, Hans, and Ruth P. Hirschland. "Muscular Fitness and Orthopedic Disability." *New York State Journal of Medicine* 54, no. 2 (1954): 212–15.

Krusen, Frank H. "The Future of Physical Medicine." *JAMA* 125, no. 16 (August 1944): 1093–97.

"Labor Notes." *Women's Wear* 11, no. 93 (October 1915): 2.

La Fetra, L. E. "The Relation of Clothing to Posture." *Proceedings of the Ninth Congress of the American School Hygiene Association*, vol. 6, 117–21. New York: American School Hygiene Association, 1917.

Laplace, Louis B., and Jesse T. Nicholson. "Physiologic Effects of the Correction of Faulty Posture." *JAMA* 107, no. 13 (1936): 1009–12.

Lardner, John. "You Can See through Her." *Los Angeles Times*, August 29, 1936.

"'Lazelle' Waists and Girdles." Display Advertisement. *Philadelphia Inquirer*, May 10, 1922.

"Learning to Stand Up No Simple Task: Took Man Centuries, Says Yale Prof." *Chicago Defender*, February 6, 1926.

Lee, Mabel. "Views of Parents on the Physical Education Program for Their Daughters." *Journal of Health and Physical Education* 4, no. 4 (1933): 12–15.

Lee, Roger I., and Lloyd T. Brown, "A New Chart for the Standardization of Body Mechanics." *Journal of Bone and Joint Surgery* 21, no. 5 (1923): 753–56.

Lehrer, Riva. *Golem Girl: A Memoir*. New York: One World, 2020.

Letendre, Andrew. "The Naked Truth about Yale's Posture Program." *Yale Alumni Magazine*, June 22, 2015. https://yalealumnimagazine.org/blog_posts/2144-remembering-yale-s -naked-posture-photos.

"Let's Get the Kinks Out of the Kids!" Display Advertisement 46. *New York Times*, February 5, 1962.

"Letters to the Editor." *New York Times Magazine*, February 2, 1995.

"Letters to the Editor." *New York Times Magazine*, February 5, 1995.

Lewin, Philip. "The Ten Commandments of Good Posture." *Hygeia* 6 (January 1928): 3–5.

Lieberman, Daniel E. *Exercised: Why Something We Never Evolved to Do Is Healthy and Rewarding*. New York: Pantheon Books, 2020.

Lieberman, Daniel E. *The Story of the Human Body: Evolution, Health, and Disease*. New York: Vintage Books, 2013.

Liptak, Adam. "Exclusion of Blacks from Juries Raises Renewed Scrutiny." *New York Times*, August 16, 2015. https://www.nytimes.com/2015/08/17/us/politics/exclusion-of-blacks -from-juries-raises-renewed-scrutiny.html.

"Little Men and Women of Today." *Boot and Shoe Recorder* (March 25, 1916): 41, 48.

Lorenz, Adolf. *My Life and Work: The Search for a Missing Glove*. New York: C. Scribner's Sons, 1936.

Lovett, Robert W. "The Occurrence and Prevention of Flat-Foot among City Hospital Nurses." *Medical and Surgical Report of Boston City Hospital* 7, no. 1 (1896): 193–201.

Lovett, Robert W. "Occurrence of Painful Affections of the Feet among Trained Nurses." *Journal of Bone and Joint Surgery* s2-1, no. 1 (August 1903): 41–60.

"Lynchburg School Has Posture Contest." *Afro-American* (Baltimore), February 3, 1934.

"Making Posture Exercises Attractive." *Hygeia* 9 (February 1931): 172.

A Manual of Camp Physiological Supports, 6th ed. Jackson, MI: S. H. Camp, n.d.

Marr, Joe. "Growing Interest in Foot Remedies." *Boot and Shoe Recorder* (January 22, 1916): 64.

Marsh, O. C. "The Ape-Man from the Tertiary of Java." *Science* 3, no. 74 (May 1896): 789–93.

Marshall, Herman W. "What Do You Know about Feet?" *Boot and Shoe Recorder* (April 15, 1922): 108–10.

Marshall, Marguerite M. "No Debutante Slouch for High School Girls." *Pittsburgh Press*, May 27, 1914.

Marx, Patricia. "Stand Up Straight!" *New Yorker*, March 29, 2021.

McCarthy, John Daly. *Health and Efficiency*. New York: Henry Holt, 1922.

McClellan, George H. "Mistaken for My Wife's Son." *American Magazine* 85, February 1918.

McDade, Phil. "UW's Good-Posture Secret." *Wisconsin State Journal* (Madison, WI), January 27, 1995.

McKenzie, R. Tait. "The Regulation of Physical Instruction in Schools and Colleges from the Standpoint of Hygiene." *Science* 29, no. 743 (March 1909): 481–84.

Mensendieck, Bess M. *The Mensendieck System of Functional Exercises*. Portland, ME: Southworth-Anthoensen Press, 1937.

Merrill, William Jackson. "Distortion of the Pelvis from Posture." *American Journal of Orthopedic Surgery* 16, no. 12 (December 1918): 492–94.

Mestre, Harold. "Seating of Women and Minors in the Fruit and Vegetable Canning Industry of California." In *Industrial Welfare Commission*. State of California, Bulletin no. 2a. Sacramento: California Printing Office, 1919.

"Military Drills in the Schools." *School Review* 24, no. 4 (April 1916): 312–20.

Miller, Kenneth D. "A Physical Educator Looks at Posture." *Journal of School Health* 21, no. 3 (March 1951): 89–94.

"A Monkey Girl: The Missing Link on Exhibition in Chicago." *Chicago Daily Tribune*, December 27, 1884.

Moore, J. Hamilton. *The Young Gentleman and Lady's Monitor, and English Teacher's Assistant*. New York: Daniel D. Smith, 1824.

Moore, Ruth C. "Somebody Ought to Tell Them." *Collier's* 90 (September 24, 1932): 18, 48–49.

Morgan, Robin. "No More Miss America! Ten Points of Protest." In *Sisterhood Is Powerful: An Anthology of Writings from the Women's Liberation Movement*, edited by Robin Morgan, 521–24. New York: Random House, 1970.

Morton, Dudley J. "Human Origin: Correlation of Previous Studies of Primate Feet and Posture with Other Morphologic Evidence." *American Journal of Physical Anthropology* 10, no. 1 (1927): 173–203.

Morton, Dudley J. "The Relation of Evolution to Medicine." *Science* 64, no. 1660 (1926): 394–96.

Mosher, Clelia Duel. "The Schematogram: A New Method of Graphically Recording Posture and Changes in the Contours of the Body." *School and Society* 1, no. 18 (May 1915): 642–45.

Mosher, Eliza M. "Faulty Habits of Posture, a Cause of Enteroptosis." *Woman's Medical Journal* 25, no. 2 (February 1915): 27–30.

Mosher, Eliza M. "Habits of Posture: A Cause of Deformity and Displacement of the Uterus." *New York Journal of Gynaecology and Obstetrics* 2, no. 13 (November 1893): 962–77.

"The Mosher-Lesley Schematograph." Advertisement. *Journal of Health and Physical Education* 4, no. 3 (March 1933): 71.

Moskowitz, Henry. "The Joint Board of Sanitary Control in the Cloak, Suit and Skirt Industry of New York City." *Annals of the American Academy of Political and Social Science* 44 (November 1912): 39–58.

Murthy, Rekha. "Posturing for an Empty Scandal? Nude Photos Taken at Ivy League and Seven Sister Schools Are Merely Pictures of the Past for Most." *Brown Daily Herald* (Providence, RI), vol. 130, no. 20, February 23, 1995.

Nelkin, Dorothy. "The Science Wars: Responses to a Marriage Failed." *Social Text* 46/47 (Spring–Summer 1996): 93–100.

"19 Girls Win in Walk Right Contest." *Afro-American* (Baltimore), December 24, 1927.

Nixon, Faye F. "Pushing Posture." *Hygeia* 13 (May 1935): 463–64.

Noel, Mary Bayley. "Improving Your Children's Posture." *Parents Magazine* 9, no. 6 (June 1934): 33–34, 75.

"Norman Graduation." *Afro-American* (Baltimore), July 11, 1931.

North v. Board of Trustees of the University, 137 Ill. 296 (1891).

Norwat, Anna M. "An Experiment with Bulletin Boards." *Journal of Health and Physical Education* 13, no. 3 (1942): 152–53, 190–91.

"Nude Photos of Yale Graduates Are Shredded." *New York Times*, January 29, 1995.

O'Connor, Liam. "How Your Shape Shapes Your Life." *Popular Science* 160, no. 5 (May 1952).

O'Donnell, Kenneth P., and David F. Powers. *"Johnny, We Hardly Knew Ye": Memories of John Fitzgerald Kennedy*. Boston: Little, Brown, 1972.

"The Orthopedic Department." *Boot and Shoe Recorder* (January 22, 1916): 75–76.

Osgood, Robert B. *Body Mechanics: Education and Practice; Report of the Subcommittee on Orthopedics and Body Mechanics*. New York: Century, 1932.

Palkovich, Ann M. "Anthropology and the Sheldon Nude Posture Photographs." Paper delivered at the Society of American Archivists, August 2014.

"Perfect Posture Goes with Brains at Girls' College." *Philadelphia Inquirer*, February 20, 1962.

Perkins, Frances. Foreword to *Industrial Posture and Seating*, prepared by the Bureau of Women in Industry, State of New York Department of Labor. Special Bulletin 104 (April 1921): 1.

Perry, Jean. "How Two Experts Shape Up on Health Care." *Daily News* (Jacksonville, NC), July 5, 1979.

Peters, Maquita. "'Power Poses' Co-Author: 'I Do Not Believe the Effects Are Real.'" *NPR*, October 1, 2016. https://www.npr.org/2016/10/01/496093672/power-poses-co-author-i-do-not-believe-the-effects-are-real.

"Photo Standalone 11—No Title." *Chicago Defender*, July 23, 1968.

"Photo Standalone 56—Charmettes Club." *New Journal and Guide* (Norfolk, VA), October 22, 1955.

"Photos of Nude Co-Eds Stir Up Detroit Furor." *Chicago Daily Tribune*, October 4, 1937.

Pilates, Joseph Hubertus. *Your Health: A Corrective System of Exercising that Revolutionizes the Entire Field of Physical Education*. New York: Joseph H. Pilates, 1934.

Plath, Sylvia. *The Bell Jar*. New York: Bantam Winstone, 1981.

Plath, Sylvia. *Letters Home: Correspondence 1950–1963*. New York: Harper & Row, 1975.

Plimpton, Pauline Ames. "Letters and Opinions." *Smith Alumnae Quarterly* (Spring 1985): 2.

Poehler, Amy. *Yes, Please*. New York: Dey Street Books, 2014.

Pohlman, Augustus Grote. "Some of the Disadvantages of the Upright Position." *American Medicine* 1, no. 9 (December 1906): 541–46.

Pontzer, H., B. M. Wood, and D. A. Raichlen. "Hunter-Gatherers as Models in Public Health." *Obesity Reviews* 19, no. S1 (December 2018): 24–35.

Porter, Roy. "Diseases of Civilization." In *Companion Encyclopedia of the History of Medicine*, edited by W. F. Bynum and Roy Porter. London: Taylor & Francis Group, 1993.

"Posteur." Display Advertisement. *Pittsburgh Post-Gazette*, October 11, 1935.

"Posture." Cambridge: Massachusetts Institute of Technology, 1928; London: Wellcome Trust, 2008. Video, 13 minutes. https://archive.org/details/Posture-wellcome.

"Posture." Worchester Film Corporation, U.S. Children's Bureau, 1926.

"Posture at Work: Influence of Occupation and Height of Furniture upon the Chest and Spine." *Iron Age* 99, no. 15 (April 1917): 930.

Posture Committee of the American Physical Education Association. "Symposium of Preventative and Corrective Physical Education." *Journal of Health and Physical Education* 2, no. 4 (1931): 11–59.

"Posture Correction Market Size, Share & Trends Analysis Report." Accessed April 11, 2023. https://www.grandviewresearch.com/industry-analysis/posture-correction-market-report.

"Posture Corrector Market: Global Industry Analysis and Forecast (2022–2029)." Accessed April 11, 2023. https://www.maximizemarketresearch.com/market-report/posture-corrector -market/146092/.

"Posture Corrector Market—Growth, Trends, COVID-19 Impact, and Forecasts (2022–2027)." Accessed July 31, 2022. https://www.mordorintelligence.com/industry-reports/posture -corrector-market.

Posture from the Ground Up. New York: Metropolitan Life Insurance Co., 1939.

"Posture Photos Return to Files." *Cornell Daily Sun* (Ithaca, NY), September 19, 1950.

"The Presidency: Back on the Course." *Time*, July 26, 1963.

Price, George. "Defective Seating, Faulty Posture Health." In *Twelfth Annual Report of the Joint Board of Sanitary Control in the Cloak, Suit, and Skirt and Dress and Waist Industries*, 35–36. New York: Joint Board of Sanitary Control, 1923.

Price, George. "Discussion." In *Proceedings of the National Safety Council, Tenth Annual Safety Congress*, 850–51. Chicago: National Safety Council, 1921.

Price, George. "Joint Board of Sanitary Control: Ten Years of Progress." In *Ten Years of Industrial Sanitary Self Control: Tenth Annual Report of the Joint Board of Sanitary Control in the Cloak, Suit, and Skirt and Dress and Waist Industries*, 3–13. New York: Joint Board of Sanitary Control, 1921.

Price, Steven D. "Letter to the Editor." *New York Times Magazine*, February 5, 1995, 14.

Primer, Ben. "Those Posture Photos." *Princeton Alumni Weekly*, June 7, 1995.

"Prison Charm School." *Ebony Magazine*, January 1960.

Prudden, Bonnie, Hans Kraus, and Marjorie Morris. *Is Your Child Really Fit?* New York: Harper, 1956.

Pryor, Alma. "Alma Pryor on Being Beautiful." *Chicago Defender*, September 6, 1975.

Pullman, M. J. *Foot Hygiene and Posture: For Adults and Children.* Los Angeles: 1933.

Ray, Marie Beynon. "Cutting a Fine Figure." *Collier's* 94 (August 18, 1934): 21, 30.

"Retail Shoe Women's Symposium." *Boot and Shoe Recorder* (May 13, 1922): 78–80.

Richardson, Frank H. "The Runabout Child and His Problems." *Hygeia* 6 (December 1928): 690–92.

"Right Posture Boy's Clothing." Display Advertisement. *Saturday Evening Post*, March 20, 1920.

Robertson, J. "Are You Norma, Typical Woman? Search to Reward Ohio Winners." *Plain Dealer* (Cleveland), September 9, 1945.

Robinson, Donald L. "Posture Program Not to Be Abolished, According to Kiphuth." *Yale Daily News* (New Haven, CT), February 10, 1956.

Rogers, James Frederick. "Constitutional and Acquired Posture." *Journal of School Health* 3, no. 1 (January 1933): 5–8.

Rogers, James Frederick. "The Long and Short of the Carriage Business." *Journal of Health and Physical Education* 3, no. 10 (December 1932): 11–59.

Rosenbaum, Ron. "The Great Ivy League Nude Posture Photo Scandal: How Scientists Coaxed America's Best and Brightest out of Their Clothes." *New York Times Magazine,* January 15, 1995.

Rugh, James T. "The Foot of the American Soldier and Its Care." *Pennsylvania Medical Journal* 22, no. 1 (January 1919): 198–205.

Rugh, James T. "Foot Prophylaxis in the Soldier." *American Journal of Orthopedic Surgery* 16, no. 8 (August 1918): 529–37.

Russell, Kelly, Shantel Gushue, Sarah Richmond, and Steven McFaull. "Epidemiology of Yoga-Related Injuries in Canada from 1991 to 2010: A Case Series Study." *International Journal of Injury Control and Safety Promotion* 23, no. 3 (July 2, 2016): 284–90.

"Samuel Higby Camp Institute for Better Posture Formed." *Women's Wear Daily* 63, no. 76 (October 1941): 16.

Sargent, Dudley Allen. "Is the Teaching of Physical Training a Trade or a Profession?" *Proceedings of the American Association for the Advancement of Physical Education,* Sixth Annual Meeting, 6–24. Ithaca, NY: Andrus & Church, 1891.

Sargent, Dudley Allen. "The Physical Proportions of the Typical Man." *Scribner's Magazine* 2 (July 1887): 13–17.

Sargent, Dudley Allen. "The Physical Test of a Man." *American Physical Education Review* 26, no. 1 (April 1921): 188–94.

Saxon, Wolfgang. "Rudolf Schlesinger, 87, Expert on the World's Legal Systems." *New York Times,* November 22, 1996.

Scalpel: University of Michigan Hospital School of Nursing. 1929. https://hdl.handle.net/2027/mdp.39015057785084.

Schaeffer, Nathan C. "Should Our Educational System Include Activities Whose Special Purpose Is Preparation for War?" *School and Society* 1, no. 9 (February 1915): 289–95.

Schereschewsky, Joseph W. "Medical Inspection of Schools." *Public Health Reports* 28, no. 35 (August 1913): 1791–1805.

Schereschewsky, Joseph W. "Some Physical Characteristics of Male Garment-Workers of the Cloak and Suit Trades." *American Journal of Public Health* 5, no. 7 (July 1915): 593–602.

Schoenfeld, Amy. "The Posture Guru of Silicon Valley." *New York Times,* May 12, 2013.

Schulenberg, D. W. "Letter to the Editor." *New York Times Magazine,* February 5, 1995, 14.

Scott, K. Frances. *A College Course in Hygiene.* New York: Macmillan, 1939.

Scurlock, Natalie. "Newest Fall Fashions Demand Good Posture." *Afro-American* (Baltimore), August 2, 1947.

Shands, A. R., Jr., Joseph S. Barr, Paul C. Colonna, and Lawrence Noall. "End-Result Study of the Treatment of Idiopathic Scoliosis: Report of the Research Committee of the American Orthopaedic Association." *Journal of Bone and Joint Surgery* 23, no. 4 (October 1941): 963–77.

Sheldon, William H. *Atlas of Men: A Guide for Somatotyping the Adult Male at All Ages.* New York: Harper, 1954.

Sheldon, William H. "The New York Study of Physical Constitution and Psychotic Pattern." In *Sensation and Measurement*, edited by Howard R. Moskowitz, Bertram Scharf, and Joseph C. Stevens, 147–55. New York: Springer, 1974.

"The Silent Epidemic." Display Advertisement 45. *Washington Post*, February 19, 1962.

Skarstrom, William. "Gymnastic Teaching." *American Physical Education Review* 13, no. 1 (January 1913): 97–104.

Slater, Diane, Vasileios Korakakis, Peter O'Sullivan, David Nolan, and Kieran O'Sullivan. "'Sit Up Straight': Time to Re-Evaluate." *Journal of Orthopaedic & Sports Physical Therapy* 49, no. 8 (August 2019): 562–64.

Sly, Margery N. "Women's Health and the Pix Controversy: Two Decades Later." Paper delivered at the Mid-Atlantic Regional Archives Conference, May 2, 2009.

Sogg, Wilton S. "Letter to the Editor." *New York Times Magazine*, February 12, 1995.

Sohrabi, Louise F. *Crooked Journey: The Story of a Woman's Fight against Scoliosis*. Alameda, CA: Rima Press, 1983.

"Something New and Vital in Shoes." Display Advertisement. *New York Times*, March 30, 1919.

Stafford, George Thomas. *Preventive and Corrective Physical Education*. New York: A. S. Barnes, 1950.

Standing Up to Life: Good Posture and Foot Health. New York: Metropolitan Life Insurance Co., 1954.

"'Stand Up Straight' Theme of Booklet Issued by H. W. Gossard—Hints to Growing Girls." *Women's Wear*, January 13, 1923.

Steinbeck, John. *Vanderbilt Clinic*. New York: Presbyterian Hospital, 1947.

Sterling, E. Blanche. "Health Studies of Negro Children: II. The Physical Status of the Urban Negro Child: A Study of 5,170 Negro School Children in Atlanta, Ga." *Public Health Reports* 43, no. 42 (October 1928): 2713–74.

Sterling, E. Blanche. "The Posture of School Children in Relation to Nutrition, Physical Defects, School Grade, and Physical Training." *Public Health Reports* 37, no. 34 (August 1922): 2043–49.

Stimson, Barbara B. "Backache." *American Journal of Nursing* 51, no. 11 (November 1951): 672–74.

Stimson, Barbara B. "The Low Back Problem." *Psychosomatic Medicine* 9, no. 3 (May 1947): 210–13.

Stoner, "Mother." "Store Life: Wait upon Your Customers with a Sprightly Step." *Women's Wear* 19, no. 19 (July 23, 1919): 53.

Sumption, Dorothy. *Fundamental Danish Gymnastics for Women*. New York: A. S. Barnes, 1929.

"Supports without Restriction, Conforms to Every Movement." *Corset and Underwear Review* 15, no. 5 (August 1920): 66.

Swartz, Nelle. "Industrial Posture and Seating." In *Proceedings of the National Safety Council, Tenth Annual Safety Congress*, 845–51. Chicago: National Safety Council, 1921.

Swarz, Lou. "Charm." *Philadelphia Tribune*, March 22, 1952.

Swift, Samuel L., M. Maria Glymour, Tali Elfassy, Cora Lewis, Catarina I. Kiefe, Stephen Sidney, Sebastian Calonico, Daniel Feaster, Zinzi Bailey, and Adina Zeki Al Hazzouri. "Racial Discrimination in Medical Care Settings and Opioid Pain Reliever Misuse in a U.S. Cohort: 1992 to 2015." Edited by Kerry M. Green. *PLoS ONE* 14, no. 12 (December 2019): e0226490.

Taylor, R. Tunstall. "Lateral Curvature of the Spine." *American Physical Education Review* 9, no. 3 (September 1904): 185–96.

Taylor, Sunaura. *Beasts of Burden: Animal and Disability Liberation*. New York: New Press, 2017.

"Television Show Launches National Posture Week." *Women's Wear Daily* 70, no. 92 (May 1945): 44.

Thomas, Leah C., and Joel E. Goldthwait. *Body Mechanics and Health*. Boston: Houghton Mifflin, 1922.

"To Breed a Brain." *Newsweek* 35, no. 19 (May 8, 1950).

Todd, Mabel Elsworth. "The Balancing of Forces in the Human Being: Its Application to Postural Patterns." In *Early Writings, 1920–1934*. New York: Dance Horizons, 1977.

Todd, Mabel Elsworth. "Principles of Posture, with Special Reference to the Mechanics of the Hip Joint." *Boston Medical and Surgical Journal* 184, no. 25 (June 1921): 667–73.

"Trupedic Shoe." *American Physical Education Review* 25, no. 3 (March 1920): 127.

United States Children's Bureau. "No. 6. Your Child's Posture." In *Lesson Material on Care of the Preschool Child: No. 1–9*. Washington, DC: Government Printing Office, 1928.

United States Public Health Service. *Studies in Physical Development and Posture*. Bulletin no. 179. Washington, DC: Government Printing Office, 1928.

United States Public Health Service. *Studies in Physical Development and Posture*. Bulletin no. 199. Washington, DC: Government Printing Office, 1931.

"Unrouged Girl Is Barnard's Best." *New York Times*, December 10, 1925.

Uribe y Troncoso, M. "Influence of Different Kinds of Handwriting on the Hygienic Posture and Deformities of School Children." *Public Health Papers and Reports* 31, pt. 1 (1905): 182–86.

"Using Your Posture for a Healthier Body." *Philadelphia Tribune*, August 21, 1990.

Voltmer, E. F. "A Plea for the Less Gifted Physically." *Journal of Health and Physical Education* 6, no. 5 (1935): 28–29, 61.

W.A.C. Field Manual: Physical Training. Washington, DC: U.S. Government Printing Office, 1943.

Wadsworth, Ruth F. *Charm by Choice*. New York: Woman's Press, 1930.

Wadsworth, Ruth F. "Sitting Pretty." *Collier's*, March 30, 1929, 22, 48.

Wallace, Alfred Russel. *Darwinism: An Exposition of the Theory of Natural Selection, with Some of Its Applications*. New York: Humboldt, 1889.

Walshe, F.M.R. "The Work of Sherrington on the Physiology of Posture." *Proceedings of the Royal Society of Medicine* 17 (1924): 4–6.

Wang, Haopeng, and Xianying Feng. "Automatic Recognition of the Micro-Expressions of Human Torso Posture in the Environment of Interrogation." In *Proceedings of the 2nd International Conference on Mechatronics Engineering and Information Technology (ICMEIT 2017)*. Dalian, China: Atlantis Press, 2017. https://doi.org/10.2991/icmeit-17.2017.4.

Ward, Henrine E. "Your Posture—Told by Miss Henrine E. Ward." *Chicago Defender*, April 27, 1935.

Warren, Samuel, and Louis Brandeis. "The Right to Privacy." *Harvard Law Review* 4, no. 5 (December 1890): 193–220.

Weinstein, S. L., and I. V. Ponseti. "Curve Progression in Idiopathic Scoliosis." *Journal of Bone and Joint Surgery (American Volume)* 65, no. 4 (April 1983): 447–55.

"What Paris Really Thinks about Corsets." *Corset and Underwear Review* 18, no. 1 (October 1921): 77–78.

Whipple, Florence C. "An Experience in Health Motivation for College Girls." *Journal of Health and Physical Education* 18, no. 9 (November 1947): 639–88.

Whitman, Armitage. "Postural Deformities in Children." *New York State Journal of Medicine* 24, no. 19 (October 1924): 871–74.

"Why Babies Move upon All Fours." *New York Times*, June 17, 1928.

"Why We Can't Stand Still." *Literary Digest* 99, no. 10 (December 8, 1928).

Wickers, Frank C., Wilton H. Bunch, and Paul M. Barnett. "Psychological Factors in Failure to Wear the Milwaukee Brace for Treatment of Idiopathic Scoliosis." *Clinical Orthopaedics and Related Research* 126 (July–August 1977): 62–66.

Williams, A. Wilberforce. "Dr. A. Wilberforce Williams Talks on Preventive Measures, First Aid Remedies Hygienics and Sanitation: Posture—A Factor in Health." *Chicago Defender*, March 5, 1921.

Wilson, Amelia. "Presto! Bad Posture Gone." *Philadelphia Tribune*, March 14, 1940.

Wolf, Naomi. *The Beauty Myth: How Images of Beauty Are Used against Women*. New York: William Morrow, 1991.

Wolfson, Theresa. "Health Education." In *Eleventh Annual Report of the Joint Board of Sanitary Control in the Cloak, Suit, and Skirt and Dress and Waist Industries*, 19–21. New York: Joint Board of Sanitary Control, 1922.

Wolfson, Theresa. "Seating Survey in the Garment Industry." *Nation's Health* 5, no. 3 (March 1923): 165–68.

"Women Who Are Inventors." *New York Times*, October 19, 1913.

"X-Ray Examinations! Urged." *Washington Post*, October 5, 1937.

Yang, Young Mo, Hee Bum Yang, Jung Soo Park, Hoon Kim, Suk Woo Lee, and Jeong Hee Kim. "Spontaneous Diaphragmatic Rupture Complicated with Perforation of the Stomach during Pilates." *American Journal of Emergency Medicine* 28, no. 2 (February 2010): 259.e1–259.e3.

Yolen, Jane. "Posture Picture on the Wall, Who's the Straightest of Us All?" *Smith Alumnae Quarterly* (Fall 1984): 19–21.

"YWCA Opens War on Beauty Contests." *New York Times*, April 18, 1924.

"Zulu Girl's Noble Carriage: It Comes of Their Habit of Carrying Heavy Burdens upon Their Heads." *Chicago Defender*, July 5, 1913.

Secondary Sources

Agocs, Steve. "Chiropractic's Fight for Survival." *AMA Journal of Ethics* 13, no. 6 (June 2011): 384–88.

Aldrich, Mark. *Safety First: Technology, Labor, and Business in the Building of American Work Safety, 1870–1939*. Studies in Industry and Society 13. Baltimore: Johns Hopkins University Press, 1997.

Alexander, Jennifer Karns. "Efficiency and Pathology: Mechanical Discipline and Efficient Worker Seating in Germany, 1929–1932." *Technology and Culture* 47, no. 2 (April 2006): 286–310.

Altschuler, Glenn C., and Isaac Kramnick. *Cornell: A History, 1940–2015*. Ithaca, NY: Cornell University Press, 2014.

Anderson, Warwick. *Colonial Pathologies: American Tropical Medicine, Race, and Hygiene in the Philippines.* Durham, NC: Duke University Press, 2006.

Aronowitz, Robert A. *Risky Medicine: Our Quest to Cure Fear and Uncertainty.* Chicago: University of Chicago Press, 2015.

Arvin, Maile. *Possessing Polynesians: The Science of Settler Colonial Whiteness in Hawai'i and Oceania.* Durham, NC: Duke University Press, 2019.

Bailey, Beth. *From Front Porch to Back Seat: Courtship in Twentieth-Century America.* Baltimore: Johns Hopkins University Press, 1988.

Bailey, Beth. *Sex in the Heartland.* Cambridge, MA: Harvard University Press, 1999.

Bailey, Courtney. "The Blessings of Providence on an Association of Archivists." September 27, 2015. Accessed July 25, 2022. https://cbaileymsls.wordpress.com/tag/j-frank-cook/.

Baker, Lee D. *Anthropology and the Racial Politics of Culture.* Durham, NC: Duke University Press, 2010.

Banner, Lois W. *American Beauty: A Social History . . . Through Two Centuries of the American Idea, Ideal, and Image of the Beautiful Woman.* New York: Alfred A. Knopf, 1983.

Barclay, Jenifer L. *The Mark of Slavery: Disability, Race, and Gender in Antebellum America.* Urbana: University of Illinois Press, 2021.

Baynton, Douglas C. *Defectives in the Land: Disability and Immigration in the Age of Eugenics.* Chicago: University of Chicago Press, 2016.

Baynton, Douglas C. "Disability and the Justification of Inequality in American History." In *The New Disability History: American Perspectives,* edited by Paul Longmore and Lauri Umansky, 33–57. New York: New York University Press, 2001.

Bederman, Gail. *Manliness and Civilization: A Cultural History of Gender and Race in the United States, 1880–1917.* Chicago: University of Chicago Press, 1995.

Belden-Adams, Kris. "'We Did What We Were Told': The 'Compulsory Visibility' and De-Empowerment of US College Women in Nude 'Posture Pictures,' 1880–1940." *Miranda: Revue Pluridisciplinaire du Monde Anglophone* 25 (2022). https://doi.org/10.4000/miranda.44430.

Bellafaire, Judith A. *The Women's Army Corps: A Commemoration of World War II Service.* Washington, DC: U.S. Army Center of Military History, 1993. https://history.army.mil/brochures/wac/wac.htm.

Bender, Daniel E. *Sweated Work, Weak Bodies: Anti-Sweatshop Campaigns and Languages of Labor.* New Brunswick, NJ: Rutgers University Press, 2004.

Bonde, Hans. "The Iconic Symbolism of Niels Bukh: Aryan Body Culture, Danish Gymnastics and Nordic Tradition." *International Journal of the History of Sport* 16, no. 4 (1999): 104–18.

Boster, Dea. *African American Slavery and Disability: Bodies, Property, and Power in the Antebellum South, 1800–1860.* New York: Routledge, 2012.

Bouk, Dan. *How Our Days Became Numbered: Risk and the Rise of the Statistical Individual.* Chicago: University of Chicago Press, 2015.

Bowler, Peter J. *Evolution: The History of an Idea.* Berkeley: University of California Press, 1984.

Bowler, Peter J. *Theories of Human Evolution: A Century of Debate, 1844–1944.* Baltimore: Johns Hopkins University Press, 1986.

Brace, C. Loring. "'Physical' Anthropology at the Turn of the Last Century." In *Histories of American Physical Anthropology in the Twentieth Century*, edited by Michael A. Little and Kenneth A. R. Kennedy, 25–53. Lanham, MD: Lexington Books, 2010.

Braff, Paul. "Saving the Race from Extinction: African Americans and National Negro Health Week." New York Academy of Medicine, February 27, 2018. https://nyamcenterforhistory .org/tag/african-american-history/.

Braun, Lundy. *Breathing Race into the Machine: The Surprising Career of the Spirometer from Plantation to Genetics*. Minneapolis: University of Minnesota Press, 2014.

Bray, Tamara L., and Thomas W. Killion, eds. *Reckoning with the Dead: The Larsen Bay Repatriation and the Smithsonian Institution*. Washington, DC: Smithsonian Institution Press, 1994.

Bronstein, Carolyn. *Battling Pornography: The American Feminist Anti-Pornography Movement, 1976–1986*. New York: Cambridge University Press, 2011.

"Brown Shoe Company, Inc. History." In *International Directory of Company Histories*, vol. 68. London: St. James Press, 2005. http://www.fundinguniverse.com/company-histories /brown-shoe-company-inc-history/.

Brumberg, Joan Jacobs. *The Body Project: An Intimate History of American Girls*. New York: Random House, 1997.

Burch, Susan. *Committed: Remembering Native Kinship in and beyond Institutions*. Chapel Hill: University of North Carolina Press, 2021.

Bynum, W. F. "Darwin and the Doctors: Evolution, Diathesis, and Germs in 19th-Century Britain." *Gesnerus* 40, no. 1–2 (1983): 43–53.

Cahn, Susan K. "If We Got That Freedom: 'Integration' and the Sexual Politics of Southern College Women, 1940–1960." In *Connexions: Histories of Race and Sex in North America*, edited by Jennifer Brier, Jim Downs, and Jennifer L. Morgan, 186–208. Champaign: University of Illinois Press, 2016.

Canaday, Margot. *The Straight State: Sexuality and Citizenship in Twentieth-Century America*. Princeton, NJ: Princeton University Press, 2009.

Chambers, Ted. *The History of Athletics and Physical Education at Howard University*. New York: Vantage Press, 1986.

Chen, Mel Y. *Animacies: Biopolitics, Racial Mattering, and Queer Affect*. Durham, NC: Duke University Press, 2012.

Clark, Constance Areson. *God—or Gorilla: Images of Evolution in the Jazz Age*. Baltimore: Johns Hopkins University Press, 2008.

Cogdell, Christina. *Eugenic Design: Streamlining America in the 1930s*. Philadelphia: University of Pennsylvania Press, 2004.

Comer, James P. *Maggie's American Dream: The Life and Times of a Black Family*. New York: New American Library, 1988.

Connolly, Cynthia A. *Saving Sickly Children: The Tuberculosis Preventorium in American Life, 1909–1970*. New Brunswick, NJ: Rutgers University Press, 2008.

Connolly, Cynthia A., and Janet Golden. "'Save 100,000 Babies': The 1918 Children's Year and Its Legacy." *American Journal of Public Health* 108, no. 7 (July 2018): 902–7.

Couturier, Lynn E. "The Influence of the Eugenics Movement on Physical Education in the United States." *Sport History Review* 36, no. 1 (2005): 21–42.

Cox, Stephen. *The Munchkins of Oz*, 3rd ed. Nashville: Cumberland House, 2002.

Craig, Maxine Leeds. *Ain't I a Beauty Queen? Black Women, Beauty, and the Politics of Race*. New York: Oxford University Press, 2002.

Creadick, Anna G. *Perfectly Average: The Pursuit of Normality in Postwar America*. Amherst: University of Massachusetts Press, 2010.

Currell, Susan. "Eugenic Decline and Recovery in Self-Improvement Literature of the Thirties." In *Popular Eugenics: National Efficiency and American Mass Culture in the 1930s*, edited by Susan Currell and Christina Cogdell, 44–69. Athens: Ohio University Press, 2006.

D'Antonio, Patricia. *Nursing with a Message: Public Health Demonstration Projects in New York City*. New Brunswick, NJ: Rutgers University Press, 2017.

Davis, Audrey B. "Life Insurance and the Physical Examination: A Chapter in the Rise of American Medical Technology." *Bulletin of the History of Medicine* 55, no. 3 (Fall 1981): 392–406.

Davis, Dana. "Noëlle Perez-Christiaens—A Posture Pioneer." Accessed July 30, 2022. https://sonomabodybalance.com/2019/11/noelle-perez-christiaens-a-posture-pioneer/.

DiPiero, Thomas. "Missing Links: Whiteness and the Color of Reason in the Eighteenth Century." *Eighteenth Century* 40, no. 2 (Summer 1999): 155–74.

Dudziak, Mary L. *Cold War Civil Rights: Race and the Image of American Democracy*. Princeton, NJ: Princeton University Press, 2000.

Duffin, Jacalyn, and Charles R. R. Hayter. "Baring the Sole: The Rise and Fall of the Shoe-Fitting Fluoroscope." *Isis* 91, no. 2 (June 2000): 260–82.

Dyreson, Mark. "Nature by Design: Modern American Ideas about Sport, Energy, Evolution, and Republics, 1865–1920." *Journal of Sport History* 26, no. 3 (Fall 1999): 447–69.

Dyson, Walter. *Howard University, The Capstone of Negro Education: A History: 1867–1940*. Washington, DC: The Graduate School, Howard University, 1941.

Emmons, Paul. "The 'Right' Angles: Constructing Upright Posture and the Orthographic View." *Proceedings of the 87th ACSA Annual Meeting* (Fall 1999): 331–36.

English, Daylanne K. *Unnatural Selections: Eugenics in American Modernism and the Harlem Renaissance*. Chapel Hill: University of North Carolina Press, 2004.

Fabian, Ann. *The Skull Collectors: Race, Science, and America's Unburied Dead*. Chicago: University of Chicago Press, 2010.

Fairchild, Amy L. *Science at the Borders: Immigrant Medical Inspection and the Shaping of the Modern Industrial Labor Force*. Baltimore: Johns Hopkins University Press, 2003.

Farrell-Beck, Jane, and Colleen Gau. *Uplift: The Bra in America*. Philadelphia: University of Pennsylvania Press, 2002.

Fear-Segal, Jacqueline, and Susan D. Rose, eds. *Carlisle Indian Industrial School: Indigenous Histories, Memories, and Reclamations*. Lincoln: University of Nebraska Press, 2016.

Fields, Jill. "'Fighting the Corsetless Evil': Shaping Corsets and Culture, 1900–1930." *Journal of Social History* 33, no. 2 (Winter 1999): 355–84.

Fields, Jill. *An Intimate Affair: Women, Lingerie, and Sexuality*. Berkeley: University of California Press, 2007.

Fissell, Mary E., Jeremy A. Greene, Randall M. Packard, and James A. Schafer. "Introduction: Reimagining Epidemics." *Bulletin of the History of Medicine* 94, no. 4 (Winter 2020): 543–61.

Flack, Robert. *A Brief History of the Family of Sonja Escherich-Eisenmenger-Weber*. Tryon, NC: 1990. https://flackology.com/genealogy/histories/Brief%20history%20of%20Sonya%20Weber.pdf.

Flamiano, Dolores. "The (Nearly) Naked Truth: Gender, Race, and Nudity in Life, 1937." *Journalism History* 28, no. 3 (Fall 2002): 121–36.

Foucault, Michel. *Discipline and Punish: The Birth of the Prison*. Translated by Alan Sheridan. New York: Pantheon Books, 1977.

Freidenfelds, Lara. *The Modern Period: Menstruation in Twentieth-Century America*. Baltimore: Johns Hopkins University Press, 2009.

Garland-Thomson, Rosemarie. *Staring: How We Look*. New York: Oxford University Press, 2009.

Gere, Cathy. "Evolutionary Genetics and the Politics of the Human Archive." In *Science in the Archives: Pasts, Presents, Futures*, edited by Lorraine Daston, 203–22. Chicago: University of Chicago Press, 2017.

Giedion, Sigfried. *Mechanization Takes Command: A Contribution to Anonymous History*. New York: Oxford University Press, 1948.

Giglio, James N. "Growing Up Kennedy: The Role of Medical Ailments in the Life of JFK, 1920–1957." *Journal of Family History* 31, no. 4 (2006): 358–85.

Gillespie, Richard. "Industrial Fatigue and the Discipline of Physiology." In *Physiology in the American Context, 1850–1940*, edited by Gerald L. Geison, 237–62. Bethesda, MD: American Physiological Society, 1987.

Gilman, Sander L. "Defining Disability: The Case of Obesity." *PMLA* 120, no. 2 (March 2005): 514–17.

Gilman, Sander L. *Fat: A Cultural History of Obesity*. Cambridge: Polity, 2008.

Gilman, Sander L. *Stand Up Straight! A History of Posture*. London: Reaktion Books, 2018.

Glick, Elisa F. "Harlem's Queer Dandy: African-American Modernism and the Artifice of Blackness." *Modern Fiction Studies* 49, no. 3 (Fall 2003): 414–42.

Goodrum, Matthew R. "The History of Human Origins Research and Its Place in the History of Science: Research Problems and Historiography." *History of Science* 47, no. 3 (September 2009): 337–57.

Gorman, Carma R. "Educating the Eye: Body Mechanics and Streamlining in the United States, 1925–1950." *American Quarterly* 58, no. 3 (September 2006): 839–68.

Gould, Stephen Jay. *The Mismeasure of Man*. New York: Norton, 1981.

Gould, Stephen Jay. "Posture Maketh the Man." In *Ever Since Darwin: Reflections in Natural History*. New York: Norton, 1977.

Goulden, Murray. "Boundary-Work and the Human-Animal Binary: Piltdown Man, Science, and the Media." *Public Understanding of Science* 18, no. 3 (2009): 275–91.

Grant, Julia. *The Boy Problem: Educating Boys in Urban America, 1870–1970*. Baltimore: Johns Hopkins University Press, 2014.

Green, Harvey. *Fit for America: Health, Fitness, Sport, and American Society*. New York: Pantheon Books, 1986.

Griswold, Robert L. "'Russian Blonde in Space': Soviet Women in the American Imagination, 1950–1965." *Journal of Social History* 45, no. 4 (Summer 2012): 881–907.

Gritzer, Glenn, and Arnold Arluke. *The Making of Rehabilitation: A Political Economy of Medical Specialization, 1890–1980*. Berkeley: University of California Press, 1989.

Grover, Kathryn. *Fitness in American Culture: Images of Health, Sport, and the Body, 1830–1940*. Amherst: University of Massachusetts Press, 1989.

Gundling, Tom. "Human Origins Studies: A Historical Perspective." *Evolution: Education and Outreach* 3 (2010): 314–21.

Gundling, Tom. "Stand and Be Counted: The Neo-Darwinian Synthesis and the Ascension of Bipedalism as an Essential Hominid Synapomorphy." *History and Philosophy of the Life Sciences* 34, no. 1/2 (December 2012): 185–210.

Haidarali, Laila. *Brown Beauty: Color, Sex, and Race from the Harlem Renaissance to World War II*. New York: New York University Press, 2018.

Haiken, Elizabeth. "Virtual Virility, or, Does Medicine Make the Man?" *Men and Masculinities* 2, no. 4 (April 2000): 388–409.

Hamilton, Diane. "Cost of Caring: The Metropolitan Life Insurance Company's Visiting Nurse Service, 1909–1953." *Bulletin of the History of Medicine* 63, no. 3 (Fall 1989): 414–34.

Hansen, Joseph, Evelyn Reed, and Mary-Alice Waters. *Cosmetics, Fashions, and the Exploitation of Women*. New York: Pathfinder Press, 1986.

Harris, James J. "The 'Tribal Spirit' in Modern Britain: Evolution, Nationality, and Race in the Anthropology of Sir Arthur Keith." *Intellectual History Review* 30, no. 2 (2020): 273–94.

Harris, Paisley. "Gatekeeping and Remaking: The Politics of Respectability in African American Women's History and Black Feminism." *Journal of Women's History* 15, no. 1 (2003): 212–20.

Harris, Ruth. *Guru to the World: The Life and Legacy of Vivekananda*. Cambridge, MA: Harvard University Press, 2022.

Hau, Michael. *The Cult of Hygiene and Beauty in Germany: A Social History, 1890–1930*. Chicago: University of Chicago Press, 2003.

Hazzard-Donald, Katrina. "Dance." In *Encyclopedia of African American History, 1896 to the Present: From the Age of Segregation to the Twenty-First Century*, edited by Paul Finkelman. New York: Oxford University Press, 2009.

Hersch, Matthew H. "High Fashion: The Women's Undergarment Industry and the Foundations of American Spaceflight." *Fashion Theory* 13, no. 3 (2009): 345–70.

Heumann, Judith. *Being Heumann: An Unrepentant Memoir of a Disability Rights Activist*. Boston: Beacon Press, 2020.

Higginbotham, Evelyn Brooks. *Righteous Discontent: The Women's Movement in the Black Baptist Church, 1880–1920*. Cambridge, MA: Harvard University Press, 1993.

Hug, P. Reginald. "Posture Queen Contests in Alabama." *Journal of Chiropractic Humanities* 15 (2008): 67–80.

Huhndorf, Shari M. *Going Native: Indians in the American Cultural Imagination*. Ithaca, NY: Cornell University Press, 2015.

Hunt-Kennedy, Stefanie. *Between Fitness and Death: Disability and Slavery in the Caribbean*. Urbana: University of Illinois Press, 2020.

Huxley, Michael R. "F. Matthias Alexander and Mabel Elsworth Todd: Proximities, Practices and the Psycho-Physical." *Journal of Dance and Somatic Practices* 3, no. 1/2 (January 2012): 25–42.

Hynes, Roger J. R. "The Most Beautiful Spines in America: The History of the Posture Queens." *Chiropractic History* 22, no. 2 (2002): 65–71.

Igo, Sarah E. *The Known Citizen: A History of Privacy in America.* Cambridge, MA: Harvard University Press, 2018.

Jacobson, Matthew Frye. *Barbarian Virtues: The United States Encounters Foreign Peoples at Home and Abroad, 1876–1917.* New York: Hill and Wang, 2001.

Jacobson, Matthew Frye. *Whiteness of a Different Color: European Immigrants and the Alchemy of Race.* Cambridge, MA: Harvard University Press, 1998.

Jain, Andrea. *Selling Yoga: From Counterculture to Pop Culture.* New York: Oxford University Press, 2014.

Jennings, Audra. "Mildred Scott: A Pennsylvania Woman at the Heart of the Early Disability Rights Movement." *Pennsylvania Legacies* 17, no. 2 (Fall 2017): 20–25.

Kasson, John F. *Houdini, Tarzan and the Perfect Man: The White Male Body and the Challenge of Modernity in America.* New York: Hill and Wang, 2001.

Kasson, John F. *Rudeness and Civility: Manners in Nineteenth-Century Urban America.* New York: Hill and Wang, 1990.

Katz, Jonathan. *The Invention of Heterosexuality.* New York: Dutton, 1995.

Katz, Michael B. *Reconstructing American Education.* Cambridge, MA: Harvard University Press, 1987.

Keller, Evelyn Fox. *The Century of the Gene.* Cambridge, MA: Harvard University Press, 2000.

Klein, Jennifer. *For All These Rights: Business, Labor, and the Shaping of America's Public Private Welfare State.* Princeton, NJ: Princeton University Press, 2003.

Kraut, Alan M. *Silent Travelers: Germs, Genes, and the "Immigrant Menace."* New York: Basic Books, 1994.

Lakoff, George, and Mark Johnson. *Metaphors We Live By.* Chicago: University of Chicago Press, 1980.

Landau, Misia. *Narratives of Human Evolution.* New Haven, CT: Yale University Press, 1991.

Lazar, J. Wayne. "Problems of Consciousness in Nineteenth Century British and American Neurology." In *Brain, Mind, and Consciousness in the History of Neuroscience,* edited by C.U.M. Smith and Harry Walker, 147–61. New York: Springer, 2014.

Leavitt, Judith Walzer. *Typhoid Mary: Captive to the Public's Health.* Boston: Beacon Press, 1996.

Lee, P. "The Curious Life of *In Loco Parentis* at American Universities." *Higher Education in Review* 8 (2011): 65–90.

Levine, Deborah I. "Corpulence and Correspondence: President William H. Taft and the Medical Management of Obesity." *Annals of Internal Medicine* 160, no. 8 (October 2013): 580–81.

Lewis, Carolyn Herbst. *Prescription for Heterosexuality: Sexual Citizenship in the Cold War Era.* Chapel Hill: University of North Carolina Press, 2010.

Lewis, Jonathan P. "Posture Pictures and Other Tortures: The Battle to Control Esther Greenwood's Body." *Quarterly Horse* 1, no. 1 (November 2016). http://www.quarterlyhorse.org /fall16/lewis.

Lindenmeyer, Kriste. *A Right to Childhood: The U.S. Children's Bureau and Child Welfare, 1912–46.* Urbana: University of Illinois Press, 1997.

Linker, Beth. "A Dangerous Curve: The Role of History in America's Scoliosis Screening Programs." *American Journal of Public Health* 102, no. 4 (April 2012): 606–16.

Linker, Beth. "Feet for Fighting: Locating Disability and Social Medicine in First World War America." *Social History of Medicine* 20, no. 1 (April 2007): 91–109.

Linker, Beth. "Spines of Steel: A Case of Surgical Enthusiasm in Cold War America." *Bulletin of the History of Medicine* 90, no. 2 (Summer 2016): 222–49.

Linker, Beth. "Strength and Science: Gender, Physiotherapy, and Medicine in the United States, 1918–35." *Journal of Women's History* 17, no. 3 (Fall 2005): 106–32.

Linker, Beth. "Toward a History of Ableness." *All of Us*, June 1, 2021. http://allofusdha.org/research/toward-a-history-of-ableness/.

Linker, Beth. "Tracing Paper, the Posture Sciences, and the Mapping of the Female Body." In *Working with Paper: Gendered Practices in the History of Knowledge*, edited by Carla Bittel, Elaine Leong, and Christine von Oertzen, 124–39. Pittsburgh: University of Pittsburgh Press, 2019.

Linker, Beth. *War's Waste: Rehabilitation in World War I America*. Chicago: University of Chicago Press, 2011.

Linker, Beth, and Emily Abel. "Integrating Disability, Transforming Disease History: Tuberculosis and Its Past." In *Civil Disabilities: Citizenship, Membership, Belonging*, edited by Nancy Hirschmann and Beth Linker, 83–102. Philadelphia: University of Pennsylvania Press, 2015.

Livingston, Julie. *Improvising Medicine: An African Oncology Ward in an Emerging Cancer Epidemic*. Durham, NC: Duke University Press, 2012.

Loschek, Ingrid. "Twentieth Century Fashion." In *Encyclopedia of Clothing and Fashion*, edited by Valerie Steele, vol. 3, 348–53. Detroit: Charles Scribner's Sons, 2005.

Lowe, Margaret A. *Looking Good: College Women and Body Image, 1875–1930*. Baltimore: Johns Hopkins University Press, 2003.

MacEwan, Charlotte G., Elizabeth Powell, and Eugene C. Howe, "An Objective Method of Grading Posture: Its Development, Routine Procedure and Applications." *Physical Therapy* 15, no. 5 (September 1935): 167–73.

Marks, Jonathan. *Tales of the Ex-Apes: How We Think about Human Evolution*. Berkeley: University of California Press, 2015.

Mauss, Marcel. "Techniques of the Body." *Economy and Society* 2, no 1 (February 1973): 70–88.

May, Elaine Tyler. *Homeward Bound: American Families in the Cold War Era*. New York: Basic Books, 1988.

McAndrew, Malia. "Selling Black Beauty: African American Modeling Agencies and Charm Schools in Postwar America." *OAH Magazine of History* 24, no. 1 (January 2010), 29–32.

McGill, Jean S. *The Joy of Effort: A Biography of R. Tait McKenzie*. Bewdley, ON: Clay, 1980.

McInerney, Joseph D. "Why Biological Literacy Matters: A Review of Commentaries Related to *The Bell Curve: Intelligence and Class Structure in American Life*." *Quarterly Review of Biology* 71, no. 1 (March 1996): 81–96.

McRuer, Robert. *Crip Theory: Cultural Signs of Queerness and Disability*. New York: New York University Press, 2006.

Meyer, Leisa D. "Creating GI Jane: The Regulation of Sexuality and Sexual Behavior in the Women's Army Corps during World War II." *Feminist Studies* 18, no. 3 (Autumn 1992): 581–601.

Meyer, Leisa D. *Creating GI Jane: Sexuality and Power in the Women's Army Corps during World War II.* New York: Columbia University Press, 1996.

Meyerowitz, Joanne. "Women, Cheesecake, and Borderline Material: Responses to Girlie Pictures in the Mid-Twentieth-Century U.S." *Journal of Women's History* 8, no. 3 (Fall 1996): 9–35.

Milam, Erika Lorraine. "Making Males Aggressive and Females Coy: Gender across the Animal-Human Boundary." *Signs* 37, no. 4 (Summer 2012): 935–59.

Millington, Brad. "'Quantify the Invisible': Notes toward a Future of Posture." *Critical Public Health* 26, no. 4 (August 7, 2016): 405–17.

Mitchell, David, and Sharon Snyder. "The Eugenic Atlantic: Race, Disability, and the Making of an International Eugenic Science, 1800–1945." *Disability & Society* 18, no. 7 (December 1, 2003): 843–64.

Mollow, Anna. "Disability Studies Gets Fat." *Hypatia* 30, no. 1 (Winter 2015): 199–216.

Montez de Oca, Jeffrey. "The 'Muscle Gap': Physical Education and US Fears of a Depleted Masculinity, 1954–1963." In *East Plays West: Sport and the Cold War*, edited by Stephen Wagg, Robert Edelman, and David Andrews, 137–62. London: Routledge, 2006.

Moran, Rachel Louise. *Governing Bodies: American Politics and the Shaping of the Modern Physique.* Philadelphia: University of Pennsylvania Press, 2018.

Morantz-Sanchez, Regina. *Conduct Unbecoming a Woman: Medicine on Trial in Turn-of-the Century Brooklyn.* New York: Oxford University Press, 1999.

Mulligan, William H., Jr. "Mechanization and Work in the American Shoe Industry: Lynn, Massachusetts, 1852–1883." *Journal of Economic History* 41, no. 1 (March 1981): 59–63.

Newcombe, Suzanne. "The Institutionalization of the Yoga Tradition: 'Gurus' B.K.S. Iyengar and Yogini Sunita in Britain." In *Gurus of Modern Yoga*, edited by Mark Singleton and Ellen Goldberg, 147–66. New York: Oxford University Press, 2013.

Noah, Trevor. *Born a Crime: Stories from a South African Childhood.* New York: Spiegel & Gau, 2016.

Nugent, Angela. "Fit for Work: The Introduction of Physical Examinations in Industry." *Bulletin of the History of Medicine* 57, no. 4 (Winter 1983): 578–95.

Nugent, Walter T. K. *Habits of Empire: A History of American Expansion.* New York: Alfred A. Knopf, 2008.

O'Brien, Elizabeth. "Pelvimetry and the Persistence of Racial Science in Obstetrics." *Endeavour* 37, no. 1 (March 2013): 21–28.

Olian, JoAnne, ed. *Children's Fashions, 1900–1950s, as Pictured in Sears Catalogs.* Mineola, NY: Dover Publications, 2003.

Owens, Deirdre Cooper. "Listening to Black Women Saves Black Lives." *The Lancet* 397, no. 10276 (February 2021): 788–89.

Owens, Deirdre Cooper. *Medical Bondage: Race, Gender, and the Origins of American Gynecology.* Athens: University of Georgia Press, 2017.

Pait, T. Glenn, and Justin T. Dowdy. "John F. Kennedy's Back: Chronic Pain, Failed Surgeries, and the Story of Its Effects on His Life and Death." *Journal of Neurosurgery: Spine* 27, no. 3 (2017): 247–55.

Parezo, Nancy J., and Don D. Fowler. *Anthropology Goes to the Fair: The 1904 Louisiana Purchase Exposition*. Lincoln: University of Nebraska Press, 2007.

Park, Roberta J. "Science, Service, and the Professionalization of Physical Education: 1885–1905." *International Journal of the History of Sport* 24, no. 12 (December 2007): 1674–1700.

Park, Roberta J. "Setting the Scene—Bridging the Gap between Knowledge and Practice: When Americans Really Built Programmes to Foster Healthy Lifestyles, 1918–1940." *International Journal of Sports History* 25, no. 11 (September 2008): 1427–52.

Patton, Clare, Marisa McVey, and Ciara Hackett. "Enough of the 'Snake Oil': Applying a Business and Human Rights Lens to the Sexual and Reproductive Wellness Industry." *Business and Human Rights Journal* 7, no. 1 (February 2022): 12–28.

Peiss, Kathy. *Hope in a Jar: The Making of America's Beauty Culture*. New York: Metropolitan Books, 1998.

Peiss, Kathy. *Zoot Suit: The Enigmatic Career of an Extreme Style*. Philadelphia: University of Pennsylvania Press, 2011.

Pelka, Fred. *What We Have Done: An Oral History of the Disability Rights Movement*. Amherst: University of Massachusetts Press, 2012.

Perlman, Robert. *Evolution and Medicine*. Oxford: Oxford University Press, 2013.

Pernick, Martin S. "The Calculus of Suffering in Nineteenth-Century Surgery." *Hastings Center Report* 13, no. 2 (April 1983): 26–36.

Peterson, Theodore. *Magazines in the Twentieth Century*. Urbana: University of Illinois Press, 1956.

Porter, Dorothy, ed. *The History of Public Health and the Modern State*. Amsterdam: Rodopi, 1994.

Potter, Claire Bond. "Taking Back Times Square: Feminist Repertoires and the Transformation of Urban Space in Late Second Wave Feminism." *Radical History Review* 113 (May 2012): 67–80.

Prendergast, Catherine, and Nancy Abelmann. "Alma Mater: College, Kinship, and the Pursuit of Diversity." *Social Text* 24, no. 1 (Spring 2006): 37–53.

Prescott, Heather Munro. *Student Bodies: The Influence of Student Health Services in American Society and Medicine*. Ann Arbor: University of Michigan Press, 2007.

Purkiss, Ava. "'Beauty Secrets: Fight Fat': Black Women's Aesthetics, Exercise, and Fat Stigma, 1900–1930s." *Journal of Women's History* 29, no. 2 (Summer 2017): 14–37.

Purkiss, Ava. *Fit Citizens: A History of Black Women's Exercise from Post-Reconstruction to Postwar America*. Chapel Hill: University of North Carolina Press, 2023.

Putney, Clifford. *Muscular Christianity: Manhood and Sports in Protestant America, 1880–1920*. Cambridge, MA: Harvard University Press, 2001.

Qureshi, Sadiah. *Peoples on Parade: Exhibitions, Empire, and Anthropology in Nineteenth-Century Britain*. Chicago: University of Chicago Press, 2011.

Rabinbach, Anson. *The Human Motor: Energy, Fatigue, and the Origins of Modernity*. New York: Basic Books, 1990.

Rainger, Ronald, Keith Ronald Benson, and Jane Maienschein. *The American Development of Biology*. New Brunswick, NJ: Rutgers University Press, 1991.

Reardon, Jenny, and Kim TallBear. "'Your DNA Is Our History': Genomics, Anthropology, and the Construction of Whiteness as Property." *Current Anthropology* 53, no. S5 (April 2012): S233–45.

Rees, Amanda. "Stories of Stones and Bones: Disciplinarity, Narrative and Practice in British Popular Prehistory, 1911–1935." *British Journal for the History of Science* 49, no. 3 (September 2016): 433–51.

Rembis, Michael. "Disability and the History of Eugenics." In *The Oxford Handbook of Disability History*, edited by Michael Rembis, Catherine Kudlick, and Kim E. Nielsen, 85–103. New York: Oxford University Press, 2018.

Reverby, Susan. *Examining Tuskegee: The Infamous Syphilis Study and Its Legacy*. Chapel Hill: University of North Carolina Press, 2009.

Rexford, Nancy. *Women's Shoes in America, 1795–1930*. Kent, OH: Kent State University Press, 2000.

Rice, Emmett A., John Hutchinson, and Mabel Lee. *A Brief History of Physical Education*, 4th ed. New York: Ronald Press, 1958.

Rider, Toby C., and Kevin B. Witherspoon. "Introduction." In *Defending the American Way of Life: Sport, Culture, and the Cold War*, edited by Toby C. Rider and Kevin B. Witherspoon, 3–10. Fayetteville: University of Arkansas Press, 2018.

Ritvo, Harriet. "At the Edge of the Garden: Nature and Domestication in Eighteenth- and Nineteenth-Century Britain." *Huntington Library Quarterly* 55, no. 3 (Summer 1992): 363–78.

Roberts, Blain. *Pageants, Parlors, and Pretty Women: Race and Beauty in the Twentieth Century South*. Chapel Hill: University of North Carolina Press, 2014.

Roberts, Dorothy E. *Killing the Black Body: Race, Reproduction, and the Meaning of Liberty*. New York: Pantheon Books, 1997.

Roberts, Samuel. *Infectious Fear: Politics, Disease, and the Health Effects of Segregation*. Chapel Hill: University of North Carolina Press, 2009.

Rogaski, Ruth. *Hygienic Modernity: Meanings of Health and Disease in Treaty-Port China*. Berkeley: University of California Press, 2004.

Rogers, Molly. *Delia's Tears: Race, Science, and Photography in Nineteenth-Century America*. New Haven, CT: Yale University Press, 2010.

Rogers, Naomi. *The Polio Wars: Sister Elizabeth Kenny and the Golden Age of American Medicine*. New York: Oxford University Press, 2014.

Rolland, John. "Mabel Todd: An Introduction to Her Work." *Contact Quarterly* (Fall 1979): 6–7.

Rose, Nikolas. "The Neurochemical Self and Its Anomalies." In *Risk and Morality*, edited by Richard V. Ericson and Aaron Doyle, 407–37. Toronto: University of Toronto Press, 2018.

Rosenberg, Charles E. *Explaining Epidemics and Other Studies in the History of Medicine*. New York: Cambridge University Press, 1992.

Rosenberg, Charles E. "Pathologies of Progress: The Idea of Civilization as Risk." *Bulletin of the History of Medicine* 72, no. 4 (Winter 1998): 714–30.

Rosenberg, Charles E. "What Is an Epidemic? AIDS in Historical Perspective." In *Explaining Epidemics and Other Studies in the History of Medicine*, 278–92. New York: Cambridge University Press, 1992.

Roth, Rodris. "Nineteenth-Century American Patent Furniture." In *Innovative Furniture in America from 1800 to the Present*, edited by David A. Hanks, 23–46. New York: Horizon Press, 1981.

Rothman, David J. *Strangers at the Bedside: A History of How Law and Bioethics Transformed Medical Decision Making*. New York: Basic Books, 1991.

Rybczynski, Witold. *Now I Sit Me Down: From Klismos to Plastic Chair: A Natural History*. New York: Farrar, Straus, and Giroux, 2016.

Rydell, Robert W. *All the World's a Fair: Visions of Empire at American International Expositions, 1876–1916*. Chicago: University of Chicago Press, 1985.

Sapp, Jan. *Genesis: The Evolution of Biology*. New York: Oxford University Press, 2003.

Sawday, Jonathan. "'New Men, Strange Faces, Other Minds': Arthur Keith, Race and the Piltdown Affair (1912–53)." In *Race, Science and Medicine, 1700–1960*, edited by Waltraud Ernst and Bernard Harris, 259–88. London: Routledge, 1999.

Schiebinger, Londa. *Nature's Body: Gender in the Making of Modern Science*. Boston: Beacon Press, 1993.

Schiebinger, Londa. "Skeletons in the Closet: The First Illustrations of the Female Skeleton in Eighteenth-Century Anatomy." *Representations* 14 (1986): 42–82.

Schlich, Thomas. "The Perfect Machine: Lorenz Böhler's Rationalized Fracture Treatment in World War I." *Isis* 100, no. 4 (December 2009): 758–91.

Schnapp, Jeffrey T. "Crystalline Bodies: Fragments of a Cultural History of Glass." *West 86th: A Journal of Decorative Arts, Design History, and Material Culture* 20, no. 2 (Fall–Winter 2013): 173–94.

Schrank, Sarah. "American Yoga: The Shaping of Modern Body Culture in the United States." *American Studies* 53, no. 1 (2014): 169–81.

Schreier, Barbara A. *Becoming American Women: Clothing and the Jewish Immigrant Experience, 1880–1920*. Chicago: Chicago Historical Society, 1994.

Schwartz, Susan E. B. *JFK's Secret Doctor*. New York: Skyhorse, 2012.

Schweik, Susan M. *The Ugly Laws: Disability in Public*. New York: New York University Press, 2009.

Scott, Holly V. *Younger than That Now: The Politics of Age in the 1960s*. Amherst: University of Massachusetts Press, 2016.

Serlin, David. *Replaceable You: Engineering the Body in Postwar America*. Chicago: University of Chicago Press, 2004.

Silver, Julie, and Daniel Wilson. *Polio Voices: An Oral History from the American Polio Epidemics and Worldwide Eradication Efforts*. Westport, CT: Praeger, 2007.

Singleton, Mark. *Yoga Body: The Origins of Modern Posture Practice*. New York: Oxford University Press, 2010.

Smith, Dale C. "Appendicitis, Appendectomy, and the Surgeon." *Bulletin of the History of Medicine* 70, no. 3 (Fall 1996): 414–41.

Smith, Frederick M., and Joan White. "Becoming an Icon: B.K.S. Iyengar as a Yoga Teacher and a Yoga Guru." In *Gurus of Modern Yoga*, edited by Mark Singleton and Ellen Goldberg, 122–40. New York: Oxford University Press, 2013.

Smith, Katharine Capshaw. "Childhood, the Body, and Race Performance: Early 20th-Century Etiquette Books for Black Children." *African American Review* 40, no. 4 (Winter 2006): 795–811.

Smith, Roger. "Representations of Mind: C. S. Sherrington and Scientific Opinion, c. 1930–1950." *Science in Context* 14, no. 4 (December 2001): 511–39.

Smith, Susan L. *Sick and Tired of Being Sick and Tired: Black Women's Health Activism in America, 1890–1950*. Philadelphia: University of Pennsylvania Press, 1995.

Smith Maguire, Jennifer. "Body Lessons: Fitness Publishing and the Cultural Production of the Fitness Consumer." *International Review for the Sociology of Sport* 37, no. 3–4 (April 2002): 449–64.

Sommer, Marianne. "Ancient Hunters and Their Modern Representatives: William Sollas's (1849–1936) Anthropology from Disappointed Bridge to Trunkless Tree and the Instrumentalisation of Racial Conflict." *Journal of the History of Biology* 38, no. 2 (2005): 327–65.

Spencer, Frank. "Sir Arthur Keith." In *History of Physical Anthropology*, edited by Frank Spencer, 560–62. New York: Garland, 1997.

Stark, Laura. *Behind Closed Doors: IRBs and the Making of Ethical Research*. Chicago: University of Chicago Press, 2011.

Starr, Paul. *The Social Transformation of American Medicine*. New York: Basic Books, 1982.

Stearns, Peter N. *Fat History: Bodies and Beauty in the Modern West*. New York: New York University Press, 2002.

Steele, Valerie. *The Corset: A Cultural History*. New Haven, CT: Yale University Press, 2001.

Stepan, Nancy Leys. *"The Hour of Eugenics": Race, Gender, and Nation in Latin America*. Ithaca, NY: Cornell University Press, 1991.

Stepan, Nancy Leys. "Nature's Pruning Hook: War, Race, and Evolution, 1914–1918." In *The Political Culture of Modern Britain: Studies in Memory of Stephen Koss*, edited by J.M.W. Bean, 129–45. London: Hamish Hamilton, 1987.

Stern, Alexandra Minna. "From Mestizophilia to Biotypology: Racialization and Science in Mexico, 1920–1960." In *Race and Nation in Modern Latin America*, edited by Nancy P. Appelbaum, Anne S. Macpherson, and Karin Alejandra Rosemblatt. Chapel Hill: University of North Carolina Press, 2003.

Stern, Alexandra Minna. "What Kind of Morph Are You? Biotypology in Transit, 1920–1960." *Remedia*, February 10, 2016. https://remedianetwork.net/2016/02/10/what-kind-of-a-morph-are-you-biotypology-in-transit-1920s-1960s/.

Stern, Alexandra Minna, Mary Beth Reilly, Martin S. Cetron, and Howard Markel. "'Better Off in School': School Medical Inspection as a Public Health Strategy during the 1918–1919 Influenza Pandemic in the United States." *Public Health Reports* 125, supplement 3 (April 2010): 63–70.

Stovall, Mary. "The 'Chicago Defender' in the Progressive Era." *Illinois Historical Journal* 83, no. 3 (Autumn 1990): 159–72.

Strings, Sabrina. *Fearing the Black Body: The Racial Origins of Fat Phobia*. New York: New York University Press, 2019.

Swazey, Judith P. *Reflexes and Motor Integration: Sherrington's Concept of Integrative Action*. Cambridge, MA: Harvard University Press, 1969.

Tenner, Edward. *Our Own Devices: How Technology Remakes Humanity*. New York: Alfred A. Knopf, 2003.

Terry, Jennifer. *An American Obsession: Science, Medicine and Homosexuality in Modern Society*. Chicago: University of Chicago Press, 1999.

Theunissen, Bert. *Eugène Dubois and the Ape-Man from Java: The History of the First Missing Link and Its Discoverer*. Dordrecht: Kluwer Academic Publishers, 1989.

Theunissen, Bert. "Marie Eugène Francois Thomas Dubois." In *New Dictionary of Scientific Biography*, edited by Noretta Koertge, 312–16. Detroit: Thomas Gale, 2008.

Thomas de la Peña, Carolyn. *The Body Electric: How Strange Machines Built the Modern American.* New York: New York University Press, 2003.

Thornton, Tamara Plakins. *Handwriting in America: A Cultural History.* New Haven, CT: Yale University Press, 1996.

Tomes, Nancy. *The Gospel of Germs: Men, Women, and the Microbe in American Life.* Cambridge, MA: Harvard University Press, 1998.

Tomes, Nancy. "The Making of a Germ Panic, Then and Now." *American Journal of Public Health* 90, no. 2 (February 2000): 191–98.

Tomes, Nancy. "Merchants of Health: Medicine and Consumer Culture in the United States, 1900–1940." *Journal of American History* 88, no. 2 (September 2001): 519–47.

Tomes, Nancy. *Remaking the American Patient: How Madison Avenue and Modern Medicine Turned Patients into Consumers.* Chapel Hill: University of North Carolina Press, 2016.

Toon, Elizabeth, and Janet Golden. "Rethinking Charles Atlas." *Rethinking History: The Journal of Theory and Practice* 4, no. 1 (2000): 80–84.

Tracy, Sarah W. "An Evolving Science of Man: The Transformation and Demise of American Constitutional Medicine, 1920–1950." In *Greater than the Parts: Holism in Biomedicine, 1920–1950,* edited by Christopher Lawrence and George Weisz, 161–88. New York: Oxford University Press, 1998.

Tracy, Sarah W. "George Draper and American Constitutional Medicine, 1916–1946: Reinventing the Sick Man." *Bulletin of the History of Medicine* 66, no. 1 (Spring 1992): 53–89.

Tracy, Sarah W. "Sheldon, William Herbert (1898–1977), Psychologist and Physician." *American National Biography.* Accessed July 25, 2022. https://doi.org/10.1093/anb/9780198606697.article.1400964.

Veder, Robin. "The Expressive Efficiencies of American Delsarte and Mensendieck Body Culture." *Modernism/Modernity* 17, no. 4 (November 2010): 819–38.

Veder, Robin. *The Living Line: Modern Art and the Economy of Energy.* Hanover, NH: Dartmouth College Press, 2015.

Veder, Robin. "Seeing Your Way to Health: The Visual Pedagogy of Bess Mensendieck's Physical Culture System." *International Journal of the History of Sport* 28, no. 8–9 (May 2011): 1336–52.

Verbrugge, Martha H. *Able-Bodied Womanhood: Personal Health and Social Change in Nineteenth-Century Boston.* New York: Oxford University Press, 1988.

Verbrugge, Martha H. *Active Bodies: A History of Women's Physical Education in Twentieth-Century America.* New York: Oxford University Press, 2012.

Vertinsky, Patricia. "Embodying Normalcy: Anthropometry and the Long Arm of William H. Sheldon's Somatotyping Project." *Journal of Sport History* 29, no. 1 (Spring 2002): 95–133.

Vertinsky, Patricia. "Physique as Destiny: William H. Sheldon, Barbara Honeyman Heath and the Struggle for Hegemony in the Science of Somatotyping." *Canadian Bulletin of Medical History* 24, no. 2 (Fall 2007): 291–316.

Vertinsky, Patricia, and Bieke Gils. "'Physical Education's First Lady of the World': Dorothy Sears Ainsworth and the International Association of Physical Education and Sport for Women and Girls." *International Journal of the History of Sport* 33, no. 13 (2016): 1500–1516.

Wailoo, Keith. *Drawing Blood: Technology and Disease Identity in Twentieth-Century America.* Baltimore: Johns Hopkins University Press, 1999.

Wailoo, Keith. *Pain: A Political History.* Baltimore: Johns Hopkins University Press, 2014.

Wald, Priscilla. *Contagious: Cultures, Carriers, and the Outbreak Narrative.* Durham, NC: Duke University Press, 2008.

Ward, Patricia Spain. *Simon Baruch: Rebel in the Ranks of Medicine, 1840–1921.* Tuscaloosa: University of Alabama Press, 1994.

Watson, Elwood, and Darcy Martin, eds. *"There She Is, Miss America": The Politics of Sex, Beauty, and Race in America's Most Famous Pageant.* New York: Palgrave Macmillan, 2004.

Weedon, Gavin, and Paige Marie Patchin. "The Paleolithic Imagination: Nature, Science, and Race in Anthropocene Fitness Cultures." *Environment and Planning E: Nature and Space* 5, no. 2 (June 2022): 719–39.

Weisz, George. *Chronic Disease in the Twentieth Century: A History.* Baltimore: Johns Hopkins University Press, 2014.

Widmer, Alexandra, and Veronika Lipphardt, eds. *Health and Difference: Rendering Human Variation in Colonial Engagements.* New York: Berghahn Books, 2016.

Wiggins, David K. "Charles Holston Williams: Hamptonian Loyalist and Champion of Racial Uplift through Physical Education, Dance, Recreation, and Sport." *International Journal of the History of Sport* 36, no. 17–18 (2019): 1531–51.

Williamson, Bess. *Accessible America: A History of Disability and Design.* New York: New York University Press, 2019.

Wilson, Daniel J. *Polio: The Epidemic and Its Survivors.* Chicago: University of Chicago Press, 2005.

Yosifon, Davis, and Peter N. Stearns. "The Rise and Fall of American Posture." *American Historical Review* 103, no. 4 (October 1998): 317–44.

Young, James Allen. "Height, Weight, and Health: Anthropometric Study of Human Growth in Nineteenth-Century American Medicine." *Bulletin of the History of Medicine* 53, no. 2 (Summer 1979): 214–43.

Index

Abbott, Grace, 76

abdominal supports, 119–28, 173. *See also* back braces; corsets; girdles

abdominal surgery, 114, 121, 294n77

ableness and ableism: class and classism intersectionality with, 8; eugenics and, 7; in evolutionary science, 22–23; fitness and physical education prejudices of, 35–36, 61, 146–48, 154, 156, 159, 163, 171, 196–97, 215–17, 220–21, 306nn38–39; gender and sexism intersectionality with, 8, 44–45; geopolitics of posture and, 7–8, 171; "good" posture standards in relation to, 51; posture assessments and, 6–7, 12, 51, 166–67, 196–97, 205, 215–22; posture commercialization and, 86–87, 96, 104–6, 112, 114, 298n17; posture photography reflecting, 12, 220, 222; in posture science, 5, 17–18, 35–36; race and racism intersectionality with, 8; research ethics and, 326n73

Aborn, Frank P., 97

ACA (American Chiropractic Association), 130–31, 157–59

Adams, Zabdiel Boylston, 96–97, 102, 298n10

Advertising Council, 170

African Americans. *See* Black people

Afro-American, 153, 154, 192

Ainsworth, Dorothy, 177–82, 187

Albert, Mildred L., 190

Alexander, F. Matthias, 48, 130, 140–41

Alexander, Jennifer, 95

Allen, Maryrose Reeves, 152–53, 189–91, 193, 208

AMA (American Medical Association), 118, 125, 133, 155, 165

American Academy of Physical Medicine and Rehabilitation, 166

American Association for the Advancement of Physical Education (later American Physical Education Association), 131–33, 154–55

American Chiropractic Association (ACA), 130–31, 157–59

American Federation for the Physically Handicapped, 218

American Indians. *See* indigenous people

American Medical Association (AMA), 118, 125, 133, 155, 165

American Orthopedic Association, 147, 217

American Posture League (APL): disbanding of, 9, 161; fitness and exercise stance of, 129; founding of, 4, 46–47, 65; gender equity goals of, 50; "good" posture standards of, 46–47, 50, 52–53, 55, 69, 71–72; Harvard Slouch posters of, 71; posture commercialization support by, 82, 83–84, 90–92, 96–97, 102, 104, 110–12, 115–16, 118–19; posture-health connection tenets of, 4–5; posture pageant involvement of,